T0310695

Legal Analytics

Legal Analytics: The Future of Analytics in Law navigates the crisscrossing of intelligent technology and the legal field in building up a new landscape of transformation. Legal automation navigation is multidimensional, wherein it intends to construct streamline communication, approval, and management of legal tasks. The evolving environment of technology has emphasized the need for better automation in the legal field from time to time, although legal scholars took long to embrace information revolution of the legal field.

- Describes the historical development of law and automation.
- Analyzes the challenges and opportunities in law and automation.
- Studies the current research and development in the convergence of law, artificial intelligence, and legal analytics.
- Explores the recent emerging trends and technologies that are used by various legal systems globally for crime prediction and prevention.
- Examines the applicability of legal analytics in forensic investigation.
- Investigates the impact of legal analytics tools and techniques in judicial decision making.
- Analyzes deep learning techniques and their scope in accelerating legal analytics in developed and developing countries.
- Provides an in-depth analysis of implementation, challenges, and issues in society related to legal analytics.

This book is primarily aimed at graduates and postgraduates in law and technology, computer science, and information technology. Legal practitioners and academicians will also find this book helpful.

Legal Analytics
The Future of Analytics in Law

Edited by

Namita Singh Malik,
Elizaveta A Gromova,
Smita Gupta
and Balamurugan Balusamy

CRC Press
Taylor & Francis Group
Boca Raton London New York

CRC Press is an imprint of the
Taylor & Francis Group, an **informa** business

A CHAPMAN & HALL BOOK

Chapman & Hall/CRC Press
Boca Raton and London

First edition published 2023
by CRC Press
6000 Broken Sound Parkway NW, Suite 300, Boca Raton, FL 33487-2742

and by CRC Press
4 Park Square, Milton Park, Abingdon, Oxon, OX14 4RN
CRC Press is an imprint of Taylor & Francis Group, LLC

Library of Congress Cataloging-in-Publication Data

Names: Malik, Namita Singh, editor. | Gromova, Elizaveta A, editor. |
Gupta, Smita (Law teacher), editor. | Balusamy, Balamurugan, editor.
Title: Legal analytics : the future of analytics in law / edited by Namita
Singh Malik, Elizaveta A Gromova, Smita Gupta, Balamurugan Balusamy.
Description: First edition. | Boca Raton, FL : Chapman & Hall/CRC Press,
2023. | Includes bibliographical references and index. |
Identifiers: LCCN 2022025397 (print) | LCCN 2022025398 (ebook) | ISBN
9781032105697 (hardback) | ISBN 9781032382074 (paperback) | ISBN
9781003215998 (ebook)
Subjects: LCSH: Technology and law. | Law--Statistical methods. |
Quantitative research. | Data mining in law enforcement.
Classification: LCC K487.T4 L433 2023 (print) | LCC K487.T4 (ebook) | DDC
344/.095--dc23/eng/20220919
LC record available at https://lccn.loc.gov/2022025397
LC ebook record available at https://lccn.loc.gov/2022025398

ISBN: 9781032105697 (hbk)
ISBN: 9781032382074 (pbk)
ISBN: 9781003215998 (ebk)

DOI: 10.1201/9781003215998

Typeset in Palatino
by Deanta Global Publishing Services, Chennai, India

Contents

Preface

Artificial intelligence is changing current visions of legal applications and changing the practice of law, for example, the computational models for legal reasoning, argumentative techniques, analyzing legal texts, and their predictive legal outcome analysis with reasons that legal professionals will be able to evaluate for themselves. Advancements in technology have revolutionized the legal field with new tools and approaches for workflows of lawyers, judges, legal researchers, and law scholars.

There is a need to explore computational legal applications, and this book makes a step ahead in it. Retrieving legal information in a smarter way, enabling cognitive computing, and collaborating between humans and computers is the priority for justice today. Self-driven cars, image recognition, speech recognition, pattern recognition, etc. are some of the innovations based on deep learning with artificial intelligence at the core. Deep learning is in the category of machine learning, wherein algorithms are used to come up with conclusions based on input data. Whereon the one hand it's advantageous to handle data as large as 560 million items at Amazon and then identify patterns in it, it is not far from gray areas; collecting quality data is resource consuming, no reasons for its conclusions, and very costly due to the amount of memory required to store input data. It is basically a resource-intensive technology, which requires powerful graphic processing units and memory.

Legal analytics technologies are examined in this book, which contribute to future predictive crime through various machine learning algorithm applications. Although technology has immense potential to do much more, but due to the lack of regulatory and governance frameworks, its use is concerning. It's time to address the urgency to enable an effective operational policing environment. Officers need to develop a new skill set to effectively understand and do risk assessments using algorithmically generated forecasts.

This book will offer to make actionable legal and compliance decisions with a focus on investigative efforts and improved outcomes. With data and technology, questions pertaining to privacy and security are on the rise; forensic data legal analytics can bring better scenarios of compliance in regard to violations as data removal, data transmission, encryption, third party misuse, AI, and analytics along with brain mapping, DNA technology, polygraph tests, etc. Technological transformation in law promises efficiency, better-informed decisions, and quantified predictions, which try to provide better professional services, to which this book is related.

Editor's Biography

Dr. Namita Singh Malik is working as Dean and Professor at the School of Law, Galgotias University, India. She has 15 years of experience in the field of law. Her teaching and research interest lies in the area of gender studies, criminal laws, forensic science, and law and technology. She has more than 25 research paper presentations and publications.

Dr. Elizaveta A. Gromova is Associate Professor and the Deputy Director for the International Cooperation and the Coordinator of the LLM Program "Law and Digital Technologies" of the Institute of Law of South Ural State University (National Research University), Russia. She is also a researcher at the International Research Projects "LegalTech: Legal Regulation of AI and Robotics, Legal Provision of the Digital Economy and Creation of the Smart Law for Smart Industry" and "Legal Regulation of the Implementation of the Components of Digital Industry in Industrial Regions." She is also an author of the monograph and textbook and numerous articles in the sphere of business law and law and digital technologies.

Dr. Smita Gupta is Associate Professor at the School of law, Delhi Metropolitan Education, affiliated with Guru Gobind Singh Indraprastha University, Delhi. She has vast teaching experience of more than 15 years in interdisciplinary areas of law and social sciences. Her area of interest lies in exploring multifaced dimensions of law with a special focus on associative links of sociology and crime.

Dr. Balamurugan Balusamy completed Ph.D. at VIT University, Vellore, and is currently working as Professor at Galgotias University, Greater Noida, Uttar Pradesh. He has 15 years of teaching experience in the field of computer science. His area of interest lies in the field of artificial intelligence, Internet of Things, big data, and networking. He has published more than 100 international journal papers and contributed book chapters.

List of Contributors

Monica Agarwal
Sharda University, Greater Noida (India)

Morozova Julia Askarovna
South Ural State University, Chelyabinsk (Russia)

Simran Bais
Maharashtra National Law University, Nagpur (India)

Swati Bansal
Sharda University, Greater Noida (India)

Maria Bazhina
Ural State Law University, Yekaterinburg (Russia) (Candidate of Juridical Science)

Paramita Choudhury
Galgotias University, Greater Noida (India)

Sayan Das
Galgotias University, Greater Noida (India)

Jayanta Ghosh
The West Bengal National University of Juridical Sciences, Kolkata (India)

Tjaša Ivanc
University of Maribor (Slovenia)

Taruna Jakhar
Institute of law, Nirma University, Gujarat (India)

Malcolm Katrak
O.P. Jindal Global University, Sonipat (India)

Swati Kaushal
Amity law School, Noida (India)

Jagdish Khobragade
Maharashtra National Law University, Nagpur (India)

Pavlos Kipouras
South Ural State University, Chelyabinsk (Russia)

Harsh Mahaseth
O.P. Jindal Global University (India)

Namita S. Malik
Galgotias University, Greater Noida (India)

Ranti Fauza Mayana
Padjadjaran University, Bandung (Indonesia)

Cocou Marius Mensah
University of Maribor (Slovenia)

Ofman Elena Mikhailovna
South Ural State University, Chelyabinsk (Russia)

Anuja Mishra
Assistant Professor, Department of Law and Governance, Central University of South Bihar, Gaya (India)

Baidya Nath Mukherjee
Christ (Deemed to be University), Delhi NCR Campus (India)

Santhi Narayanan
Sharda University, Greater Noida (India)

Oksana Vitalievna Ovchinnikova (Oksana V. Ovchinnikova)
South Ural State University (National Research University) (Russia)

Disha Pandey
Galgotias University, Greater Noida (India)

Saumya Pandey
Banaras Hindu University, Varanasi (India)

Tatyana Pavlovna Pestova (Tatyana P. Pestova)
South Ural State University (National Research University) (Russia)

Ahmad Ramli
Padjadjaran University, Bandung (Indonesia)

Snizhana Romashkin
Stashis Scientific Research Institute for the Study of Crime Problems, National Academy of Law Sciences (Ukraine)

Tisni Santika
Pasundan University, Bandung (Indonesia)

Surya Saxena
The ICFAI University, Dehradun (India)

Rusman Galina Sergeevna
South Ural State University, Chelyabinsk (Russia)

Bhupinder Singh
Christ (Deemed to be University) Delhi NCR Campus (India)

Chandrika Singh
Himachal Pradesh National Law University, Shimla (India)

Vijoy Kumar Sinha
The West Bengal National University of Juridical Sciences, Kolkata (India)

David W. Tushaus
Missouri Western State University (USA)

Niteesh Kumar Upadhyay
Galgotias University, Greater Noida (India) and Research Advisor South Ural State University (Russia)

Dulatova Natalya Vladimirovna
South Ural State University, Chelyabinsk (Russia)

Vera Aleksandrovna Zadorozhnaia
South Ural State University (National Research University) (Russia)

Sergey Vasilievich Zuev (Sergey V. Zuev)
South Ural State University (National Research University) (Russia)

1

Artificial Intelligence, Automation, and the Legal System

Disha Pandey and Namita Singh Malik

CONTENTS

Introduction

Automation and artificial intelligence (AI) have often been used interchangeably, but they are fundamentally different. We see automation every day everywhere around us, like the auto-generated marketing messages and e-mails to customers and the CCTV cameras watching us. "Automation" can be described as manually configured software that follows pre-programmed rules. It prevents humans from performing repetitive, monotonous tasks and helps them improve performance, speed, and quality of their work by reducing errors. On the other hand, AI uses technologies, systems, or processes that are designed to mimic human beings and make decisions like them. However, both AI and automation together have helped professionals work easier and more accurate, starting from collecting information to execution in cases.

Digital transformation is seen almost in all industries now, with law not being an exception. The legal industry is also focusing on achieving practical, economical, and efficient outcomes. It would not be appropriate for a lawyer of the 21st century to manually do tasks when he could use technology. Therefore, the legal field is gradually paving its way toward "legal automation". Legal automation is the execution of legal tasks, processes, and workflows through machines wholly or partially. The term "legal automation" includes the design, execution, management, and automation of legal tasks, processes, and decision making. It helps legal professionals automate their routine manual tasks using software. The basic advantage of legal automation is to increase the productivity of law firms and reduce their costs and time consumption on low-value work. Its advantages include easier access to data, less duplication of work, better client experience, and high-value work with

DOI: 10.1201/9781003215998-1

less headcount. For example, automation tools help lawyers to attract a whole new client base, particularly in the fields of family law and estate planning law. But the discussion brings us to questions like – what is the scope of automation in the legal domain? Can automation replace legal professionals? Can machines outperform human performance in the range of activities requiring cognitive capabilities? Is it possible to train all legal professionals in AI? What could be the possible challenges faced in automating the legal field? What are the effects of automation on legal professionals, etc.?

Origin of Artificial Intelligence

Around the 1700s, philosophers often contemplated human thinking being artificially mechanized and manipulated by intelligent machines. Philosophers, logicians, and mathematicians started taking interest in AI when they considered manipulation of symbols mechanically, leading to the invention of the programmable digital computer, the Atanasoff–Berry Computer (ABC), in the 1940s. This further inspired the scientists to create an "electronic brain" or an artificially intelligent being. However, the question arose whether the machine would use its own intelligence to render commands or follow algorithms and commands? To answer this question Sir Alan Turing proposed a test called the "Turing test". This test successfully measured a machine's ability to replicate human actions to an extent which is indistinguishable from human behavior. It was after this that AI projects were developed in a manner which allowed the performance of tasks in human-like creativity. As per Turing, an AI machine showed intelligence if the responses submitted by the same were indistinguishable from real human responses. Later, during a summer conference, John McCarthy, a computer and cognitive scientist, coined the term "artificial intelligence". Thus, from the 1950s, many scientists, programmers, logicians, and theorists have helped increase the modern understanding of AI. With each new decade came new findings about AI that added to people's fundamental knowledge of AI, which became a tangible reality for present and future generations.

Artificial Intelligence and Automation Transforming the Legal Landscape

The recent technological progress has changed the concept of what machines can do. Automation and AI together have redefined the roles of machines by leading to unimaginable progress in businesses and economy and meeting new demands of society. Specifically speaking of the two, "automation" has made work quicker, easier, accurate, and cost effective by helping humans to do away with repetitive and monotonous tasks and "artificial intelligence" is designed to perform tasks involving human intelligence to learn from experiences so that they can self-select responses. The use of software to reduce human effort is not new to industries, but AI has provided whole new possibilities. More clearly, automation reduced human effort only to a limited extent. It is only when AI and automation are combined, one will be able to reduce not just human effort but also totally remove the need for such intervention altogether. This combination is known as *automation continuum*.

Recently, like other industries, the legal field has also witnessed a shift from manual to automated working. Today's legal professionals can carry out tasks faster and with more accuracy. These include collecting data, documenting, and predicting litigation outcomes. AI has remarkably increased productivity in these areas as it can work autonomously. A lot of law firms and lawyers have started using existing AI to gain maximum efficiency and quality. They have gained maximum advantage by using AI and automation, combinedly making it more effective and competitive. But the scope of AI remains unexplored in the legal field, also posing some challenges which will be discussed further.

Scope of Automation in the Legal Domain

A law professional's job includes a combination of many tasks like documentation, contract drafting, applying legal knowledge to the facts, articulation, judgment, execution, etc. These tasks may involve repetition, monotony of the procedure of compliance, etc. Legal automation can free up the professional's time from low-value work and enable them to focus on more complex cases. This does not mean that legal automation can handle every work a lawyer does, as automation lacks the cognitive capabilities. However, it can do several everyday tasks of the lawyers which are low value, four to five times faster than manually done. For lawyers in law firms, it might mean automating the document review process and arranging the data and for in-house lawyers, it might mean automating creation of routine contracts. The use of technology is growing all the time, and the categories of work addressed by legal automation are also rapidly increasing. Some categories include:

- Setting up tasks, taking follow-ups, and managing internal and external communications are some of the routine tasks where maximum automation is seen.
- Mapping and analyzing important data.
- The advent of secure electronic signatures has led to the framing of automated digital agreements, which has saved a lot of manual documentation.
- Legal teams are choosing to deploy contract automation workflows rather than scaling expensive headcount.
- Often legal teams struggle with thousands of PDFs and word contracts having unstructured data. AI contract reviewers can in no time tell them what is there in those contracts.
- Automated workflows can help to stitch various manual processes to help make common workflows scalable like intake of new queries or arranging legal matters according to their priorities.
- A lot of time and risk can be saved by automating contract management like contract renewal, etc.
- Knowledge automation can help us gather legal know-how and precedents easily and in a faster manner.
- Automated billing management helps cut down costs on outside counsels, particularly for many in-house teams.

While automation isn't something new, the adoption of legal automation is still an infancy for legal professionals.

Benefits of Legal Automation

Will a lawyer who has spent lakhs on obtaining a law degree and got trained for years want to waste his time copying files and doing monotonous, repetitive tasks? Certainly NOT. Thanks to legal automation for saving the precious time of the lawyers and helping them invest in more productive and significant tasks. Apart from this, it has also helped in several other ways like:

- Legal professionals indulge in both high- and low-value work. Automation has helped them invest their time in more strategic and commercial work hence saving a lot of time spent on paperwork.
- Drafting contracts based on a format may need a lot of head count. Automating contract creation has helped reduce a lot of hiring for the legal teams. Also, this ensures a lot of accuracy and is cost effective.
- A lawyer may have to perform many tasks on a routine basis, which is basically a lot of duplication of work. For example, drafting the same non-disclosure agreement (NDA) on an everyday basis. Automation will lead to the generation of such contracts through a template, thereby saving a lot of low-value work.
- If low-value work is automated, this will do a lot of cost saving, which will in turn save a lot of client's money.
- Legal professionals face a huge outflow of information on an everyday basis. Managing all this raw data can be a tedious process, sometimes demanding a huge investment of time. Tasks like due diligence reviews, automated contract renewals, and signing through electronic signatures have made a lawyer's work much easier.

As legal professionals are experiencing more and more benefits of automation, this is encouraging them to learn and adopt automation for maximum tasks.

Will AI and Automation Replace the Legal Professionals in Future?

As discussed earlier, technological advancements are seen in almost all fields. In the 21st century, AI has changed the manner law is practiced by automating the lawyering process to an extent, and legal professionals are successfully adopting the new tools to enhance their professional offerings. However, the impact of AI cannot be generalized owing to the different areas of practice and firm structures in the field. Another major challenge is the awareness and acceptability of AI and automation by legal professionals who still trust the traditional methods of lawyering to be reliable. Not surprisingly, humans trust another human's intelligence more than machines and face-to-face interactions remain the most prevalent way of communication. Lately, there has been a huge outcry about AI replacing legal professions or almost taking over all tasks done by humans. A better understanding of AI in the legal system would bring us to the fact that AI will not replace legal professionals; rather it will enhance a lawyer's productivity and efficiencies, also increasing the relative value of exclusive human skills. The former CJI, Hon'ble Justice SA Bobde, in his address at the launch of the AI portal "SUPACE" said,

> A futuristic judiciary is not an impossible dream now …. These assistive tools must give us all the information necessary to arrive at a correct decision in the case, according to the demands of the laws, justice and the Constitution. But we will not allow it to tell us what to decide … Where we rely upon judicial technology to aid a judicial decision, it must be subject to the final consideration of a human.

This can be better understood by understanding how AI and automation work in the legal field. Although the definitions of AI would vary, the term mostly means the automation of intelligent behavior through the computer process. Modern AI systems use algorithms to detect patterns in data; based on these they make assessments that resemble those made by humans. But the data required to perform such tasks is mostly huge. A good example can be the classification of e-mail as spam. After being trained with data from millions of mails, the AI can filter the key characteristics of a spam mail and segregates the mails showing similar characteristics.

More advanced AI systems perform functions using deep learning (a subset of machine learning where artificial neural networks – algorithms built around the neural structure of the human brain – learn from data) and legal analytics to make probabilistic inferences in a matter based on similarly decided matters. Though AI is rapidly developing, these intelligent machines have their own limitations as there are many tasks which demand a set of skills and high cognition which only humans can possess. Many legal problems cannot be solved through automated processes but would require a lawyer's involvement. Legal analytics may help locate precedents according to the present case in question, but only a lawyer can decide which precedent would best advance a client's position. Machines cannot determine the precedents on which the courts are likely to rely. Also, precedents can be overruled. Similarly, AI can draft legal contracts like non-disclosure agreements or non-compete clauses, but they may not necessarily be applicable to all employees.

Further, having access to the right data is the most important thing in litigation. The right investigation of a particular matter and collection of relevant evidence can only be done by a lawyer as this would also involve skills to extract information, tact, and ethics. In litigation, it is also important for the lawyers to build a rapport with the clients as the clients trust the lawyers with the outcome of the case. The lawyers need to explain to the clients the reasons for their act and are also accountable for justifying the conclusions reached, which certainly machines cannot do. Also, negotiations and mediation proceedings between parties can only be done by lawyers as AI and automation are still not helpful in satisfying such tasks.

Sometimes, the clients do not have a clear picture of laws and the legal system, and hence, they often want to resort to illegal, misguided, and unethical means of winning a case. This is where the lawyer balances the illegitimate desires of the clients and ethical legal conduct. An intelligent machine may determine the client's unlawful course of conduct but can never stop the client or make him understand the desired conduct. Core values such as truth and transparency would be compromised without a lawyer's involvement.

Last but not the least, AI and automation could benefit lawyers if it is accessible to every legal professional. According to recent studies, most professionals and law firms do not want to invest huge amounts in buying technology and the others lack awareness and are not trained to use these technologies.

Use of AI in the Legal Field – Major Challenges

While the use of AI has shown promising results in the legal field, there are major issues involved with it which abstain legal professionals from completely relying upon the use AI. Some such issues are:

- Laws cannot be static; they are required to be amended according to the current needs of society. Legal professionals need a high level of articulation, creativity, interpretation, and innovative skills to argue or decide a case. Artificial intelligence cannot ever replicate humans as they function according to the manner they are programmed and the data entered into them; for example, if machines deliver judgments instead of judges, they would just be based on past cases and would lack the articulation, interpretation, experience, and most importantly, the empathy factor used by judges while deciding cases. Also, the predictions made by the machines can also be based on biased data as machines function on a large amount of data entered into them.

- A major issue faced is the reluctance of legal professionals towards using AI as most of them resort to the traditional ways as automation does not feel very comfortable to them. They also fear unemployment as AI is getting more prevalent with each passing day. Moreover, the scenario of the developing countries is not at par with the developed countries when it comes to the use of AI in the legal field as developing countries still use outdated technologies and machines.

- Use of AI is also expensive as the machines also need to be updated with changing times. These intelligent humans incur huge installation and maintenance charges, which are not affordable by most of the organizations.

- As AI can mimic the human brain and is trained to perform functions surpassing human intelligence, it may also commit mistakes. Hence, it becomes important to know the legal status of AI. Can AI be held liable for the mistakes it commits? Can AI be held under civil, criminal, or tortious acts? Can AI enforce if there is a violation of its right? The legal personality of AI is still not clear under any of the laws. This creates an issue when the legal status of machines and robots comes into question, as the liability of AI is not made clear under any of the laws.

- AI can only function as an intelligent machine after a huge amount of data is entered into its system. But the question arises regarding the security of data as the client's confidential information cannot be put at risk. In most cases, the personal information of the clients has faced privacy and security issues.

- Application of AI also faces certain IPR issues as AI has now started creating paintings, music, software, etc. So, if the work done by AI was to be copyrighted, certain rights and duties must be granted to AI first. This would require changes in the IPR legislation of most countries. Similarly, if AI is given legal status, it would enjoy certain rights under the industrial and employment laws. The lack of clarity on these issues makes the application of AI in the legal field even more challenging.

- Another challenge faced is whether AI can be penalized for its wrongdoing. Since AI has yet not got legal status, it cannot be held liable in its own capacity, nor can it enforce for violation of its right as it does not have rights at all. The bigger question is, who can be held responsible in this case – the developer, programmer, retailer, or the end customer? What would be the nature of liability – joint or contributory? These issues raise several questions pertaining to adapting and adopting AI in the legal field.

The Future Ahead

Artificial intelligence is no more a dream but a reality. Today, AI is assisting legal professionals by reducing the time taken to perform numerous everyday tasks related to trials, decision making, and execution. It has also helped legal professionals to increase their productivity and efficiency by investing more time in high-value work, which in turn has reduced the pendency of cases. However, with the enhanced automation and AI promising sustainability, productivity, and efficiency, there come several legal, institutional, and socio-economic challenges that the legal professionals should be aware of so that they can make the maximum utilization of the benefits of AI for economic prosperity and growth. AI will soon change the way people look at lawyering and help them to explore new horizons. It will bring a new wave of innovation and revolution for mankind. At the same time, professionals will have to balance the pros and cons of AI. Since AI is the capability of a machine to mimic intelligent human behavior, we need to adopt an approach which efficiently regulates the functioning of AI systems.

Suggestions

- In the absence of a regulatory framework, it is mandatory to have contracts between the AI user and the AI developer, making clear the scope of services offered and liabilities, clearly outlining roles and responsibilities for determining the liability between the parties.
- Countries should try to enact comprehensive legislation covering data protection and privacy considerations clearly covering the liability clause, etc.
- The government should enact mechanisms giving prior information to the individuals regarding data extraction and the purpose for doing so.
- With the advancements in the technology where AI and legal automation is becoming the future, we need to parallelly upgrade the laws and the literary framework in the country,
- As AI is growing every day, we are unaware of the future advantages and dangers; therefore, we must have technological regulators and comprehensive legislations to monitor its growth.
- It is important to decide what can be the scope of legal automation and what are the tasks which should be compulsorily left to humans to perform. Roles like prediction, documentation, and contract drafting can be automated, whereas areas like judgment including empathy and creativity should be the areas of humans as they can better satisfy their clients.

Lastly, technology can no doubt improve the way legal professionals work and can solve disputes in an unimaginable way, which were previously not possible. While there may be several challenges in the application of AI to the legal system, it can aid in magical transformation.

Bibliography

Adam, A. (1996), 'Constructions of gender in the history of artificial intelligence', *IEEE Annals of the History of Computing*, 18(3), 47–53.

Albus, J. S. (1983), 'Robotics: Challenges to present-day technology: Robots must be produced in great numbers with more advanced capabilities, and we must lose our fear that they are taking over our jobs', *IEEE Potentials*, 2(Fall), 24–27. https://doi.org/10.1109/MP.1983.6499635.

Aly, A., Griffiths, S. and Stramandinoli, F. (2017), 'Metrics and benchmarks in human-robot interaction: Recent advances in cognitive robotics', *Cognitive Systems Research*, 43, 313–323.

Anthony, E. D. (2020), 'The future of law firms (and lawyers) in the age of artificial intelligence', *Revista Direito GV*, 16, 1–12. https://doi.org/10.1590/2317-6172201945.

Asimov, I. (1950), *I, Robot*, Gnome Press, New York.

Banakar, R. (2014), *Normativity in Legal Sociology: Methodological Reflections on Law and Regulation in Late Modernity*, Springer International Publishing, Switzerland.

Bellman, R. (1978), *An Introduction to Artificial Intelligence: Can Computers Think?*, Boyd & Fraser Pub. Co., San Francisco, CA.

Bonnefon, J.-F., Shariff, A. and Rahwan, I. (2016), 'The social dilemma of autonomous vehicles', *Science*, 352(6293), 1573–1576.

Bostrom, N. (2014), *Superintelligence: Paths, Dangers, Strategies*, 1st edn, Oxford University Press, Oxford, UK.

Calo, M., Froomkin, M. and Kerr, I. (2016), *Robot Law*, Edward Elgar Publishing Limited, Northampton, MA.

Calo, R. (2017), 'Artificial intelligence policy: A primer and roadmap', *UC Davis Law Review*, 51, 399–435.

Cerka, P., Grigienė, J. and Sirbikytė, G. (2017), 'Is it possible to grant legal personality to artificial intelligence software systems?', *Computer Law & Security Review*, 33, 685–699.

Charniak, E. and McDermott, D. (1985), *Introduction to Artificial Intelligence*, Addison-Wesley Longman Publishing Co., Inc., Boston, MA.

Davies, C. R. (2011), 'An evolutionary step in intellectual property rights – Artificial intelligence and intellectual property', *Computer Law & Security Review*, 27(6), 601–619.

de Graaf, M. M. and Allouch, S. B. (2013), 'Exploring influencing variables for the acceptance of social robots', *Robotics and Autonomous Systems*, 61(12), 1476–1486. https:doi.org/10.1016/j.robot.2013 .07.007.

Dietterich, T. G. and Horvitz, E. J. (2015), 'Rise of concerns about AI: Reflections and directions', *Communications of the ACM*, 58(10), 38–40.

Doyle, B. (2016), 'Do robots create jobs? The data says yes!', in *Proceedings of ISR 2016: 47st International Symposium on Robotics*, pp. 1–5.

Dyrkolbotn, S. (2017), 'A typology of liability rules for robot harms', in Aldinhas Ferreira, M., Silva Sequeira, J., Tokhi, M., Kadar, E., and Virk, G. (eds), *A World with Robots*, Intelligent Systems, Control and Automation: Science and Engineering, vol. 84, Springer International Publishing, Cham, pp. 119–133. https://doi.org/10.1007/978-3-319-46667-5_9.

Ess, C. (2014), *Digital Media Ethics, Digital Media and Society*, 2 edn, Polity Press.

Ferreira, M. I. A. and Sequeira, J. S. (2017), 'Robots in ageing societies', in Aldinhas Ferreira, M., Silva Sequeira, J., Tokhi, M., Kadar, E., and Virk, G. (eds), *A World with Robots*, Intelligent Systems, Control and Automation: Science and Engineering, vol. 84, Springer International Publishing, Cham, pp. 217–223. https://doi.org/10.1007/978-3-319-46667-5_17.

Frey, C. B. and Osborne, M. (2013), *The Future of Employment: How Susceptible Are Jobs to Computerisation?* Technical report, Oxford Martin School, University of Oxford, Oxford.

Gerke, S., Minssen, T. and Cohen, G. (2020), 'Ethical and legal challenges of artificial intelligence-driven healthcare', in *Artificial Intelligence in Healthcare*, pp. 295–336. https://www.ncbi.nlm.nih.gov/pmc/articles/PMC7332220/.

Gilligan, C. (1982), *In a Different Voice*, Harvard University Press, Cambridge, MA.

Goertzel, B. (2007), 'Human-level artificial general intelligence and the possibility of a technological singularity', *Artificial Intelligence*, 171(18), 1161–1173. https://doi.org/10.1016/j.artint.2007.10.011.

Gogoll, J. and Müller, J. F. (2016), 'Autonomous cars: In favor of a mandatory ethics setting', *Science and Engineering Ethics*, 23(3), 681–700. https://doi.org/10.1007/s11948-016-9806-x.

Grinbaum, A., Chatila, R., Devillers, L., Ganascia, J.-G., Tessier, C. and Dauchet, M. (2017), 'Ethics in robotics research: CERNA mission and context', *IEEE Robotics & Automation Magazine*, 24, 139–145. https://doi.org/10.1109/MRA.2016.2611586.

Gupta, P. (2019), 'Artificial intelligence: Legal challenges in India', *ResearchGate*, 3(1), 133–141. https://www.researchgate.net/publication/335967041.

Gupta, S. (2020), 'Artificial intelligence in legal profession', *Law Audience Journal*, 2(3), 1–27. https://www.lawaudience.com/artificial-intelligence-in-legal-profession/#google_vignette.

Haddadin, S. (2014a), 'Competitive robotics', in *Towards Safe Robots*, Springer Tracts in Advanced Robotics, vol. 90, Springer, Berlin and Heidelberg, pp. 217–252. https://doi.org/10.1007/978-3-642-40308-8_9.

Haddadin, S. (2014b), 'Towards the robotic co-worker', in *Towards Safe Robots*, Springer Tracts in Advanced Robotics, vol. 90, Springer, Berlin and Heidelberg, pp. 195–215. https://doi.org/10.1007/978-3-642-40308-8_8.

Hassler, S. (2017), 'Do we have to build robots that need rights?', *IEEE Spectrum*, 54(3), 6.

Haugeland, J. (1985), *Artificial Intelligence: The Very Idea*, Massachusetts Institute of Technology, Cambridge, MA.

http://www.oxfordmartin.ox.ac.uk/downloads/academic/future-ofemployment.pdf

https://lawreview.law.ucdavis.edu/issues/51/2/Symposium/51-2_Calo.pdf

https://scholar.law.colorado.edu/articles/1234/?utm_source=scholar.law.colorado.edu%2Farticles%2F1234&utm_medium=PDF&utm_campaign=PDFCoverPages

Jones, S. E. (2006), *Against Technology: From the Luddites to Neo-Luddism* (1st ed.), Routledge. https://doi.org/10.4324/9780203960455.

Kauffman, M. E. and Soares, M. N. (2020), 'AI in legal services: New trends in AI-enabled legal services', *SpringerLink*, 223–226. https://doi.org/10.1007/s11761-020-00305-x.

Kumar, N. (2021), 'Legal AI: An automated versus autonomous future', *Forbes*, pp. 7–9.

Kurzweil, R. (1990), *The Age of Intelligent Machines*, MIT Press, Cambridge, MA.

Kyriakidou, M., Padda, K. and Parry, L. (2017), 'Reporting robot ethics for children-robot studies in contemporary peer reviewed papers', in Aldinhas Ferreira, M., Silva Sequeira, J., Tokhi, M., Kadar, E., and Virk, G. (eds), *A World with Robots*, Intelligent Systems, Control and Automation: Science and Engineering, vol. 84, Springer International Publishing, Cham, pp. 109–117. https://doi.org/10.1007/978-3-319-46667-5_8.

Legg, M. and Bell, F. (2019), 'Artificial intelligence and the legal profession: Becoming the AI enhanced lawyer', *University of Tasmania Law Review*, 38(2), 34–59. https://www.lawsociety.com.au/sites/default/files/2020-11/Artificial%20Intelligence%20and%20the%20Legal%20Profession%20-%20Becoming%20the%20AI-Enhanced%20Lawyer.pdf

Makridakis, S. (2017), 'The forthcoming artificial intelligence (AI) revolution: Its impact on society and firms', *Futures*, 90, 46–60.

Manyika, J., Chui, M., Miremadi, M., Bughin, J., George, K., Willmott, P. and Dewhurst, M. (2017), *A Future that Works: Automation, Employment, and Productivity*, Technical report, McKinsey Global Institute. https://www.mckinsey.com/featured-insights/digital-disruption/harnessing-automation-for-a-future-that-works.

Markovic, M. (2019), 'Rise of the robot lawyers?', *Arizona Law Review*, 61, 326–349. https://arizonalawreview.org/pdf/61-2/61arizlrev325.pdf.

Murphy, R. and Woods, D. D. (2009), 'Beyond Asimov: The three laws of responsible robotics', *IEEE Intelligent Systems*, 24(4), 14–20.

Nilsson, N. J. (1998), *Artificial Intelligence: A New Synthesis*, Morgan Kaufmann Publishers Inc., San Francisco, CA.

Pagallo, U. (2013), *The Laws of Robots*, Springer Netherlands, Dordrecht. https://doi.org/10.1007/978-94-007-6564-1.

Paul-Choudhury, S. (2017), 'A robot tax is only the beginning', *New Scientist*, 233(3115), 25.

Remus, D. and Levy, F. S. (2016), 'Can robots be lawyers? Computers, lawyers, and the practice of law', *SSRN Electronic Journal*.

Rich, E. and Knight, K. (1990), *Artificial Intelligence*, 2nd edn, McGraw-Hill Higher Education.

Sahota, N. (2019), 'Will A.I put lawyers out of business?', *Forbes*, pp. 3–6.

Schellekens, M. (2015), 'Self-driving cars and the chilling effect of liability law', *Computer Law & Security Review*, 31(4), 506–517.

Shekhar Sarmah, S. (2019), 'Artificial intelligence in automation', *RESEARCH REVIEW International Journal of Multidisciplinary*, 4(6), 14–17. https://www.researchgate.net/publication/336085049.

Surden, H. (2019), 'Artificial intelligence and law: An overview', *Georgia State University Law Review*, 35(4), 1305–1337.

Tay, B., Jung, Y. and Park, T. (2014), 'When stereotypes meet robots: The double-edge sword of robot gender and personality in human–robot interaction', *Computers in Human Behavior*, 38, 75–84.

Wang, W. and Siau, K. (2019), AI, machine learning, automation, robotics, future of work and future of humanity: A review and research agenda', *Journal of Database Management*, 30(1), 61–79. https://www.researchgate.net/publication/333423274.

Winston, P. H. (1992), *Artificial Intelligence*, 3 edn, Addison-Wesley Longman Publishing Co., Inc., Boston, MA.

Yu, P. K. (2020), 'Artificial intelligence, the law-machine interface, and fair use automation', *Alabama Law Review*, 72(1), 187–238. https://papers.ssrn.com/sol3/papers.cfm?abstract_id=3665489.

2

Moving One Step Further: The Integration of AI-Based Mechanisms into the Indian Judicial System

Harsh Mahaseth and Saumya Pandey

CONTENTS

Introduction

Artificial intelligence (AI) is expected to stride towards the path of sustainable, effective, and speedy justice delivery with zero interference in the decision making. The Apex Court in India is exploring these newer technologies with an intent to adopt these futuristic technologies in its day-to-day process (*ThePrint*, 2019). Though the Court is exploring newer cognitive tools, it is made evident that these technologies will only be used to assist judges in their decision-making process and will never replace human judges and human discretion.

A significant part of lawyers' and judges' time is spent in legal research, analysis of facts, in search of appropriate legal provision, and doing such other similar set of work that mainly involves analyzing and deep research skills. Any assistance or aid in doing such work can prove miraculous in saving the Court's time and speedy justice delivery, which would subsequently reduce the burden of the Indian judicial system. Nowadays, there is a rapidly growing consensus that AI can transform and revolutionize human existence across the globe in a tremendous manner. The Pandemic has alleviated the talk on digitization and expediating the use of digital and electronic media in our judicial system so that the Pandemic does not take a toll on justice delivery. Taking a cue from this, it is expected that the deployment of AI in the Indian judicial system might improve the Court's functioning processes and bring in administrative efficiency. These cognitive tools can be deployed to assist in the decision-making process.

DOI: 10.1201/9781003215998-2

With the increasing use of AI in justice delivery, there is also lingering fear of its unethical use and a serious threat to privacy. Further, these futuristic technologies should not cause any harm to our constitutional and legal principles, which are inevitable for a democracy like India. In the context of the Judicial system, where a lot of information is involved, which sometimes are too sensitive that it is advisable not to make them public, in such instances, the use of AI or machine learning can pose serious concern, and hence, it's very important that a strict compliance is ensured and these competing interests of disclosure and privacy are balanced.

Background of the Situation in India

There are nearly 30 million cases pending in India with many more being filed every day. As mentioned by Justice VV Rao – it might take as many as 320 years for the Judiciary to clear backlogs (Ghosal, 2020). This pendency rate will only grow with time as the population is increasing and the disputes will increase accordingly. There are multiple causes for the delay like lack of infrastructure, loopholes in the law itself, inefficient method of police investigation, redundant court processes and voluminous paperwork, and one very prominent cause for the delay is the vacancy of judges. The number of judges per million people in India is 20.91, which is abysmally low and a great cause for concern (Deka, 2021).

As per the National Judicial data grid, which monitors the courts, there are around 37.7 million cases pending in India, with more than 3.7 million cases pending before the higher courts, district, and taluka courts, which is a cause for great concern (Vishwanath, 2021).

The problem of Shortage of Judges Which Leads to Delay in Delivery of Justice

The inadequate number of judges leads to a delay in justice delivery. It is such a sorry state of affairs that against the sanctioned strength of 1,080, less than 416 judges are appointed in high courts across the nation, which creates the largest vacancy and eventually delays justice delivery. It is quite disappointing to mention that more than 660,000 cases remain pending for more than 20 years and 131,000 cases for more than 30 years, making speedy redressal very critical for the Judiciary (Krishnan, 2020).

Another thing that has escalated this problem is the sudden closure of the courts due to the Pandemic, which was eventually resumed through virtual court hearings but it had a serious impact on lower courts that are not lashed with technology and were not ready for this shift towards technology.

Apart from all these issues, one very prominent reason for the delayed justice and backlogs is the inadequacy of judges. The Judiciary is going through a critical phase, and it is more important than ever to overhaul this defective system and fill the vacancy of judges but it is critical to note that the problem with the judiciary is more nuanced. The entire appointment process is made by the State Government and this whole appointment process is clearly very different from the collegium system of the higher courts. While the system involves a rigorous appointment process from holding competitive exams to promotions solely relying on Annual Confidential Reports and seniority, sadly, there are many loopholes that have crept in with the passage of time, which somehow have led to the hiring and promotion of incompetent people. Consequently, it will not be wrong to point out that we lack in talent at the lower judiciary.

"In the 2014 Delhi Judicial Services exam, for example, out of 1,000 applicants, only 115 qualified for the 85 posts available" (Ghosal, 2020), and this was not limited to just this, but it was more surprising to find out that 68 out of those 885 failed were serving as judges at the lower Judiciary in other states, respectively. The 115 qualified were seen to be in relation of judges that had been appointed prior. This clearly reflects a problematic state of the judicial system.

It is almost improbable for the lower court judges to reach the Supreme Court or High Court via promotion. The selected judicial members from the respective bars to the High Courts is quite high in comparison to their elevation from subordinate Judiciary.

The "*S.P. Gupta case and the Advocates on Record Association vs the Union of India*" has made it quite explicit that the service year spent at lower Judiciary will not count in for promotion to Supreme Court. It made it mandatory for all of the Supreme Court promotions to be determined solely on the basis of the candidate's High Court judicial experience, making the chance of appointment to the apex judiciary practically impossible.

Inadequacy of judges has worsened the situation and led to overburdening of the Court. It is crystal clear that with the passage of time, this problem will only aggravate as the number of litigants will increase with the growing population. Hence, it is very urgent for the Court to shift its focus towards newer technologies.

The Need for Assistance: Which Can Be Done through AI and What Can AI Do in This Manner?

The ethical and responsible use of artificial intelligence can be embedded in the judicial process to enhance its efficiency and provide administrative assistance. The Indian Judiciary, which has successfully created infrastructure for online functioning of the Court, has already realized the potential of these technologies, which can help the Court in significantly reducing the pendency. AI can be deployed in two major areas, enhancing the administrative process's efficiency and by augmenting the decision-making process, both of which would significantly ease the court process and lower the burden on the Court.

To increase the administrative efficiency, the steps can be taken to deploy AI, which runs on task-specific and precisely designed algorithms to do specific and routine tasks such as scheduling hearings and creating cause lists, to complex tasks like discovery and review of evidentiary documents. Likewise, other tasks that involve complexity could also be achieved through careful programming and algorithms, for instance, smart e-filling, prioritization of cases or notifications, and to help in tracking of cases.

AI can be smartly designed and deployed in research work, which can eventually help the judges decide the cases as precedents play a major role in the Indian Judiciary. The decision of the Supreme Court is binding on all other courts and hence a lot of time is spent on research before any decision is delivered. The integration of AI in the judicial system will increase administrative efficiency and help in the speedy delivery of justice. Further, computational tools can be adopted to enhance the justice delivery such as those used in traffic challans and motor vehicle compensation claims.

Supreme Court Portal for Assistance in Court's Efficiency (SUPACE)

The Supreme Court has launched an AI interface, Supreme Court Portal for Assistance in Court's Efficiency (SUPACE), to assist the Court in administrative work and to ease their workload. It is designed to help judges decide cases efficiently by bringing in all facts and data, which will automatically reduce the time taken by a judge in deciding a case. SUPACE will be a blend of AI and human intelligence. In 2019, under the chairmanship of Justice L. Nageswara Rao, an AI Committee was formed as a strong need was felt for the introduction of cutting-edge new age technology into the judicial system to reduce its burden and enhance its efficiency.

The AI committee of the Supreme Court has well recognized the potential of new age technology in processing information effectively. SUPACE is a blend of AI and human intelligence. It will not have any effect on decision making but it would play a crucial role in bringing in all relevant facts and processing all information as the primary duty of judges is to carefully look into all facts available on the facts in issue in any particular case and for that they have to check for precedents on that line which is a tedious process. The role of AI gets pertinent here more than ever as it helps in processing information, comparison of evidence, contradiction, and there is a research option available that helps in finding judgments at the click of a button.

> SUPACE is one-of-a-kind solution that is fully customizable and behaves uniquely like its user. The AI within it adapts and adopts user behaviour based on incremental usage of the platform. This is one of the first examples of mass customization in the world.
>
> **(Mehra, 2021)**

SUPACE is believed to be a fully customizable app and behaves uniquely like its user. The AI involved in this portal adopts and adapts according to its user's requirement and functions accordingly. It is said to have the potential of augmenting the efficiency of the legal researchers and judges to work on cases, extract relevant information, read case files, manage teamwork, and draft case documents. This is believed to be one of the brilliant examples of mass customization in the world.

The Supreme Court has adopted machine learning and AI to deal with huge chunks of cases pending in Indian courts. SUPACE is a blend of human intelligence and futuristic technologies which will assist the judges in their routine work by reducing their workload and increasing their efficiency. It has been clarified by the honorable chief that the decision-making process will be sole responsibility of the judicial minds, with only difference being that AI will provide assistance with any knowledge, be it of law or facts, which a judicial mind needs for forming their final decision. The only crucial difference SUPACE is expected to create is speedy justice delivery by making all information and case precedents readily available. The Supreme Court has laid down a strong foundation for this futuristic technology in accelerating court processes and enhancing efficiency. The ethical and responsible use of these cutting-edge new age technologies can be embedded in the judicial set up for its advancement and efficiency enhancement without causing any intrusion in the decision-making process.

Supreme Court Vidhik Anuvaad Software (SUVAS)

The Indian Judiciary has already created basic infrastructure under its e-court project to overhaul the judicial system and is now looking to deploy machine learning and artificial intelligence in its day-to-day court process. Recently, in the year 2020, all the 17 branches of Supreme Court have become paperless. The Court has also developed few other tools with the help of these futuristic technologies such as SUVAS and SUPACE to ease the court process.

SUVAS is a tool developed to help in translation. In fact, SUVAS was the pioneer of a task-based algorithm which was employed in judicial system. The main function of SUVAS is to process natural language, expediting translation of court orders and judgments. On November 25, 2019, SUVAS – "Supreme Court Vidhik Anuvaad Software" was launched at the commemoration of Constitution Day (*Press Release- Supreme Court of India*, 2019).

> SUVAS is a machine-assisted translation tool trained by Artificial Intelligence. This Tool is specially designed for Judicial Domain and has the capacity and capability to translate English Judicial documents, Orders, or Judgments into nine vernacular language scripts and vice versa. This is the first step towards the introduction of Artificial Intelligence in the Judicial Domain.
>
> (*Press Release- Supreme Court of India*, **2019**)

SUVAS is relied on currently for translating Supreme Court judgments into nine vernacular languages – *"Assamese, Bengali, Hindi, Kannada, Marathi, Oriya, Tamil, Telugu, and Urdu"*. In the coming time, this software will not be restricted to these few languages but will be extended to translate other languages. Currently, cases related to Labour laws, Rent Act, Land Acquisition, Service law, Compensation, Criminal law, Family Law, Civil procedure code, Personal Law, Religious and Charitable Endowments, Mortgage, Eviction under the Public Premises (Eviction) Act, Land Laws and Agriculture Tenancies, Consumer Protection Act are being translated with the help of these software in preferred languages. The use of software will sooner be relied upon by the High Courts and district courts as well after the amazing response received at the Apex Court.

SUVAS is based upon neural machine translation (NMT) technology, making it a typical translation software "to translate words from one language to another. In the case of translation, each word in the input sentence (e.g English) is encoded as a number to be translated by the neural network into a resulting sequence of numbers representing the translated target sentence (e.g Bengali)".

This is certainly an amazing head start for legal-technological advancement, which will help ease the Court's functioning and make the orders and judgments readily available in one's vernacular language. The AI-enabled software will provide assistance to the persons seeking justice and simultaneously make courts more efficient.

How Can AI Help the Judiciary and the Challenges Ahead?

The integration of artificial intelligence in the Indian judicial system is paving way for faster justice delivery. AI will automate and complement several tasks. The legal professionals will be able to devote their time in solving more critical issues innovatively and the judges will be able to conduct trials at a faster speed and take swift decisions, which will reduce the burden of the Court eventually. Artificial intelligence has the potential to ease the judicial process and assist legal professionals in developing advanced legal reasoning, augment decision making, and assist in the interpretation of laws.

However, the foremost thing expected from the AI committee of the Supreme Court is to chart out the long, medium, and short-term usage and goals they need to accomplish by integrating AI in the Indian judicial system. They need to clearly establish the ethical rules for deployment of AI in the Indian judicial system so that it doesn't create any sort of ruckus at any point of time. To achieve the two-pronged objective of administrative efficiency and assistance in the decision-making process with zero interference, it is highly important that the deployment of AI takes place in a phased manner.

The Pandemic has brought major changes in the functioning of Judiciary, causing most of the judicial work to shift online. It necessitated the digitization of the courts through the e-court project, created virtual courts, and paved way for online dispute resolution. With this advancement taking place, AI has also been gaining recognition. It is significant to deploy futuristic technologies such as AI to increase administrative efficiency by automating the daily process of the registry.

To do so, we need to develop task-specific, narrowly customized algorithms that can be used to help run the administrative system more efficiently, from simple operations like scheduling hearings and compiling cause lists to more sophisticated duties like discovery and document review (Brickell, 2020). Interventions at the level of smart e-filing, intelligent filtering/prioritization of cases or notifications, and case monitoring are all small chores that can benefit from AI.

In India, the preliminary work in the use of AI has already commenced. SUVAS was the pioneer of such task-specific algorithms, designed by the Supreme Court's AI Committee. It relies on natural language processing (an ML process), easing and expediting translation of judicial orders and rulings (Sengupta et al., 2021). Additionally, as was announced last year, the Supreme Court (SC) AI Committee is also working on a composite new tool named SUPACE (Supreme Court Portal for Assistance in Court Efficiency) (Shekhar, 2021), which will target different processes like data mining, legal research, projecting case progress, etc. (Shanthi S, 2021). There is also in-house software being piloted in the 17 benches of the Supreme Court to make them paperless (Singh, 2020).

AI-designed bots are becoming increasingly ubiquitous across different sectors like insurance, banking, e-commerce, etc. Bots are convenient and interactive tools for providing common information to a user in a conversational format. With respect to the justice system, legal robotics can play a crucial role in serving as intelligent and dynamic repositories of FAQs, which aids the public's understanding of laws. This would be extremely useful for common citizens and potential litigants in getting basic inputs on a prospective legal case, and making better-informed decisions, inter alia whether litigation is needed or not. In addition to providing better information, legal robotics can also improve access to legal services. Accessing these or even grappling with a potential legal situation can be a daunting conundrum for a common person. Intelligent algorithms (or bots) can be useful in furnishing basic legal information to potential litigants and readily connecting them with legal aid services or pro-bono lawyers. Basic legal services like drafting and conveyancing, legal analyses and interactive breakdown on laws, etc., can be some modes for mainstreaming access to such services, without the trouble of locating and paying for expensive lawyers.

Given the growing interest of governments and public institutions globally, its application and governance have become a crucial talking point to harness the transformative powers of artificial intelligence. In order to develop trust in AI-driven technology in the Indian court system, a thorough legal, regulatory, and ethical framework is required. However, the integration of AI poses significant challenges as well. It is imperative to ensure that this process of integrating AI into the judicial system is concomitant with our country's constitutional and ethical principles (Sengupta et al., 2021). AI can perpetuate biases either intentionally or unintentionally posing threat to the fairness and impartial judicial system. These technologies are designed on heavy datasets which carry enormous possibility of biasness as present in the original dataset. In fact, there always exists a chance of personal biasness as can be induced by the developer of the algorithms. On the contrary, as advocated by the NITI Aayog, this might be avoided by establishing an oversight organization to establish norms, rules, and benchmarks for the usage of artificial intelligence across sectors, which will be necessary for public sector procurement. Field experts from computer science, artificial intelligence, legal experts, sector specialists, and representatives from civil society, humanities, and social science are likely to make up the body (Sharma, 2020). The main work of this body would be to maintain vigilance and to prohibit any kind of biasness from creeping in.

Further, the use of AI in the judicial system can be two-fold. Firstly, they could be a support system in decision making, and secondly, they could surpass the human mind and render decision on their own. The Indian Judiciary has repeatedly clarified that the use of AI in the judicial system will be limited to rendering support to the judges and they will never be used to supplement the human mind. Therefore, the use of AI in the judicial system is merely in the form of cognitive tools.

Also, it is pertinent to note that it has been found in research that the use of computer was always meant to reduce the efforts of the decision-making process and not to merely act as a support system in improving the quality of their decision, so it would not be a completely false belief if it is said that there are possibilities that the integration of AI in the judicial system will meddle with the decision-making process. And if happens so, then it will not improve adjudication but will only worsen the situation (Dymitruk, 2019).

Another doubt which arises is the chances of malfeasance on the part of the developer of these algorithms who can use suboptimal technologies to supplant human judgment considering the newness of this technology. Hence, it is essential for the judges to keep their eyes and ears open while deciding cases and shouldn't shy away from deviating from these algorithms if it seems justifiable to their conscience.

An excessive reliance on such systems may result in legal issues being decided by computer programs despite there being an impression that all principles of the human adjudicating process are obeyed. As a result, AI-driven technology can completely alter the very nature of how we identify judicial processes. These tools have the capacity to become prescriptive, potentially overshadowing case-specific reasoning, and instead reduce judicial decisions to purely statistical and algorithmic outcomes.

Adding on to this is the problem of Judiciary reaching a point of stagnation. Indian Constitution is a vivid example of how law should keep changing with the time so as to keep with the changing societal beliefs, values, and dynamics. In an AI-centric judiciary, while fostering the rule of stare decisis, in pursuit of consistency, it is a plausible fall-out that the precedents become stagnant. For instance, the right to privacy has been read into Article 21 of the Indian Constitution, after almost three decades of conflicting jurisprudence. Had the same case been decided by an AI adjudicatory tool, designed and trained on the same conflicting jurisprudence, it would have foreseeably reinforced the same principle and legal disposition. Therefore, as AI increasingly becomes enmeshed within the justice system, and even aids in the judicial decision-making process, it is vital to retain humans within the loop. Human oversight and discretion are needed to complement the efficiency of intelligent decision-making tools, to prevent any unfavorable value lock-ins. Moreover, there are opinions that the AI-driven judicial system may fundamentally alter its constitutional role especially as an institutional checks and balances against executive and legislative overreach (Michaels, 2020, p. 88).

Also, it would not be wrong to ask if an AI-driven technology can act like a human judge of constitutional courts whose job is much more than rendering decision. It is noteworthy that the working of constitutional courts is complex and serious in nature, requiring a weighing of law and facts, tempered with reasoned discretion, to balance competing interests.

It is seen quite often that a judge has to act rather innovatively in deciding a case involving crucial interpretation of legal principles and has to differ with her predecessor to balance the present need of the society. This ability of judges to deal with such intricate and complex issues does not come handy but it is acquired through severe hard work and persistence, which again is a result of year long experience gained due to the exposure they get by holding that particular position. Though it has been clarified by the Chief Justice that the use of AI will be limited to offering assistance, but the possibility of AI meddling with the judicial decision making with the increasing use of AI cannot be completely ruled out; hence, it is vital than ever to retain human mind within the judicial loop under tough screening so as to not lead to any kind of interference in the decision making and keep the sanctity of the Court intact.

A complete transference of judicial functions over to AI will certainly face the challenge of how this technology will perform the entire spectrum of roles and obligations that are presently required of human judges, but with the adoption of these technologies in the judicial system, they are sure to increase the institutional efficiency and reduce the burden on Judiciary by providing them assistance and saving their time by doing clerical work which is usually the most erratic process.

The foremost exercise that needs to be undertaken by the Judiciary is to clearly determine the road lying ahead for AI-enabled judiciary. It is no doubt that the road ahead is full of doubts related to privacy infringement and pose serious concern of data protection, but it is also contended that with great self-regulation by developers and external regulation by parliament through statue, rules, and laws, these concerns can be easily resolved with the passage of time. It is surely expected to create an everlasting impact on the Judiciary in the longer run by easing down the court process and removing pendency of cases. It is supposed to overhaul the entire judicial system by increasing the overall efficiency of the justice system.

Conclusion

There is a need to amplify justice delivery. The use of AI-based mechanisms will help in improving institutional efficiency. It will also provide legal assistance in improved decision making, transforming the justice system in India. AI-based mechanisms are efficient tools for intelligent legal analytics and research. It can ease the access to justice as legal robotics can play a crucial role in serving as intelligent and dynamic repositories of FAQs, which aids the public's understanding of laws.

References

Brickell, J. (2020, January 31). AI-enabled processes: And you thought E-discovery was a headache! *New York Law Journal.* https://www.law.com/newyorklawjournal/2020/01/31/ai-enabled -processes-and-you-thought-e-discovery-was-a-headache/.

Deka, K. (2021, February 8). On India's judiciary: Bogged by a backlog. *India Today*. https://www.indiatoday.in/magazine/nation/story/20210208-bogged-by-a-backlog-1763840-2021-01-30.

Dymitruk, M. (2019). Ethical artificial intelligence in judiciary. *Internet of Things – Digital Edition of Proceedings of the 22nd International Legal Informatics Symposium 2019*. https://jusletter-it.weblaw.ch/issues/2019/IRIS/ethical-artificial-i_a54e474060.html__ONCE&login=false.

Ghosal, S. (2020, January 8). Why India's courts are struggling to find judges. *Business Standard India*. https://www.business-standard.com/article/specials/why-india-s-courts-are-struggling-to-find-judges-116061600898_1.html.

Krishnan, M. (2020, June 29). 3.7 million cases pending in courts for over 10 years: Data. *Hindustan Times*. https://www.hindustantimes.com/india-news/3-7-million-cases-pending-in-courts-for-over-10-years-data/story-ytI7P0rm5Plwe5r8ubNVyJ.html.

Mehra, S. (2021, April 7). AI is set to reform justice delivery in India. https://indiaai.gov.in/article/ai-is-set-to-reform-justice-delivery-in-india.

Michaels, A. (2020). Artificial intelligence, legal change, and separation of powers. *University of Cincinnati Law Review*, *88*(4), 1083.

Press Release- Supreme Court of India. (2019, November 25). https://main.sci.gov.in/pdf/Press/press%20release%20for%20law%20day%20celebratoin.pdf.

Sengupta, A., Jauhar, A., & Misra, V. (2021). *Responsible AI for the Indian Justice System – A Strategy Paper*. Vidhi Centre for Legal Policy. https://vidhilegalpolicy.in/research/responsible-ai-for-the-indian-justice-system-a-strategy-paper/.

Shanthi S. (2021, May 29). Behind SUPACE: The AI portal of the Supreme Court of India. *Analytics India Magazine*. https://analyticsindiamag.com/behind-supace-the-ai-portal-of-the-supreme-court-of-india/.

Sharma, Y. S. (2020, November 17). NITI Aayog wants dedicated oversight body for use of artificial intelligence. *The Economic Times*. https://economictimes.indiatimes.com/news/economy/policy/niti-aayog-wants-dedicated-oversight-body-for-use-of-artificial-intelligence/articleshow/79260810.cms.

Shekhar, S. (2021, May 7). Supreme Court embraces artificial intellegence, CJI Bobde says won't let AI spill over to decision-making. *India Today*. https://www.indiatoday.in/india/story/supreme-court-india-sc-ai-artificial-intelligence-portal-supace-launch-1788098-2021-04-07.

Singh, A. (2020, May 26). Supreme Court develops software to make all its 17 benches paperless. *The Economic Times*. https://economictimes.indiatimes.com/news/politics-and-nation/supreme-court-develops-software-to-make-all-its-17-benches-paperless/articleshow/75989143.cms.

ThePrint. (2019, November 27). 'AI can improve judicial system's efficiency'—Full text of CJI Bobde's constitution day speech. *ThePrint*. https://theprint.in/judiciary/ai-can-improve-judicial-systems-efficiency-full-text-of-cji-bobdes-constitution-day-speech/326893/.

Vishwanath, A. (2021, March 27). Pandemic impact: Record pendency of cases at all levels of judiciary. *The Indian Express*. https://indianexpress.com/article/india/pandemic-impact-record-pendency-of-cases-at-all-levels-of-judiciary-7247271/.

3

Using Artificial Intelligence to Address Criminal Justice Needs, Problems, and Perspectives

Niteesh Kumar Upadhyay and Snizhana Romashkin

CONTENTS

Introduction

Every significant technological breakthrough has the potential to either advance or harm society. The technology measurement and analysis capacities of artificial intelligence can very well help in resolving some of the biggest transportation, humanitarian assistance, and public health problems, including supporting significant changes in clinical diagnosis and treatment, integrating transport and communal growth, and exacerbating the dangers of climate change.[1]

The criminal justice system, including courts and court management, is overburdened with lakhs of pending cases, and it requires new solutions, especially which are AI-based in order to reduce the pendency in the courts and delay in justice. Most of these AI-based technology solutions and companies are owned by private enterprises, and delivering the justice delivery system in the hands of private players will be a huge debate that we need to take into consideration first before its implementation.[2]

In the recent past, we have seen the use of AI for many functions including surveillance by police, online dispute resolution, victim and criminal profiling, and also extensively used in crime prevention. The AI-based devices are used by law enforcement agencies to

DOI: 10.1201/9781003215998-3

penetrate deeply into the preparatory stage of the crime and also AI is also used to inquire about the crime which has already been committed. AI assists the law enforcement agencies by analyzing a large amount of data including live videos and in deciding the probability of any crime and crime prevention. AI reads the trends, looks, and personal details of people who are more prone to become a victim or commit a crime, and after analyzing the pattern it suggests the action that can be taken to prevent the crime.[3]

The idea of artificial intelligence was lifted from the scientific realm to conversations in the upper levels of science, business, and government. However, specialists only just commenced to consider the effect of intelligent machines on human rights, and since then, people really do not seem to conform about what the result means.[4]

Currently, humans might see major technical differences in the way vulnerability analyses are carried out. Assessments are given between algorithm-based, big data analysis (accompanied by artificial intelligence). It is argued that the process establishes a new boundary of precision, to the point that it can also eradicate certain types of bias.

Though artificial intelligence is commonly used within the criminal justice system, the use of "clever" technological tools in law enforcement is still at the initial development stage and has already met with resistance, evaluating their pros and cons. In most cases, such AI tools are used as "preliminary risk assessment" methods in almost every country. Criminal justice algorithms applied as "risk assessments" or "proven-based methods" are intended to forecast the conduct of defendants.

It is quite often that these specialized methods are used to set bail, to determine sentences, guilt, or innocence. However, in most cases, these methodologies are hidden from the public. The hidden methodologies create suspicion in the mind of the people about their objectivity and impartiality. Many non-supporters of AI say that these AI tools work with machine learning, and the data fed into these algorithms are encoded with human bias and discrimination will remain in AI also.[5] The main argument for such kind of behaviour by AI is that human beings have common sense and knowledge to understand the world, which enables the humans to adapt and decide as to what is just and unjust during any new situation, whereas AI completely lacks such common sense and hence cannot be much relied.[6] The biggest concern about utilizing AI in the criminal justice system is that humans are not sure that AI will share the principle of respect for human rights.[7]

Artificial Intelligence and the Definitional Dilemma

Let us first determine the definition of artificial intelligence. According to the definition provided by Marshal S. Willick, he determines AI as "the capability of a device to perform functions that are normally associated with human intelligence, such as reasoning, learning and self-improvement". The author also indicates that the definition always includes four categories, i.e., AI technology, AI simulation, AI modelling, and AI theory.[8]

In 1955, theoretical physicist McCarthy characterized AI's objectives as "to create machines that behave as though they were intelligent". As of today, we know that this definition is not acceptable to the demands of modern technology.[9]

Recently, Stanford University, in its report, defines artificial intelligence as "a science and a set of computational technologies that are inspired by – but typically operate quite differently from – the ways people use their nervous systems and bodies to sense, learn, reason, and take action".[10]

Stuart Russell and Peter Norving combine the numerous specialized definitions of artificial intelligence into four distinct categories of thought systems and human attitudes and define them as systems that think like humans; act like humans; think rationally, and act rationally.[11] As All sop CJ observed, the scale of automation may differ from "decision-support" to "human-in-the-loop" and the complete absence of persons starting the decision-making process.[12]

Use of Artificial Intelligence Technology in the Criminal Justice System

Michael E. Donohue's article notes that the methods that have been used to calculate the risk that the defendant would again break the law were first used to determine how the criminals were to be released in the 1930s. After that, the scientists developed four generations of indicators for evaluating recidivism risk, called the "Correctional Offender Management Profiling Tool for Alternative Sanctions" (COMPAS), the "Level of Service Inventory-Revised" (LSI-R), and these tools depended on the services of a professional assessment officer. The officer will then gather data (social network, family background, and neighbourhood) on the suspect and perform an interview and produce a risk ranking.[13]

Artificial intelligence is now widely used for radiological images and helps medical staff to predict the cause and manner of death of any person. Artificial intelligence is not only used for radiological images but it is also used for forensic science including DNA analysis.[14] In the past decade, artificial intelligence was used extensively for the purpose of video and image analysis, which will help the criminal justice system. Artificial intelligence works very efficiently in image and video analysis and also chances of error are less, which makes it more favourable for this work than human beings.[15] Traditional software that assisted the criminal justice system was limited to predetermined features such as eye shape, eye colour, and facial attributes or demographic information for pattern analysis.[16] The modern AI algorithms can learn to develop and determine by itself the complex facial recognition feature to identify weapons, objects and accidents, and crimes; detect patterns of anomalous behaviour; predict the behaviour of a crowd, and protect critical infrastructure.[17]

Artificial Intelligence and Risk Assessment

AI technology has optimum potential to provide help to the law enforcement officers in informing about any terrorist attack or any other dangerous attack against civilians or civilian and state property. AI technology in robots and drones can help in risk assessment and prevention by performing safety surveillance.[18] Risk assessment algorithms prepare details of criminal profiles, behaviour patterns, and movements and will try to find out the likelihood of committing the crime and will further inform law enforcement officers, which can lead to the prevention of the commission of the crime. The risk assessment algorithms prepare a recidivism score and help law enforcement. These risk assessment tools are also useful for the judiciary in terms of deciding and determining the type of rehabilitation facilities suitable for a particular criminal, whether bail is to be granted to them or not and how the low recidivism scale criminals are treated leniently, whereas criminals accused with a high recidivism score bear harsh punishment and measures. In the last few years, many countries around the globe have invested on risk assessment tools in their criminal justice systems and hope to see more AI tools related to risk assessment in future.[19] Some of the crime prevention and risk assessment tools that are widely used are *ShotSpotter,*[20] *Hikvision,*[21] *PredPol,*[22] etc.

There is another side of these risk assessment tools also, which can be termed as discriminatory toward a particular community or group. As machine learning uses statistics to find patterns of data, if historical crime data is fed into AI, it will create bias as it will pick patterns in associated crime and the chance of it picking the same community or class or nationality of people will be high and on a higher recidivism score.[23]

Artificial Intelligence and Crime Prevention

As we have already gone through AI as a tool for risk assessment, now in this part, we will discuss how AI helps in the prevention of crime and criminal activities. With the advent of AI technology, surveillance of areas prone to crime or terrorist attack is always under a scanner and any suspicious or malevolent activity is reported to law enforcement.[24] Video analytics with integrated facial recognition and voice recognition can detect criminals and civilians at multiple locations with the help of multiple AI-based cameras and circuits and help movement and pattern analysis, identify crime and criminals, take a record of criminals' licensed number plates and other details like tattoo and facial expressions, help law enforcement in suspect identification, etc. AI-like shot-spotter not only detects the location of gunshots but also determines shot timings and the number of firearms used and estimates the class of weapon and the calibre of the gun used, and all of these factors will help law enforcement officers to catch the criminals and help during the investigation.[25] AI generates a massive volume of data, which helps it to detect crimes which otherwise may go unnoticed, hence helping law enforcement in ensuring larger public safety, crime prevention, and robust criminal justice systems.[26]

AI can also perform functions like intruder alarms or alarms at places in order to deter the criminals before the commission of the crime itself and also inform law enforcement that a crime might be about to take place. For example, person "A" is going to kill person "B" by using a gun and the AI, after looking at the past records, movement history, and threat analysis, informs law enforcement officers about the same or alarms the person "B" that he might face a deadly attack.[27]

Interpol in Europe manages the International Child Sexual Exploitation image database (ICSE DB), which helps to identify and fight child sexual abuse including paedophiles.[28] This AI works by recognizing and analyzing furniture, carpets, curtains, and room accessories in the background of an abusive image or video, even background noise in the video, and if things match with perpetrators of crime or victims, it gives an alert and helps prevent child sexual abuse by the paedophiles.[29] Child sexual abuse by paedophiles is also controlled by a new technique that we know as chatbots; these chatbots act as a real person during chat and find about perpetrators of webcam and chat-based sexual abuse. Sweetie is a virtual character used by the Dutch children's rights organization Terre des Hommes for curbing webcam sex and webcam sex tourism. Sweetie posing as a Filipino child identifies abusers in chatrooms and online forums and finds out people who contacted the virtual character and asked Sweetie for webcam sex.[30]

AI will help in many ways in crime prevention by uncovering the criminal network, modus operandi of work and area and location of the crime and identifying probable victims and criminals.

Artificial Intelligence and Evidence Gathering

AI algorithms are not limited to risk assessment and crime prevention but are also really helpful in evidence gathering as well. AI can find out not just traditional evidence like DNA, semen, saliva, and blood but can also gather digital evidence that relates to the crime. AI can track the use of mobile phones, the Global Positioning System, and computers to gather additional evidence at any crime scene, which will be very useful for our justice delivery system.[31]

Forensic scientists can use various AI tools, which are made for crime scene investigation, and these AI can help in the discovery, recovery, recording, capturing, analysis, storage, and transmission of evidence through lab-on-chip technology. AI will be more reliable in evidence gathering than human beings as the chances of bias because of undue influence and coercion by criminals, politicians, and bureaucrats will not be there. As we have seen in many cases, because of the involvement of high-profile people involved in crime, tampering with evidence happens and will create a lot of problems for law enforcement and justice delivery mechanism, but these problems will not be there if AI is used for the purpose of gathering evidence.[32] AI can perform all kinds of simple and complex tasks like identification of victims and criminals, fingerprint capture, audio and video recording, face, number plate, and weapon detection.[33]

Artificial Intelligence as a Tool for Justice Delivery Mechanism

The Ministry of Justice in Estonia is developing a robot judge who will adjudicate petty disputes of valuation less than 7000 euros among parties. This AI robot will collect the documents uploaded by both parties, and after perusal of the documents, it will decide the matter on merits and accordingly will decide which party is liable to pay and how much and the same is to be paid to whom. In the Estonian justice system, the appeal from the decision of a robot judge can be filed with a human judge.[34]

Petty crimes (including incidents like theft, criminal damage, and anti-social behavior) in Europe happen at volume, also known as "volume crime", and affects almost everyone including civilians, business owners, and infrastructure owners. The P-REACT project is designed as a petty crime incident surveillance platform.[35] This AI algorithm is embedded with video and audio sensors for surveillance of petty crime incidents and will provide an alert detection after going through the video and audio analysis.

AI methods are also widely used in private dispute resolution, especially in online transactions. Many countries worked and still working on the suggestions to use AI methods for public services, for instance, the "automatic online conviction" process in the United Kingdom, Estonian "robot judge" for small claims.[36]

Some other element that needs to be considered is the effect of artificial intelligence and similar techniques on harm assessment. Most AIs are designed to learn about their environment by accessing a big quantity of data.[37]

The key challenge of introducing co-robotics in the judiciary is to facilitate the functioning of communication between humans and machines. The use of AI in the judiciary can help mitigate the impact of external factors such as fatigue and emotional instability. After all, intelligent decision making may reflect various human-made, functional systems that arise from the legal system, AI training information, or cyber security web development itself.[38]

There are two approaches to the challenge of co-robotics in the judiciary. In order to retain human control, one has to either enable the functioning of communication between humans and machines or strictly separate them from each other.[39]

As Thomas Julius Buocz also in his research mentions that it is especially important to distinguish between transparency regarding the general functioning of a machine and an explanation of a specific machine-generated decision. The main questions which arise are how to avoid problems between human judges and co-robotics decisions and what happens if the results are different.[40]

As has been described by Matilda Claussén-Karlsson in her research that such human mistakes may be the same in the AI mechanism. In order to comply with John Stuart Mill's Harm Principle, the State is entitled to monitor and control the individual's behaviour at risk of injury or death to everyone else. Legal moralists claim that the state can, even in the absence of injury, use criminal law to prevent immoral wrongdoings. Moral standards could be said to be a non-judicial element that influences the legislation. If information technology can benefit humans, it is important not to limit the advantageous possible outcomes that arise. There are still consequences of significant harm.[41]

In addition, in certain cases, the various defined actors may be considered to be morally wrong when they do not participate in the AI commission of a crime. However, where the action is neither harmful prima facie nor distant, the moral error should not be used as a sufficient excuse for the State to interfere in the autonomy of the person.

It must be recognized that the use of artificial intelligence is meant to be efficient, particularly customer-based and network-based, in order to provide access to the law. As a counter-example, there is the expanding and controversial use of AI to support judgments based on re-offending and historical information; the objective of this program is to stimulate punishment and close the gap that exists within the judicial system.[42]

Artificial intelligence and Automation of Prisons

AI can help in providing vocational training to jail inmates that could help them re-enter society successfully after their jail period is over. Artificial intelligence techniques also pose serious questions related to human rights like the right to privacy and data protection, equality, prohibition of discrimination, and effective remedy and access to public services and social rights. All human rights and freedoms are interrelated, and hence all kinds of human rights are impacted by the use of artificial intelligence.

Since the year 2018, the Punjab Police has been using the artificial intelligence system, namely Punjab Artificial Intelligence System (PAIS), which was developed by Staqu. This AI is well equipped with basic features such as face search, text search, and video search and has a database of around one lakh criminals housed in different jails in the State of Punjab.[43] AI in jails can prevent jailbreaks, smuggling of drugs, weapons, mobile phones, etc. As we all know, many criminals operate even from behind bars and AI can keep surveillance of such criminals and can prevent the crimes in jails and protection homes. By using AI in prisons, crimes like murder inside prisons, torture by police officials, and gang fights inside jails can also be controlled. AI tools can be used in jails to give an alarm in case anything illegal or suspicious is flagged by the AI device.[44]

Artificial Intelligence as an Important Tool Used for Digital Forensic

AI can perform a very important function in current times, and that is of digital forensics. Nowadays, every person irrespective of whether he is poor or rich carries cell phone(s), laptops, and other devices like memory cards, etc., which may contain very important data related to the commission of any crime, and AI can help recover that data even after the data is deleted or destroyed. AI can perform the job of extracting, analyzing, and storing the digital evidence found in any electronic device. AI can, with ease and in a fraction of minutes, find any data which is incriminating and can store it, and the same can also be transferred to a lab or court as per need.

Major Challenges of AI in the Sphere of Criminal Justice

It is necessary to understand that the use of AI is intended to be beneficial, user-based, and network-based in order to increase access to justice.[45]

The essence of crime has been revolutionized by the development of AI. For example, cybercrimes can only be committed technically by measuring the means and ends of a crime, such as malicious software or data loss. Cybercrimes are common offences that can be pared back or repeated using modern technologies such as fraudulent activity or money laundering. Digital money trafficking, drone monitoring, data security, "dark web" marketplaces for counterfeit products, new technology-enabled abuse, and intimidation are all problems where algorithmic solutions are candidates for necessary and proportionate actions to maintain the rule of law.[46]

The suggestion is whether a potential alternative would be a framework to implement AI criminal responsibility within such a program that recognizes only the state of the actus reus when determining a felony and that appears to be inappropriate in the context of the basic concepts of criminal law. We agree with the view that in such a case, where mens rea is excluded, it would be similar to involuntary acts that exclude liability.[47]

"New offenses" regarding the illegal use of AI and robotics must be carried out. On the one hand, AI systems may become "victims" of crime, and it is likely that new concepts and guidelines will be required to cope with similar circumstances.[48]

For example, artificial intelligence can be disrupted by third parties in such a way as to prevent such systems from achieving their objectives and/or cause them to commit a crime. One might think of people who purposefully interrupt the software of self-propelled vehicles, thereby causing an accident that was completely beyond the control of the developer and the consumer of the car.[49]

Also, there are various documented cases of AI which have gone erroneous in the criminal justice system. The application of AI in this way often occurs in two separate parts: probability scores assessing whether or not a convict is likely to recur in order to recommend rehabilitation and bail or so-called "predictive police" using knowledge from various data points to forecast when or where crime will arise and to direct effective law enforcement action.[50] The usage of equity-scoring for accusers is advertised as the avoidance of the perceived moral mistake of judges in their sentence and bail decisions. Yet predictive policing initiatives seek to allow effective use of often-limited police services to prevent violence, though there is also a significant likelihood of misappropriation.[51]

Another challenge may be discussed in the context of criminal responsibility. As a result, there are several types of AIs, but a few similar characteristics will be shared by them: unaccountability, unpredictability, and autonomy. Such aspects are also the major reasons for the question of liability. Unpredictability, along with discretion, restricts possible claimants to those who have an obligation to act and, as a result, responsibility cannot be defined in certain cases for actors who should be liable.[52] The major reason for this issue is the absence of a valid trigger when the AI operates efficiently without requiring any real individual. When the AI operates fully autonomously, there is no defined causal chain between the complainant and the AI until the implementation or usage of the AI itself is dangerous.[53]

Protecting Privacy in an AI-driven World

Artificial intelligence gathers a lot of data about people, their movements, buying choices, and their interaction with the outer world, and almost everything is under surveillance. AI surveillance and tracking pose a serious threat to the privacy of the public at large. A well-reported case of such a privacy breach is the case of a pregnant teenage girl who received coupons for products related to pregnancy from a retail chain and the most shocking part of the whole event is that she never disclosed it to anyone, not even to her own family, that she is pregnant.[54] The retail chain used a big data analysis technique to identify the customer who was pregnant and required pregnancy-related products. The big data analyzed the purchasing pattern of the teenage girl and found that she is buying skin care and health supplements which pregnant women take, and this way the AI predicted her pregnancy. This way AI can be really dangerous to the privacy of people, which is one of the fundamental rights given to citizens in most of countries.[55]

Artificial Intelligence and Discrimination

As we are aware that human beings are full of bias, which could be based upon various grounds related to race, religion, caste, colour, sex, and personal and political preferences, and it is a notion that AI will not have such kind of discrimination, but in reality, AI machines can also foster discrimination because of machine learning and human intervention at the stage of development of AI. AI learns from the data that is fed into it; however, if the data fed is biased in any way, AI will also show and reflect bias in its nature. For example, an automated analysis tool for job applications may be biased in choosing men over women or choosing people of one particular state over other. One more such example of bias is word embeddings, which will associate certain words to a particular gender, like nurse will be a female and doctor will be male; also, the word "commander" will be associated with the male gender.[56]

During the last few years, chatbots are used by almost all types of industries, including weather transport, hotel, IT, and even the judiciary. These chatbots can easily reply to any question that you have in your mind and are pre-programmed based on AI. Many chatbots are used in civil as well as online dispute resolution systems all around the globe. AI chatbots will learn by themself and possess a chance of being racist and adopt a discriminatory set of values for a particular class, caste, sex, and gender.[57]

Facial recognition AI can also create discrimination and can be very stereotypical at times, and one such example is when Faception AI classified and depicted a stereotypical representation of white-collar crime offenders as rich and classy, and the criminal wears nice sunglasses and terrorists were depicted covering their heads having long facial hair, which also one can relate to a particular religious community.[58] These facial recognition systems not just discriminate but also can be grossly erroneous at times; for example, in June 2017, during the 'Union of European Football Associations' (UEFA) champions league final match, around 92% possible match alerts given by a facial recognition AI were incorrect and almost the same level of incorrect recognition was done during the London street carnival with around 98% facial recognition alerts being wrong.

Many of the AI techniques for surveillance and traffic control can be used for dual purposes and at times can be used for classification, discrimination, dehumanization, polarization, and preparation, and maybe with the advancement in AI technology, it can also be used for the extermination of people. China has used artificial intelligence facial recognizing technology to track the movement of Uighurs (the Muslim minority group in the western region of Xinjiang). Machine learning helps automated facial recognition webcams to collect data on Uighurs and can easily differentiate Uighurs and non-Uighurs. AI recognizes patterns or traits of Uighurs by analyzing millions of pictures of Uighurs already fed in its system and can easily distinguish this ethnic group from others. This data gathering may be used for crime or genocide against the Uighur community. Many AIs like this perform a dual function at times, which can prove to be really dangerous for the human rights of any community.[59]

Artificial Intelligence and Opacity

In the criminal justice system, not just the decision or judgment is important but also the way how the judgment is reached and the reasons behind any judgment. AI in the criminal justice system will not be able to explain how it had come to any conclusion as most of these AI algorithms are protected behind a veil of IPR secrecy. When AI use deep learning algorithms, the process itself becomes very complicated and beyond the understanding of human beings.

Artificial Intelligence in the Criminal Justice System and Cybercrime

Artificial intelligence is used for keeping, sharing, and storing a large amount of data which is very prone to data theft by cybercriminals. Data theft can happen from two places, one is from the data which is stored in AI and the other is by using AI data stored in any other computer system, computer network, or computer storage. Artificial intelligence can bring a lot of new crimes into the picture or at least can introduce a lot of new ways of doing such crimes. For example, AI can generate audio or video data mimicking a voice which looks like a real person and performs social engineering to extract money or even to do aggravated forms of crime like murder, kidnapping, extortion, etc. Google has already attempted to show that AI can be used to create phone call operators that are indistinguishable from humans. Many new kinds of crimes can be committed using chatbots, including crime related to privacy infringement, data theft sextortion, etc.

Serious threats are also created by AI related to deepfakes by which any video can be created in which a person is acting or saying anything that in reality is not true or he has not done any such thing as mentioned in the video. These videos will shape public and societal opinions about any person and can be very annoying and threatening to humans and their reputations. These videos will also decrease the value our court system places on video evidence and can be really harmful to the criminal justice system. It is highly probable that in the near future AI will be capable of guessing passwords from the millions of usually opted passwords and can hack any system, transfer money, data, and details about anything.

AI devices can decide what to do and what not to do with the help of machine learning. Machine learning will remove the human intervention in any task performed by an AI device and hence will be more vulnerable to cybercrime. Cybercriminals may hack AI systems to perform tasks which are illegal or that humans have never imagined that AI is capable of performing. AI designers can also face problems due to zero-day attacks on their computer systems.[60] AI developers can also, because of their criminal intent, develop a bug in any new AI device, which may breach the law of the land. In future, AI-based devices can play a larger role in curbing cybercrime and also generating a new kind of threat by playing a more active role in cybercrime done on social media platforms, etc. The capability of AI devices to do cyber-social engineering sounds realistic, and it's not science fiction.[61] The AI-based cybercrime or hacking of AI devices requires AI to curb it, so we have to develop a countermeasure AI device against these cybercrimes.

Conclusion

The introduction of artificial intelligence in justice is currently debatable and is accompanied by different approaches, from the active use of artificial intelligence in resolving various categories of disputes (copyright disputes, commercial disputes) to criminal liability for using artificial intelligence algorithms to predict court decisions.

The dual presence of AI assistance and a human judge raises the possibility of a secret renunciation of decision-making authority. In order to prevent this, it is important to discuss in time how AI's assistance could be incorporated into the judiciary. The crucial question in this discussion is whether to let the AI and the human judge work alongside one another or to separate them institutionally. In this regard, one of the key issues is facilitating functioning communications between humans and AI.

The use of AI in the judicial system may help in reducing the effect of different factors including fatigue and mental instability. Nevertheless, AI decision-making can expose specific human-made systemic assumptions that derive from the legal system, the AI training data, or the AI software itself.

The law enforcement will be best equipped to respond to events, prevent threats, schedule actions, divert money, and analyze and assess illegal behaviour with the use of artificial intelligence and advanced policing technologies, paired with computer-aided intervention and actual social surveillance camera videos. Technologies have the potential to be a regular function of the law enforcement sector, to provide investigative assistance and to make it much easier for law enforcement officers to keep the public safe.

The new technologies, conversely, are not yet capable of delivering machine learning and artificial intelligence with a skill set that is wide enough for the work of a judge. It is not so because the requisite analytical challenges are too difficult to be dealt with by the AI, but because the job of a successful judge consists of a combination of abilities, including science, grammar, reasoning, imaginative problem solving, and social skills.

Researchers also criticize AI on the aspect that rather than giving full control to AI devices, it is much better if it provides support to human actions during its application. This is because a human decision goes through the process of the OODA loop, which simply means to observe, orient, decide, and act, and if AI is performing all four stages, the chances of violations of law are added. The most suitable solution to such a problem is that AI should perform functions of observation and orientation, but the later functions of decision making should be performed by a human controlling it and later AI can execute the final act.

Before we try to make AI more impactful in the criminal justice system, some kind of consensus needs to be developed about the code of conduct of AI and how law enforcement should use AI during the process of trial, execution, arrest, etc.

Data is the new oil of the world and can be used for various purposes from making a profit, crime, stalking, cyber crimes, etc. Hence, we need a minimum standard for the use of algorithmic tools by the criminal justice system, especially the by police. It is imperative that AI should respect human rights and administrative law principles at all points of time and in all situations.

A formalized system of scrutiny is also required with AI, be it a commission, board, task force, or response team in order to control the negative impact of AI. These boards should not only contain lawyers but should also have experts from multidisciplinary fields including engineers, scientists, academicians, senior police officers, forensic experts, doctors, and ethical hackers. The approach of work of this formalized system should be collaborative and should represent experts and stakeholders from different disciplines. The board will decide which AI can be implemented and which cannot and what care the criminal justice system requires before relying on the work done by AI. The board should also consider the chances of error and bias and should take all factors into consideration before using AI for any criminal justice system-related work. After due implementation of the above suggestions, we can make use of AI in the criminal justice system a reality.

Notes

1. Christopher Rigano, "Using Artificial Intelligence to Address Criminal Justice Needs", National Institute of Justice, accessed December 17, 2021, https://www.ncjrs.gov/pdffiles1/nij/252038.pdf.
2. Tim Wu, "Will Artificial Intelligence Eat the Law? The Rise of Hybrid Social-Ordering Systems", 119 *Columbia Law Review*, no. 1 (2019), https://scholarship.law.columbia.edu/cgi/viewcontent.cgi?article=3602&context=faculty_scholarship.

3. Ales Zavrsnik, "Automated Justice: Social, Ethnical and Legal Implications", Institute of Criminology, accessed December 18, 2021, http://inst-krim.si/en/automated-justice-social-ethical-and-legal-implications/.

4. Kiel Brenna-Marquez and Stephen Henderson, "Artificial Intelligence and Role-Reversible Judgment", 109 *Journal of Criminal Law and Criminology*, no. 137 (2019), pp 137–164.

5. Katie Brigham, "Courts and Police Departments are Turning to AI to Reduce Bias, but Some Argue It'll Make the Problem Worse", *CNBC*, accessed November 20, 2021, https://www.cnbc.com/2019/03/16/artificial-intelligence-algorithms-in-the-criminal-justice-system.html.

6. Ben Dickson, "AI's Struggle to Reach "Understanding" and "Meaning"", accessed July 13, 2021, https://bdtechtalks.com/2020/07/13/ai-barrier-meaning-understanding/.

7. Tania Sourdin, "Judge v. Robot? Artificial Intelligence and Judicial Decision-Making", 41 *UNSW Law Journal* 1114 (2018), pp 1115–1122.

8. M. S. Willick, "Artificial Intelligence: Some Legal Approaches and Implications", 4(2) *AI Magazine*, November 1983, p. 24, https://doi.org/10.1609/aimag.v4i2.392.

9. Jonas Schuett, "A Legal Definition of AI", accessed December 18, 2021, https://ssrn.com/abstract=3453632.

10. Matilda Claussén-Karlsson, "Artificial Intelligence and the External Element of the Crime", accessed December 18, 2021, https://www.diva-portal.org/smash/get/diva2:1115160/FULLTEXT01.pdf.

11. A. Padhy, "Criminal Liability of the Artificial Intelligence Entities", 8(2) *Nirma University Law Journal*, https://nulj.in/index.php/nulj/article/view/124, pp 15–20.

12. Monika Zalnieriute and Felicity Bell, "Technology and the Judicial Role", in Gabrielle Appleby and Andrew Lynch (eds), *The Judge, the Judiciary and the Court: Individual, Collegial and Institutional Judicial Dynamics in Australia*, UNSW Law Research Paper No. 19–90, https://ssrn.com/abstract=3492868 or http://dx.doi.org/10.2139/ssrn.3492868.

13. Michael E. Donohue, "A REPLACEMENT FOR JUSTITIA'S SCALES? MACHINE LEARNING'S ROLE IN SENTENCING", *Harvard Journal of Law & Technology*, no. 32 (Spring 2019), pp 657–664.

14. Christopher Rigano, "Using Artificial Intelligence to Address Criminal Justice Needs", National Institute of Justice, accessed November 30, 2021, https://www.ncjrs.gov/pdffiles1/nij/252038.pdf.

15. Harry Surden, "Artificial Intelligence and Law: An Overview", *Georgia State University Law Review* 1305 (2019), pp 1306–1311.

16. Supra note 14.

17. Christopher Rigano, "Using Artificial Intelligence to Address Criminal Justice Needs", National Institute of Justice, accessed November 30, 2021, https://www.ncjrs.gov/pdffiles1/nij/252038.pdf.

18. Vladimir Murashov, Frank Hearl, and John Howard, "Working Safely with Robot Workers: Recommendations for the New Workplace", *Journal of Occupational and Environmental Hygiene*, no. 3 (2016), https://www.ncbi.nlm.nih.gov/pmc/articles/PMC4779796/.

19. Daniel Faggella, "AI for Crime Prevention and Detection- 5 Current Applications", accessed December 30, 2021, https://emerj.com/ai-sector-overviews/ai-crime-prevention-5-current-applications/.

20. The AI algorithm uses smart city infrastructure to find about fired gunshots in any area so that law enforcement officers can reach to the spot as soon as possible.

21. This is a Chinese AI tool by which surveillance can be done of any area. The Hikvision camera can scan car license number plates, facial recognition of criminals and victims, identify objects like bobby traps, etc.

22. PredPol is used to find out future spots of crime by using data related to past crime and nature of crimes. This AI uses past data and observe crimes which took place earlier and predict where future crimes will take place. This AI software is used in many parts of the United States of America.

23. Karen Hao, "AI is Sending People to Jail—and Getting it Wrong", *MIT Technology Review*, accessed January 20, 2020, https://www.technologyreview.com/2019/01/21/137783/algorithms-criminal-justice-ai/.

24. Anonymous, "Justice of the Future: Predictive Justice and Artificial Intelligence", Council of Europe, accessed December 18, 2021, https://www.coe.int/en/web/cepej/justice-of-the-future -predictive-justice-and-artificial-intelligence.

25. Anonymous, "The ShotSpotter Solution", accessed December 30, 2021, https://www.shotspotter.com/technology/

26. Michael E. Donohue, "A Replacement for Justitia's Scales?: Machine Learning's Role in Sentencing", *Harvard Journal of Law & Technology* 32 (2019): 657.

27. Andrew Guthrie Ferguson, "Big Data and Predictive Reasonable Suspicion", *University of Pennsylvania Law Review* 163 (2015), pp 329–334.

28. Claudia Peersman, "Catching Child Abusers with Artificial Intelligence", Elsevier, accessed October 30, 2021, https://www.elsevier.com/connect/catching-child-abusers-with-artificial -intelligence.

29. Anonymous, "Microsoft has Created a Tool to Find Pedophiles in Online Chats", *Technology Review*, accessed November 20, 2021, https://www.technologyreview.com/2020/01/10/130941 /microsoft-has-created-a-tool-to-find-pedophiles-in-online-chats/.

30. Anonymous, "Microsoft has Created a Tool to Find Pedophiles in Online Chats" *Technology Review*, accessed November 30, 2021, https://www.technologyreview.com/2020/01/10/130941 /microsoft-has-created-a-tool-to-find-pedophiles-in-online-chats/.

31. Katalin Ligeti, "Artificial Intelligence and Criminal Justice", accessed December 21, 2021, http://www.penal.org/sites/default/files/Concept%20Paper_AI%20and%20Criminal%20Justice _Ligeti.pdf.

32. Iria Giuffrida, Fredric Lederer, and Nicolas Vermerys, "A Legal Perspective on the Trials and Tribulations of AI: How Artificial Intelligence, the Internet of Things, Smart Contracts, and Other Technologies Will Affect the Law", *Case Western Reserve Law Review* 68, no. 68 (2018), pp 747–752.

33. Lizzi Goldmeier, "How Artificial Intelligence is Revolutionizing Investigation for Law Enforcement", BriefCam, accessed November 10, 2021, https://www.briefcam.com/resources /blog/how-artificial-intelligence-is-revolutionizing-investigation-for-law-enforcement/

34. Michael E. Donohue, "A Replacement for Justitia's Scales?: Machine Learning's Role in Sentencing", *Harvard Journal of Law & Technology* 32 (2019), pp 657–662.

35. Anonymous, "Successful Final P-React Project Review Meeting, Athens, Greece-9th and 10th of June 2016", accessed December 10, 2021, http://p-react.eu/#:~:text=The%20P%2DREACT %20project%20will,alert%20detection%20and%20storage%20platform.

36. Eric Niiler, "Can AI Be a Fair Judge in Court? Estonia Thinks So", *Wired*, accessed June 30, 2021, https://www.wired.com/story/can-ai-be-fair-judge-court-estonia-thinks-so/.

37. Matilda Claussén-Karlsson, "Artificial Intelligence and the External Element of the Crime", accessed December 18, 2021, https://www.diva-portal.org/smash/get/diva2:1115160/FULLTEXT01.pdf.

38. Cary Coglianese and Lavi M. Ben Dor, "AI in Adjudication and Administration", *Brooklyn Law Review* 86 (2021), https://scholarship.law.upenn.edu/cgi/viewcontent.cgi?article=3120&context =faculty_scholarship.

39. Nigel Stobbs, Dan Hunter, and Mirko Bagari, "Can Sentencing Be Enhanced by the Use of Artificial Intelligence?", accessed December 10, 2021, https://eprints.qut.edu.au/115410/10/CLJ prooffinal25Nov2017.pdf.

40. Thomas Julius Buocz, "Artificial Intelligence in Court: Legitimacy Problems of AI Assistance in the Judiciary", *Retskraft – Copenhagen Journal of Legal Studies* 2 (2018), pp 41–46.

41. M. Veale, "Algorithm Use in the Criminal Justice System Report", *The Law Society*, accessed June 30, 2021, https://www.lawsociety.org.uk/topics/research/algorithm-use-in-the-criminal -justice-system-report.

42. Maxim Dobrinoiu, "The Influence of Artificial Intelligence on Criminal Liability", *Universitatea Nicolae Titulescu*, no. 1 (2019).

43. Swati Sudhakaran, "How AI Can Be Used in Policing Reform Clinical Justice System", *ThePrint*, accessed March 28, 2021, https://theprint.in/tech/how-ai-can-be-used-in-policing-to-reform -criminal-justice-system/384786/.

44. John Morison and Adam Harkens, "Re-engineering Justice? Robot Judges, Computerized Courts and (Semi) Automated Legal Decision-Making", accessed December 20, 2021, https://www.cambridge.org/core/journals/legal-studies/article/reengineering-justice-robot-judges-computerised-courts-and-semi-automated-legal-decisionmaking/E153E0FB25BB155B971AA38284EC7929.

45. Gabriel Hallevy, "The Criminal Liability of Artificial Intelligence Entities – From Science Fiction to Legal Social Control", *Akron Law Journal*, no. 4 (2016), https://ideaexchange.uakron.edu/cgi/viewcontent.cgi?article=1037&context=akronintellectualproperty.

46. M. Veale, "Algorithm Use in the Criminal Justice System Report", *The Law Society*, accessed June 30, 2021, https://www.lawsociety.org.uk/topics/research/algorithm-use-in-the-criminal-justice-system-report.

47. Maxim Dobrinoiu, "The Influence of Artificial Intelligence on Criminal Liability", *Universitatea Nicolae Titulescu*, no. 1 (2019).

48. Monika Zalnieriute and Felicity Bell, "Technology and the Judicial Role", accessed December 10, 2021, https://www.austlii.edu.au/au/journals/UNSWLRS/2019/90.pdf.

49. Katalin Ligeti, "Artificial Intelligence and Criminal Justice, AIDP-IAPL International Congress of Penal Law", accessed November 20, 2021, penal.org/sites/default/files/Concept%20Paper_AI%20and%20Criminal%20Justice_Ligeti.pdf.

50. Laviero Buono, "Artificial Intelligence (AI) and the Criminal Justice System", accessed November 20, 2021, https://eucrim.eu/events/artificial-intelligence-ai-and-criminal-justice-system/.

51. Anonymous, "Report of Access Now: HUMAN RIGHTS IN THE AGE OF ARTIFICIAL INTELLIGENCE", accessed December 10, 2021, https://www.accessnow.org/cms/assets/uploads/2018/11/AI-and-Human-Rights.pdf.

52. A. Padhy, "Criminal Liability of the Artificial Intelligence Entities", *Nirma University Law Journal*, no. 8 (2019), https://nulj.in/index.php/nulj/article/view/124.

53. Morgan Livingston, "Policy Memo: Preventing Racial Bias in Federal AI", accessed March 30, 2021, https://www.sciencepolicyjournal.org/article_1038126_jspg160205.html.

54. Anonymous, "Algorithms in the Criminal Justice System: Risk Assessment Tools", accessed December 31, 2021, https://epic.org/algorithmic-transparency/crim-justice/.

55. Kashmir Hill, "How Target Figured Out a Teen Girl Was Pregnant Before Her Father Did", *Forbes*, accessed November 18, 2021, https://www.forbes.com/sites/kashmirhill/2012/02/16/how-target-f igured-out-a-teen-girl-was-pregnant-before-her-father-did/.

56. Anonymous, "AI in the Criminal Justice System", *Epic*, accessed December 20, 2021, https://epic.org/algorithmic-transparency/crim-justice/.

57. Anonymous, "Law Enforcement Chatbots", accessed January 21, 2021, https://aiethics.princeton.edu/wp-content/uploads/sites/587/2018/10/Princeton-AI-Ethics-Case-Study-4.pdf.

58. Peter Brown, "Artificial Intelligence: The Fastest Moving Technology", accessed March 10, 2021, https://www.law.com/newyorklawjournal/2020/03/09/artificial-intelligence-the-fastest-moving-technology/.

59. Paul Mazur, "One Month, 5,00,000 Face Scans: How China is Using A.I. to Profile a Minority", *NY Times*, accessed July 31, 2021, https://www.nytimes.com/2019/04/14/technology/china-surveillance-artificial-intelligence-racial-profiling.html.

60. Jake Frankenfield, "Zero-Day Attack", accessed November 20, 2021, https://www.investopedia.com/terms/z/zero-day-attack.asp.

61. Marc Wilczek, "Cybercrime: AI's Growing Threat", accessed December 10, 2021, https://www.darkreading.com/risk/cybercrime-ais-growing-threat-/a/d-id/1335924.

4

Artificial Intelligence Inroads into the Indian Judicial System

Santhi Narayanan, Monica Agarwal, and Swati Bansal

CONTENTS

Introduction

"A futuristic judiciary is not an impossible dream now".

– Justice Bobde, at the inaugural ceremony of SUPACE, AI Tool for Supreme Court

The impact of artificial intelligence (hereinafter referred to as "AI") is widening its spread to the work realm of socio-economic institutions. It is a system wherein computers are configured for undertaking tasks to act and think like human beings. These systems are usually powered by either deep learning or machine learning. Certain authors argue that no technology as much as AI has stoked the dystopian fears of society (Dixon, 2021). Law is an inherently favorable area where AI is applicable as it relies on a system of logic wherein the system of precedents can be applied to a case, and the inference is drawn based on it. Thus, its effect extends to the law-making process and its application (Buchholtz, 2020).

DOI: 10.1201/9781003215998-4

AI is a computer process that an ordinary person would consider intelligent (Irons & Lallie, 2014), and some of the exciting activities include speech recognition and problem solving (Habeeb, 2017). Judge Herbert B. Dixon Jr. (Ret.) defines AI as "AI combines algorithms and machine learning as a substitute for the human brain to predict, analyze, forecast, and decide" (Dixon, 2020). In addition, AI is a branch of computer science that studies the relation between computation and cognition (Gruber et al., 2020).

However, to date, neither the scientific community nor the international organizations have defined "artificial intelligence" (Ad hoc Committee on Artificial Intelligence, 2020). The feasibility study report of the Ad hoc Committee of Artificial Intelligence states, "AI can be used as a 'blanket term' for various computer applications based on different techniques, which exhibit capabilities commonly and currently associated with human intelligence".

Application and Importance of AI

AI is used to model legal ontology and occupies an essential role in the legal informatics system to cover computational models of legal reasoning, automatic classification of legal texts, lawbots to tackle repetitive and small legal tasks and due diligence reports. Due diligence reports are a very mechanical job, and hence, with the help of AI, the reports can be generated by feeding a set of parameters to dissect the data. The main focus of AI has been on managing the documents, including patent portfolios, customized documents for filing, and due diligence reports. The increasing acceptance of AI can help companies embark upon digitizing their documents and saving legal and space constraints for maintaining the documents. In addition, the AI helps to provide support in case of giving legal advice. However, this needs to be under the supervision of experts and in a transparent manner. Hence, in the recent past, AI solutions in the legal sector cover three main areas: automation in practice, document analysis, and legal research (Yu & Ali, 2019).

Using AI in Judiciary

AI is progressing at a fast rate, and the countries' legal systems cannot afford to ignore this emerging revolution in technology. The rapid technological changes and computational power available in AI have made the Judiciary embrace it. AI helps to change the way judicial decisions are taken. A decision can be taken with the help of AI by evaluating the case files, in which both the procedural aspects and the analogous judicial decisions were made earlier. In this process, a dataset is analyzed to determine the statistical correlation between the given data and the associated judicial decisions to come to a judicial decision. As the algorithm starts processing more data, the accuracy level of the system is also said to improve, and predictive judicial decisions based on AI are plausible. Whether this predictive judicial decision based on algorithms would lead to a better decision or undermine the existing judicial principles remains to be seen. AI frameworks can hone the decision-making capabilities only over some time as the decision-making capacity depends on the mechanical analysis of data over some time (Chahal, 2018). Judicial data derived by the use of AI can help improve the transparency in the judicial system through consistency and predictability of the application of the law. The judicial systems worldwide have adopted automation and are still in early developmental stages and are being evaluated for their robustness. Some of the countries, the United States, Canada, the United Kingdom, and Europe, have initiated some level of prediction systems (Arias, 2020).

Arguments for the Use of AI

While administering justice in individual cases, the Judiciary provides justice to the parties of the case and broadly sets the standards of justice for society. The work of the Judiciary is basically to process the information brought before it through reasoning and set procedures to conclude in the form of information. It is not in every case a judge uses complex procedures to conclude; sometimes, it requires only a simple assessment or an issue of admissibility, etc. Hence, the requirement of technology is not similar for all cases. However, in cases where repeatedly there are predictable outcomes under similar situations, AI can be promoted. The arguments that favor AI utilization to give judicial decisions are that in the existing system, sometimes the judicial decisions lack transparency and are biased or delayed. The main advantages of using AI in the Judiciary are efficiency, consistency, and access to justice (Levendowski, 2018). Using AI for decision making can be quick, cheap, and less subject to bias, making the system efficient, fairer, and accessible to the ordinary person (Volokh, 2018). Another author said that there is no right to a human decision but merely a right to a calibrated decision by a machine (Huq, 2020). Certain other authors compare AI to a self-driving car and terms laws as self-driving too, and the advances in AI would help identify the rules applicable to real-life situations (Casey & Niblett, 2016; 2017). Tools are being developed to support judicial activities and to build analytical tools for decision making.

Automation: Several repetitive processes like docket management and due diligence reports can be automated. These tools help to draft documents in minutes, which require days of manpower to draft. Further, document analysis is a time-consuming process, and AI solutions help to improve this time-consuming process by automating the review of documents. Also, it helps to mark relevant parts of the documents to reduce time wastage in discovering the facts for a given legal case. In addition to this, these tools can accurately bill the productive invoiceable hours and simplify the billing system. For example, a tool known as "Smart Contract Risk Identification Software (SRIC)" helps lawyers review and analyze the contracts by classifying the document into clause-based categories and helps the lawyers assess the risk and severity of the given clause. With frequent document review, the tool gets smarter and recognizes facts and specific patterns. The document review can help the lawyers to prepare their cases accordingly. Also, AI tools are developed to serve as personal legal assistants like Siri or Cortana but with a knowledge of law. The AI chatbots can provide preliminary legal advice to many people simultaneously and then get them on board to help them get on board as full-time paying clients for the legal practice.

Simplification of legal language: The client who comes in for consultations cannot understand the technical terms and legal jargon. AI can help the client to understand the legal jargon before they meet the lawyer. AI assistant tools can also help generate a case file and give opinions to seek a lawyer or not in the given case.

Organizing: AI can be used to scan the innumerable number of documents filed during a court and come with Venn diagrams and graphs to ease the work of the judge who spends a considerable amount of time making different connections and evidence through the several documents before it. The graphs or Venn diagrams can help the judges reduce the time of decision making.

Dispute resolution: In future, a clause might be inserted in the contract that the matter can be resolved through alternate dispute resolution or decisions made by AI. Further, from voluminous documents, AI can be used to calculate the loss suffered by the parties. However, in this case, the parties can go to the courts, which helps cut the costs of protracted litigation.

Conducting trials: AI can help with repetitive tasks and legal research, thereby helping the judges to make decisions in a faster and effective manner, which in turn helps reduce the pendency of cases.

Consideration of small value pecuniary matters: Decision in criminal matters requires emotions, empathy, and the human element; however, in civil matters of small pecuniary value like traffic tickets, AI can be used to frame punishments in case of violations.

Deciding territorial jurisdiction of the cases: When one inputs the data into the system regarding the information regarding where parties had filed the case, AI can identify the territorial jurisdiction of the concerned matter. This identification reduces the time taken by the court to decide the jurisdiction. As the judges are so involved in administrative work, this can help reduce the burden on the judges.

Arguments against the Use of AI

If AI has arguments for improving lives, it has also raised concerns about the negative aspects that can affect human life if the tools are in the wrong hands, so examining these risks is a requirement.

Unemployment and declining role of people: It is a general belief that there could be layoffs as the machines would outperform human beings. Also, there is a general sentiment that AI could take the role of people engaged in judicial services by introducing AI into the judicial system. The traditional way of maintaining files and records becomes obsolete over time, but digitizing these data requires human supervision. Therefore, further interventions are required for processing, authenticating, and utilizing the data, as highlighted in the report titled "Just and Equitable AI Data Labelling: Towards a Responsible Supply Chain" by Aapti Institute, 2021. In addition, this report also indicates that creating robust datasets for AI will create job opportunities, and the potential has also been recognized by the NITI Aayog 2018 in the National Strategy for AI.

Need for a regulatory framework: There is a lack of a proper regulatory framework to take care of the implementation of AI. Law attempts to code the policies driven by moral and ethical principles. Voluntary codes exist in relation to the usage of AI like the UK Data Ethics Framework, etc. The Commission for the Efficiency of Justice (CEPEJ) of the European Council has also laid down certain ethical principles for the use of AI in the administration of justice. However, these are only in broad terms; hence, its compliance and enforcement are big challenges.

Lack of digital literacy: This would be a challenge to utilizing services as the services offered by the AI in courts. In a country like India, digital accessibility and the appropriate digital infrastructure, especially in the remote parts of the country, pose a challenge to deploying AI-based solutions.

Authentication of evidence: The authentication of evidence during court proceedings using AI tools is still a sore point as AI is still not without flaws, and the integrity of such tools is also a matter of deliberation.

Data protection issues: As AI technology becomes powerful, these tools can be utilized for nefarious activities causing harm to the public. The data protection issues raised in the Cambridge Analytica scandal are one of the significant persisting concerns of the utilization of AI.

Ethical and human rights issues: The ethical and human rights issues will pose fresh challenges to technology. Even though AI is capable of processing with speed and can process much beyond human beings, AI systems might remove human errors but still cannot be trusted to be fair and neutral in their approach as the codes could contain bias and prejudice. AI operates on material and data. There could be categories of data incorporating an inbuilt bias, i.e., the technology can be coded to negatively impact specific demographic-based characteristics and discriminate on its basis. The AI programs may carry inherent bias as data is fed into the system by humans; there can be specific categories set that may be inherently biased. Over a while, the AI system may behave in unexpected ways.

AI in India

In India, the first task force was set up under N. Chandrasekharan to propose measures to improve the collaboration of AI in different fields (Artificial Intelligence Taskforce Report, 2018). The subsequent strategy discussion paper titled "National Strategy for Artificial Intelligence" by NITI Aayog in 2018 threw light on the ethical, data security, and primary and ethical challenges of AI. The paper identified barriers such as (a) research expertise in the application of AI, (b) very high resource costs and less awareness regarding the adoption of AI, (c) privacy and security issues, (d) an enabling ecosystem, and (e) no enabling collaborative approach to the adoption of and application of AI. In the Government Artificial Intelligence Readiness Index, 2019 compiled by Oxford Insights and the International Development Research Centre, India ranks 17th, whereas the top five countries are Singapore, the UK, Germany, the USA, and Finland. This index scores various governments based on the preparedness to use AI to deliver their public services.

AI in Indian Judiciary

The wheels of justice need to keep rotating. If a spoke gets loose, the entire machinery is taken for a toss and leads to justice not reaching the needy at a time when needed. Innumerable reports indicate the delay in the disposal of cases by the Indian Judiciary, with many cases dragging on for decades. The judicial systems are blamed for the delay. However, the delay is sometimes not only due to the large number of cases being filed in the courts but also because of the lack of appropriate systems in place. The system is facing huge pressure due to the sheer size of cases filed in the courts, and the delay in hearing the cases has unfortunately affected millions of litigants in receiving the right kind of justice.

The National Judicial Data Grid (NJDG) shows 10,512,502 cases pending in India's district and taluka courts as of July 28, 2021. Out of this pendency of cases, the cases which await record are 418,076 (16.89 %), and unattended cases are 487,255 (19.69 %). This points to the need of using more technology in the justice delivery system. Surmounting this considerable backlog of cases is a critical challenge for the Indian judicial system. The pandemic has aggravated the situation. Therefore, it is a wake-up call for the Indian Judiciary to adopt artificial intelligence to help manage this huge caseload that the courts are currently facing. The research from the Vidhi Centre for Legal Policy (2021) states that integrating AI into judicial decision-making processes would help improve the administrative efficiency of the courts and expedite the judicial processes and improve access to justice for the ordinary person. The report further noted, "The pandemic has led to a surge in discussion around increasing digitization through the eCourts Project, creation of virtual courts, and the potential of online dispute resolution. Within this conversation, AI has also become an increasing talking point". The Indian Judiciary has also not been far behind in adopting technology to improve its functioning.

AI Tools Adopted into the Indian Judicial System

E-Court's Project

With a vision to transform the Indian Judiciary, the e-Courts Project was implemented based on the "National Policy and Action Plan for Implementation of Information and Communication Technology (ICT) in the Indian Judiciary – 2005". This is a pan-India project for the district courts across the country. The objectives of this project are to (a) ensure that the delivery of citizen-centric services is provided in a timely and efficient manner as per the e-courts litigant charter, (b) developing and installing decision support systems in the courts, (c) providing transparency and accessibility of information to various stakeholders through automating the processes, and (d) increasing the productivity and efficiency of the judicial system in qualitative and quantitative terms and making it accessible, reliable, transparent, affordable, and cost effective. The first phase of the project began in 2007 and concluded in 2015 with the district courts being computerized and the staff training for the usage of the digital platforms. Phase II of the project focuses more on delivering services to litigants, lawyers, and stakeholders. There is an increasing emphasis on making the services available in local languages and further improvisation of the National Judicial Data Grid (NJDG) to include more information to the public, governments, and courts. In 2013, the e-courts project was launched. The National Judicial Data Grid now has more than 2,852 district and taluka courts, providing their case status, pending cases, disposed of cases, and orders/judgments. The data is analyzed from time to time and utilized for both policy and decision making for speedy justice delivery. Also, this portal acts as a repository of orders/judgments of the district courts. The analytical tools incorporated help in analyzing data in real time for effective dashboards on the court and case management. This portal also helps to serve as a judicial performance indicator.

SCI-Interact

The Supreme Court developed software called "SCI-Interact", in 2020 to convert the entire 17 benches to paperless mode. This software comprises five components, namely scanned copies of pending cases, e-filing of new cases, IT hardware, Multiprotocol Label Switching (MPLS) networks with dual redundancy, and security audit, everything on one platform to reduce human touch, help in quick decision making, and speedy disposal of cases.

Legal Information Management & Briefing System (LIMBS)

LIMBS is a web-based application introduced by the Department of Legal Affairs. This application can help monitor cases from high courts and tribunals and help track the life cycle of the cases.

Supreme Court Vidhik Anuvaad Software (SUVAS)

Supreme Court Vidhik Anuvaad Software (SUVAS) is an AI system introduced by the Supreme Court in 2019, which would help to translate the judicial order in English to nine vernacular languages and vice versa, which will solve the issue of translation challenges of orders/judgments/documents. It reduces the time taken for the case to be pending due to challenges faced in translating the documents of the given case. The nine languages are – Assamese, Bengali, Hindi, Kannada, Marathi, Odia, Tamil, Telugu, and Urdu. The cases related to Labour Law, Rent Act, Consumer Protection, Agricultural Tenancies, Ordinary Civil, Personal Law, Religious and Charitable Endowment, Simple Money and Mortgage Matters, and Family Law are getting translated.

Supreme Court Portal for Assistance in Courts Efficiency (SUPACE)

The Supreme Court Portal for Assistance in Courts Efficiency (SUPACE) project was inaugurated on April 6, 2021, by the then Chief Justice of India, S.A. Bobde. This tool seeks to reduce the workload of the judges of the Supreme Court. This system will process the information and provide the data needed for the judges to decide by filtering the information provided by the parties by the inbuilt legal research tools. The system will not make decisions of its own but would assist the decision with the necessary data in a convenient format to reach a decision. Thus, the judges will be able to get across the crucial data of multiple cases that they are hearing simultaneously in an easy and accessible manner without wasting time on legal research. The portal has been implemented as a pilot project in the High Court of Bombay and Delhi dealing with criminal matters. This assists the courts in improving efficiency and pendency in the judicial process, which can be automated using AI.

Analysis of the Implementation of AI in the Indian Judiciary

It is a technological and human challenge to implement AI in the Indian judicial system due to the complexities present in the system. In the Indian scenario, there is a requirement for digitalization of the documents, primarily non-commercial documentation, which can help people to save many legal costs. The Judiciary has to convert the entire court records like case files, pleadings, submissions, judgments, and all the orders online issued by all courts. The judge spends much time analyzing the case and looking up and analyzing case precedents to avoid ambiguity in decision making. This costs much time for the judges and delays in delivering judgments. AI can help provide them with tools to make this process easier. Using AI, the judges can identify that no two defendants, whichever part of the country they belong to, have different jail terms for a similar offense in a similar situation. The AI can translate voluminous documents, help understand the de facto meaning of the words, and nuance the language used in the documents.

In India, now, the number of tech-savvy judges in the Indian courts is on the lower side. Hence, the interest in AI in the Judiciary is much less when compared to the judicial systems abroad. Further, the R&D investments for AI in the Indian legal space are comparatively low compared to the US, where big companies are targeting lawyers as their potential customers.

The usage of AI for decision making has been questioned on the grounds of not being able to incorporate the human touch. The chairman of the Indian Supreme Court's AI committee Justice L. Nageswara Rao, opined, "It is implausible that the AI system will make human lawyers or judges redundant". Former Judge B.N. Srikrishna, in his book *"India 2030: Rise of a Rajasic Nation"*, also states,

> AI will not only help organize cases, but it will also bring references into the judgment at a speed not seen so far. Technology will ensure that those who do not have access to justice due to distance will not be excluded anymore.

The area of judicial decision making is a very complex one that sees an amalgamation of legal expertise with cognitive and emotional competence and serves an essential social standing. The issue of "justice" is sensitive as an individual's case might involve deep feelings, which can impact the expectations of various legal stakeholders shaping the understanding and development of the legal system. The social dynamics of the case does not come into consideration. It needs to be said that the legal acumen and empathy used by the judges to make decisions can never be replaced. The use of AI decisions will hinder the ability to change the law or question the law with the changing social circumstances and the development of law. However, AI can help study the patterns and correlations between the various cases examined as precedents for a particular case, which would take much time to decipher by a person in an ordinary course. However, AI tools should be configured to give results compatible with the fundamental rights enshrined in our Constitution.

The Way Forward

Human involvement in the process of decision making has a significant value and is not just an opinion factory, and it creates a beneficial dialogue between law and society (Michaels, 2019). Law is created, monitored, and updated by human society. The argument presented before the judge in a given case is not just for the benefit of the litigant's case but to persuade the judge. This helps to develop the law, setting precedents for future cases potentially. The arguments raised sometimes create dilemmas and seek a relook at the existing law. Then, it is up to the judges to make decisions to put the law on the right course to meet the needs of justice. Dissent between judges in a judgment is entirely plausible. However, the dissent opens the available legal choices and allows the public to debate, which helps in the development of the law. Using AI in judicial decision making can result in the overvaluing of the uniformity provided by the judgments.

Further, it would discourage the public legal debates forming the voice of the society. The legal community not only monitors the law, but it helps spread the notion that we are governing ourselves by laying down the laws. If there is no due consideration for the arguments, then no quality argument would be developed, which might have been helpful to change the due course of the law. The judges respond to the lawyers, and the lawyers respond to the litigants' needs, spreading the power to shape the law across the legal community and society. The lawyers in the present system do get a chance to address the judges' concerns whenever there is a need. It helps us understand that we are governed by the laws that we have created and gives us the power to control ourselves rather than external authority. Codes are used to develop tools, and these codes tend to be covered by intellectual property rights. This code should be made public, and everyone has a right to examine it free of cost so that it can be examined by legal experts and computer scientists to help in the improvement of their ability to make decisions.

It has been well documented in the literature that there could be bias inbuilt into AI. The question that remains to be seen is whether the AI tools can change in the changing circumstances and then ensure the separation of powers between the legislature, executive, and Judiciary. A further issue could also arise when the virtual tool and the real-life interaction may face which has never been faced before and lead to several alterations in various branches of law. In addition, the level of transparency gets reduced over a while. The service providers may not want to make a transparent system to enforce and protect their interests.

The legal boundaries are likely to become blurred between organizations, individuals, and non-humans. As these boundaries get blurred, auditing of information of these tools can become difficult. Hence, there should be a framework of safe operations as the organizations may not undertake stringent measures to ensure that the systems work according to rules without compromising personal autonomy. It can give rise to new inequalities caused by restricted access to these technologies. Law dispensation should be in a manner that should provide unrestricted access to technologies to every ordinary person and reduce inequalities.

Regulations have a critical underlying role to play while implementing and incorporating AI technologies to deal with the bad actors seeking to exploit the opportunities (Székely et al., 2011). Proper legal frameworks should come into existence to keep pace with the technological developments due to the spread of such applications and services. The framework needs to identify the impact of these technologies at an individual level or at the state level to develop a framework for governance across the country, keeping in mind the cultural and geographic diversity. Also, the individual protection, efficiency, and consistency of the use of regular communications between the stakeholders and lawmakers lay down the ethical principles on which the system works. The concerned stakeholders should consider ethical impact assessment to assess the impact of the emerging technologies. For this, steps should be taken in the initial stage so that as the technologies roll out, it garners extensive use for the benefit of the ordinary person. Keeping in view that technology and law cannot be silos now and, in the future, legal education needs to incorporate the latest technological changes in its curriculum. Proper forums for emerging technologies need to be set up wherein the stakeholders, government, researchers, and experts are made part, who can review and suggest policy frameworks compatible with the changing needs. To handle the ethical implications arising out of AI is to establish efficient control mechanisms to ensure that they do not result in unethical practices affecting society (Székely et al., 2011).

The people in the departments seeking to employ and develop AI tools should exercise appropriate judgment and reasonable care and take deliberate steps to avoid and minimize unintended bias while developing and deploying the tool. It is not yet feasible to explain how the result can be achieved using AI tools (Reiling, 2020). The operational methods for using AI should be auditable and need to have transparent data sources, design procedures, and documentation. The safety and security of these systems should be tested from time to time across the various stages of the output and has to be continuously monitored, evaluated, and optimized for better and appropriate outputs. Thus, caution needs to be exercised while building tools to support people-centric administration.

Conclusion

It is pertinent to note that AI cannot interact with clients to understand the client's issues or argue on behalf of the client in a court of law. However, AI can help in the repetitive tasks undertaken by the lawyer or assist him in developing the brief of the case. Hence, it is expected that the next generation of lawyers will acquire specific technical skills to handle technological developments. The lower-cost services enabled by AI in the judicial system can lead to increased access to legal services (Branting et al., 2021). There could be ethical questions about mitigating risks and adverse outcomes; however, AI has vast potential to better the lives of people. Future research in AI has to be with a multidisciplinary approach.

The responsible use of AI would result in efficiency enhancement to be embedded in the legal and judicial processes. The Judiciary now has set on a path of transformation by introducing AI into its functioning of the Indian judicial system. This can help to improve the efficiency of justice delivery. This, in turn, can help in the ease of living of the citizens, ultimately resulting in ease of doing business, paving the way for the more foreign direct investment required for the Indian economy's growth. Future research in AI should be multidisciplinary, and the lawyers have to gear up for future changes considering the ethical considerations.

Acknowledgment

The authors wish to thank Mr. Sanjeev Kumar Singh, Advocate Supreme Court of India, Mr. Shighra, Advocate Supreme Court of India, and Mr. Aman Naqvi, Advocate Delhi High Court, for their expertise and assistance throughout all aspects of our study and for their help in writing the manuscript.

Bibliography

Aapti Institute. (2021). *Just and equitable data labelling*. https://uploads.strikinglycdn.com/files/7d492f74-a51f-423b-bf5d65c9f88eee06/AI_Data_Labelling_Report_DIGITAL_25FEB1033.pdf.

Ad hoc Committee on Artificial Intelligence. (2020). Feasibility study on a legal framework on AI design, development and application. https://rm.coe.int/cahai-2020-23-final-eng-feasibility-study-/1680a0c6da.

Arias, P. C. (2020). Artificial intelligence & machine learning: A model for a new judicial system? *Revista Internacional Jurídica y Empresarial*, 3, 81–91.

Branting, L. K., Pfeifer, C., Brown, B., et al. (2021). Scalable and explainable legal prediction. *Artificial Intelligence and Law, 29*, 213–238. https://doi.org/10.1007/s10506-020-09273-1.

Buchholtz, G. (2020). Artificial intelligence and legal tech: Challenges to the rule of law. In Thomas Wischemeyer & Timo Rademacher (Eds.), *Regulating artificial intelligence* (pp. 175–198). Cham: Springer.

Casey, A. J., & Niblett, A. (2016). Self-driving laws. *University of Toronto Law Journal, 66*(4), 429–442.

Casey, A. J., & Niblett, A. (2017). Self-driving contracts. *Journal of Corporation Law, 43*, 1.

Chahal, Y. (2018, June 9). *India's 'Unregulated' tryst with artificial intelligence: Looking into future without a law?* https://www.livelaw.in/indias-unregulated-tryst-with-artificial-intelligence-looking-into-future-without-a-law/.

Dixon Jr, H. B. (2020). What judges and lawyers should understand about artificial intelligence technology? *Judges' Journal. -American Bar Association, 59*(1), 36–38.

Government Artificial Intelligence Readiness Index 2019, https://www.oxfordinsights.com/ai-readiness2019.

Granado, D. W. (2019). Artificial intelligence applied to the legal proceedings: The Brazilian experience. *Revue Internationale de droit des données et du numérique, 5*, 103–112.

Gruber, J., Handel, B. R., Kina, S. H., & Kolstad, J. T. (2020). *Managing intelligence: Skilled experts and AI in markets for complex products* (No. w27038). Cambridge, MA: National Bureau of Economic Research.

Habeeb, A. (2017). Artificial intelligence. *Research Gate, 7*(2).

https://www.indiatoday.in/india/story/supreme-court-india-sc-ai-artificial-intellegence-portal-supace-launch-1788098-2021-04-07.

https://indiaai.gov.in/research-reports/responsible-artificial-intelligence-for-the-indian-justice-system.

Huq, A. Z. (2020). A right to a human decision. *Virginia Law Review, 106*, 611.

Irons, A., & Lallie, H. S. (2014). Digital forensics to intelligent forensics. *Future Internet, 6*(3), 584–596.

Kauffman, M. E., & Soares, M. N. (2020). AI in legal services: New trends in AI-enabled legal services. *SOCA, 14*, 223–226. https://doi.org/10.1007/s11761-020-00305-x.

Levendowski, A. (2018). How copyright law can fix artificial intelligence's implicit bias problem. *Washington Law Review, 93*, 579.

Michaels, A. C. (2019). Artificial intelligence, legal change, and separation of powers. *University of Cincinnati Law Review, 88*, 1083.

National Judicial Data Grid, https://njdg.ecourts.gov.in/njdgnew/index.php.

National Strategy for Artificial Intelligence, NITI Aayog (2018). https://indiaai.gov.in/research-reports/national-strategy-for-artificial-intelligence.

Reiling, A. D. (2020). Courts and artificial intelligence. In *IJCA* (Vol. 11, p. 1). https://heinonline.org/HOL/LandingPage?handle=hein.journals/ijca11&div=18&id=&page=.

Report of Task Force on Artificial Intelligence, https://dipp.gov.in/whats-new/report-task-force-artificial-intelligence.

Ret, J. H. B. D. J. (2021). Artificial intelligence: Benefits and unknown risks. *The Judges' Journal, 60*(1), 41–43.

Székely, I., Szabó, M. D., & Vissy, B. (2011). Regulating the future? Law, ethics, and emerging technologies. *Journal of Information, Communication and Ethics in Society, 9*, 180–194.

Vidhi Centre for Legal Policy. (2021, April). Responsible AI for the Indian Justice System – A strategy paper. https://vidhilegalpolicy.in.

Volokh, E. (2018). Chief justice robots. *Duke Law Journal, 68*, 1135.

Yu, R., & Alì, G. S. (2019). What's inside the black box? AI challenges for lawyers and researchers. *Legal Information Management, 19*(1), 2–13.

5

The Role of Language Prediction Models in Contractual Interpretation: The Challenges and Future Prospects of GPT-3

Malcolm Katrak

CONTENTS

Introduction

Imagining human communication without natural language (NL) would be impossible. By enabling humans to communicate and express thoughts, NL became a *sine qua non* for human communication. Considering its importance, technology has adapted to understand and process NL in the same manner as human-to-human communication. This has been described as "natural language processing (NLP)", a phenomenon which began in the 1950s, essentially combining the fields of computer science and linguistics (Johri et al., 2021). Endowed with the ability to understand text and words, computer programs have been able to perform tasks such as speech recognition, language translation, entity recognition, natural language generation, etc. A recent technological development in NLP is the language model, Generative Pre-trained Transformer (GPT)-3. An autoregressive language model (LM), GPT-3 can, *inter alia*, generate code, write and summarize text, answer queries, and philosophize. Such technology will have a ripple effect on almost all fields that work with language. One such field is the legal profession, where language, often described as "legalese", is given utmost importance. It has been recurrently said that legal language is a barrier that non-lawyers are barely able to cross (Burton, 2018). The difficulty normally

DOI: 10.1201/9781003215998-5

faced by consumers in understanding standard form contracts that they give their assent to constitutes an example of the same (Wilkinson-Ryan, 2014). Arbel and Becher (2022) are among the first, and the strongest, proponents of the GPT-3 technology. They suggest the use of such technology in mobile applications called "smart readers", which have the ability to simplify, personalize, construct, and benchmark contracts. It is true that GPT-3 is a skillful model; however, it still has its shortcomings, such as occasionally generating incoherent texts, being unable to solve some queries, and, most importantly, training data bias leading to racist and sexist texts (Johnson, 2021). These shortcomings are visible when applied to contractual and statutory interpretation; one such example is the anti-consumer bias in the training data (Kolt, 2022). Nevertheless, the technology is worth understanding to appreciate the progress in the field of linguistics, especially in relation to the development of NL. To that end, Chapter 2 provides a conspectus of the development of NL models and the GPT technologies. Chapter 3 delves into the impact GPT-3 may have on certain businesses, specifically the legal profession. Chapter 4 embarks on analyzing the consumer issues pertaining to standard form contracts. Chapter 5 concludes.

Linguistic Chameleons: Natural Language Models

As mentioned earlier, NLP involves computer manipulation of natural language. NLPs often use language models (LMs) to carry out probability analysis of the sequence of words in a sentence (Jozefowicz et al., 2016). Consider the following sentence: "he took the water bottle, opened its cap to …". An LM would assign a higher probability that the subsequent term would be "drink", "sip", or "have a sip". The LM does this through the n-gram approach, which entails assigning probabilities using n-size (Jurafsky & Martin, 2008). The n-grams may be *unigram*, wherein each term is evaluated independently, *bigram*, wherein one preceding word is evaluated, or *trigram*, wherein two preceding words are evaluated. Subsequent technological developments have resulted in large neural network LMs being formed. These neural networks are nothing but structures akin to those of neurons in the human brain, where one is connected to another, and the input from one neuron is sent to the others (Elman, 1990). Essentially, these neural networks contain nodes which are connected to each other for easier passing of data. NLP neural networks use word embeddings, which allow words to be represented by numerical values. One example of this is Google's Word2vec algorithm, which uses a neural network model to do a probability analysis. The advantage of neural networks is the inherent temporal memory that allows the LM to find connections and dependencies. For instance, a recurrent neural network (RNN) passes information in a chain-like structure between the network, allowing each word to be processed separately (Phi, 2018). This inevitably speeds up the computational process and drastically reduces the cost. The process of training RNNs, however, has its own set of problems. Consider the sentence, "My aim was to become a soccer player. I represented my school during various sporting events … Carrying forward my childhood dream, I now play professional". The information provided in the sentence would imply the word "soccer" should be the succeeding word. To narrow down the profession, a specific context is required from the essay. Going back only a few words would provide limited context, giving rise to the problem of long-range dependency (LRD) (Goodfellow et al., 2016). LRD makes the RNN ineffective when the distance between relevant information and the space where it is required is large. To curb this, a long-short term memory

(LSTM), which acts like a memory cell, was developed (Hochreiter & Schmidhuber, 1997). The LSTMs work as gates, where the input gate allows the flow/storing of information, the forget gate allows the information to be deleted so that new information can be stored, and the output gate controls the output of the information (Sak et al., 2014).

Recent developments in relation to neural networks include the rise of *attention* mechanisms (Niu et al., 2021), which allow the RNNs to focus on the input with varying levels of attention. The information can be hidden in any word, and, therefore, it is necessary to pay close "attention" to every word of the input. This "attention" mechanism is at the heart of the current technology, namely, transformers (Kobayashi et al., 2020). Vaswani et al. (2017) were the first to conceptualize transformers, a type of neural network architecture based solely on attention mechanisms. What makes a transformer model unique is its feature of "self-attention". This helps the neural networks remove ambiguity and learn semantic roles. Consider the following sentences: "I waited on tables to pay for my brother's hospital costs" and "the only place to sit was on the table, and so I waited on that table for hours". For a human, it is easy to decipher that the phrase "waited on tables" implies "serving food or drinks" whereas "waited on that table" in the second sentence implies "staying on that table". The similarly worded phrases entail two very different things, something a human can easily identify by looking at the context. *Self-attention* helps neural networks to decipher phrases that are ambiguous (Gehring et al., 2017). Transformer models have now been used to generate predictive text, including computer codes, music, written sonnets, and translations. Given that there is unlimited data for an NLP to work on, it is necessary to train these models for specific purposes. This is where the development of pre-training in LMs is essential (Qiu et al., 2020). Pre-training, a recent development in LM, allows the model to be fine-tuned for specific tasks. Essentially, pre-training has two stages. First, an LM is trained on a generic dataset; this training may be termed "pre-training the model". Second, the model is initialized, which can then further train on a specific dataset to carry out the relevant task for which the model was intended (Sarker, 2021). While datasets are available for the LMs to train; these LMs can either be taught by example or provided data without any example. The former can be counted as supervised learning, where the algorithm is taught by example and by iteratively making predictions on the data. As the input name suggests, supervised learning models require human intervention for labeling the input data. Meanwhile, unsupervised learning requires limited human intervention as the data is not required to be labeled. The algorithms, using unsupervised learning, detect hidden patterns in unlabeled data.

The Rise of GPT Models

In the last few years, OpenAI, an artificial intelligence and deployment company, has developed the Generative Pre-trained Transformer (GPT) models to assist in language prediction (Brown et al., 2020). GPT models have been shown to have increased levels of performance in text summarization, language translation, textual entailment, etc. GPT-1, the first GPT model, had a semi-supervised learning for natural language (Radford et al., 2018). This allowed for unsupervised language modeling with a large corpus of text, eventually followed by supervised fine-tuning. According to OpenAI researchers, supervised learning requires large, cleaned datasets, which are relatively expensive to work with. Therefore, unsupervised learning helps, especially with the availability of raw data. This was indicated by the results of GPT-1, a semi-supervised model that performed better than the supervised specific-domain models (Radford et al., 2018). More importantly, GPT-1 performed incredibly well on zero-shot learning, which entails learning by a model without the opportunity to refer to any previous examples. With the substantial results evidenced by GPT-1's performance, NLP models with larger datasets could transform NL. Keeping this in mind, OpenAI scaled up the data for GPT-2, a large-scale unsupervised language model (Radford et al., 2019). GPT-2 was trained on a dataset of eight million web pages with 1.5 billion parameters. This was more than ten times the data that GPT-1 was trained on. With one simple human written sentence, GPT-2 was able to generate written samples that seem indistinguishable from NL (Martin, 2019). As far as zero-shot training is concerned, GPT-2, like its predecessor, was able to perform substantially well without any domain-specific data being provided. More importantly, GPT-2 was able to perform with an accuracy of 92–96%, equivalent to humans, on several language tasks.

In 2020, OpenAI came up with GPT-3, an autoregressive language model trained on 175 billion parameters (Brown et al., 2020). To put it in context, GPT-3 has ten times more parameters than its closest competitor, Microsoft's Natural Language Generator Model, and 100 times more parameters than its predecessor, GPT-2. For training GPT-3, researchers used the Common Crawl dataset, which constituted nearly a trillion words. The Common Crawl dataset contains petabytes of data collected over the years and includes webpage data, text extracts, and metadata extracts. Researchers also added Wikipedia pages and a large corpus of books to train the model. Owing to its large capacity, GPT-3 was able to clearly outperform GPT-2 on several parameters. Some of the things that GPT-3 can do is write codes, design web templates, write symphonies, autocomplete images, answer queries, etc. Large LMs like GPT-3 recognize patterns in the available text data to allow for predictions. GPT-3 learns from these patterns and applies them to its performance of different tasks. The process of providing examples for GPT-3 to learn and predict is called in-context learning. In most cases, GPT-3 is provided with input descriptions and examples of the task to be completed, whereas the final task, similar to that of the input description, must be completed by the model. Learning from a few previous prompts without the addition of previous training data is called "few-shot learning" (Brown et al., 2020). Thus, the essential difference is the number of examples available in the training dataset. "Few-shot learning" entails a few examples, whereas "zero-shot learning" entails the absence of examples. Consider the following example of a translation prompt:

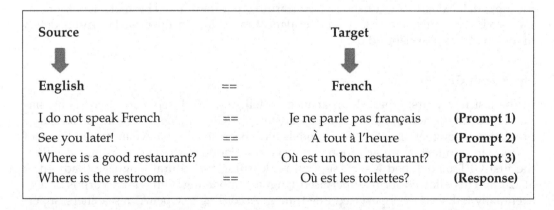

Source		Target	
English	==	French	
I do not speak French	==	Je ne parle pas français	**(Prompt 1)**
See you later!	==	À tout à l'heure	**(Prompt 2)**
Where is a good restaurant?	==	Où est un bon restaurant?	**(Prompt 3)**
Where is the restroom	==	Où est les toilettes?	**(Response)**

Sample 1: English text into French prompt (translation) (GPT-3)

As seen above, these prompts provide GPT-3 with valuable instructions on how to perform the new tasks. Another example of GPT-3's prowess is its query-solving ability. Take, for instance, Marv, a factual chatbot, which gives sarcastic replies. The first few prompts are provided by the user. Thereafter, the model will have the ability to answer the succeeding queries.

Prompt 1-

User: How many pounds are in a kilogram?

Marv: This again? There are 2.2 pounds in a kilogram. Please make a note of this.

Prompt 2-

User: When did the first airplane fly?

Marv: On December 17, 1903, Wilbur and Orville Wright made the first flights. I wish they'd come and take me away.

Response-

User: Why is the sky blue?

Marv: I'm not sure. I'll ask my friend Google.

Sample 2: Conversation with a sarcastic factual chatbot (GPT-3)

Evidently, there are many applications for GPT-3. The performance shown by GPT-3, in some regards, indicates that language models are here to stay and transform businesses in a positive way.

Mapping the Implications of GPT-3 in the Field of Law

Inevitably, GPT-3 can impact the functioning of several businesses by making predictions to generate human-like texts. As of 2021, according to OpenAI, more than 300 applications are using the GPT-3 model. A conspectus of the companies that use GPT-3 technology reveals an array of businesses from customer service chatbots to gaming applications. Consider the following applications/companies – (1) Viable, an application which helps companies structure customer feedback, essentially by using GPT-3 to identify themes, emotions, and sentiment from surveys and obtaining a summary within seconds; (2) Fable Studio, a virtual reality entertainment company that has used GPT-3 to power "Virtual Beings", a project aimed at creating interactive stories; and (3) Algolia, a search-as-a-service platform that provides consumers optimal search and discovery experiences across the digital platform, which has used GPT-3 in its recent "Algolia Answers" product for fast and relevant search results. These instances are a testament to the growing reach and proliferation of GPT-3 across the board. As such, it follows that, before long, the use of GPT-3 will permeate into the legal profession.

GPT-3 in the Legal Profession: Pushing the Envelope for the Applications of AI in Law

Research on AI in law has been on the block for over 70 years. Throughout these years, research on the application of AI in law has developed exponentially, especially after law firms began utilizing AI for different services (Kauffman & Soars, 2020). According to Yu and Alì (2019), AI systems in law can be grouped into three macro areas. The first of these is "unstructured data analysis and due diligence", which includes document analysis, contract analysis, and discovery. For instance, companies like Kira Systems, LawGeex, and eBrevia provide services that review documents, evidence, and contracts, drastically reducing manual review time and increasing accuracy. The second group is "legal research and analytics", where AI applications perform swift legal research and predict litigation outcomes. LexisNexis's Lex Machina is famous for performing such outcome analysis, while also helping with crafting winning strategies. Lastly, the third group constitutes practice management applications, which assist lawyers in drafting documents and briefs. With the increase of data, advancement of algorithms and glaring inefficiencies in the legal practice, language models appear to be the next step in AI research related to law. Generally, LMs have been used for sentiment analysis, forming textual summaries, extracting legal content, information retrieval, answering queries, and predicting case outcomes. Recently, researchers have been taking a lot of interest in making datasets and evaluating pre-trained models on these datasets (Hendrycks et al., 2021a). Probably, the most important research on LM in law is carried out by fine-tuning "Bidirectional Encoder Representations from Transformers", or simply "BERT", a pre-trained model, on legal documents. Elwany et al. (2019) have noted that a large legal lexicon adds valuable improvements to BERT's performance on almost all parameters, i.e., classification tasks, accuracy, and training speed.

As far as GPT-3 is concerned, it can perform most functions that current domain-specific law AI technologies are able to perform. As soon as the GPT-3 model was released, developers started to use the API to perform legal functions. Some of the results, which were posted by the developers on Twitter, have been reproduced in Table 5.1.

Apart from developers posting their views about GPT-3 on Twitter, Hendrycks et al. (2021b) carried out possibly the only study that empirically evaluates the performance of GPT-3 on a range of 57 different subjects from STEM to social sciences, including law. They

TABLE 5.1

Researchers Using GPT-3 to Perform Domain-Specific Tasks

Researcher	Input Data	Instruction	Output by GPT-3
Daniel Gross (Gross, 2020)	26 US Code § 2801 (Imposition of Tax)	Plain English translation of the section	"If you get money from someone who is not living in America anymore because they gave up citizenship, you have to give extra taxes on it".
Francis Jervis (Jervis, 2020)	Plain English sentences into legalese (with two prompts) E.g., "The people I am suing are my landlords"	Write it like an attorney	"The defendants are, in some manner, the owners and lessors of the real property where the plaintiff resides"
Michael Tefula (Tefula, 2020)	Contract legalese into plain English (with two prompts) E.g., "The sale of all or substantially all of the assets of the company or sale of shares involving change in control will be treated in the same way as liquidation."	Plain English translation of the contractual provision	"If the company is sold or the new owner takes control the proceeds of the sale will be distributed as in the liquidation clause above."

(Gross, 2020; Jervis, 2020; Tefula, 2020)

found that GPT-3 was able to assess language in greater detail than its previous models. However, GPT-3 showed lopsided performance and knowledge gaps with respect to certain subjects. For instance, the model was able to achieve a 69% accuracy for US foreign policy and only a meager 26% accuracy for college chemistry. For GPT-3, the lowest accuracy has been felt in the STEM fields, where domain-specific understanding of calculus and mathematics is required. Interestingly, under the humanities group, law was subdivided into professional law, jurisprudence, and international law. The model was able to perform extremely well on factual questions of international law and jurisprudence with extremely high accuracy, whereas on the professional law front, the model showed a very low accuracy percentage, just above the random chance mark. One reason for the deviation in performance was the presence of hypothetical application questions posed under the professional law category. The researchers noted that it was necessary to see whether the open-ended learning models were able to apply legal knowledge to the hypothetical scenarios. This assumes relevance in the context of contractual interpretation, which hinges on situational readings of the bare terms of the contract. The same shall be dealt with at length in the subsequent chapter.

GPT-3 as an Aid to Contractual Interpretation: Tipping the Scales against the Consumer?

Almost all consumer contracts are standard form contracts, which are boilerplate. Consumers have little bargaining power when it comes to standard form contracts, primarily due to either the lack of alternatives or all competitors offering similar terms (Radin, 2013). In most cases, the vendor often has bargaining power, which makes it difficult for the consumer to negotiate favorable terms. Therefore, there is a tendency on part of the consumers to disregard information by not reading the contracts or to decide on intuition. Considering this, most firms strategically place one-sided clauses in contracts, which affect the readability of the contract itself (Benoliel & Becher, 2019). The legalese in these contracts is also often difficult to understand for not only the consumers but also the judges. Eisenberg (1995), in his paper aptly titled "Text Anxiety", points to the oral arguments in the case of *Gerhardt v. Continental Insurance Co.* (1966), where the judges read out clauses of the insurance policy in question and said, "I don't know what it means. I am stumped. They say one thing in big type and in small type they take it away". Eisenberg (1995) then follows this with a question – when the consumer knows he will probably be unable to understand dense text, why should he read it? Another cause of non-readership is the rational apathy that the consumers may face when it comes to the reading of standard form contracts. Plaut and Bartlett's (2012) study indicates that 17.6% of the participants pointed out that "I (consumer) really don't care" showcasing apathy or indifference. This, of course, boils down to the cost involved in reading, knowing, understanding, and acknowledging the terms of the contract, which the consumer is consciously unwilling to take. Consumers' beliefs also play an important role in the reading of contracts. Consumers may believe that no one reads the contractual terms and, therefore, following others' behavior makes sense. In turn, too much interrogation regarding the substance of contracts may imply distrust towards the other party (Hillman & Rachlinski, 2002). More importantly, consumers who inquire about policies and terms are often ridiculed and termed as "nudnicks" – a Yiddish term loosely translated as "a pest" (Arbel & Shapira, 2020). Arbel and Shapira argue that nudnicks are those consumers who complain, post online reviews, and, occasionally, file lawsuits against the sellers. Sellers, unfortunately, may tend to depict the nudnick as a troll who is out for revenge. That said, Arbel and Shapira suggest that many consumers, who may be termed as "nudnicks", do care about the services provided by the seller. The problem arises when the seller portrays them as idiosyncratic or someone who is out to take revenge. Furthermore, research in the field of psychology shows that the concept of attention can be voluntary and involuntary. In most cases, the aim of the consumers is to maximize efficiency and, therefore, they will put in fixed energy into the product attributes rather than focusing their attention on the attributes unimportant to them (Becher, 2007). Thus, most empirical studies indicate that even among consumers who read contracts, most of them skim through the clauses in the standard form pertaining to product attributes, price, functionality, instead of navigating through, *inter alia*, dispute settlement, indemnity, and force majeure boilerplates. Ultimately, reading of standard form is a rarity among consumers and most empirical evidence does indicate the same. As Ben-Shahar (2009) points out, "Dedicated readers can expect only heartache, which is a very poor reward for engaging in such time-consuming endeavor".

To curb the problem of non-readership, academicians, in the past, have given certain suggestions. Ben-Shahar (2009) has argued that using non-legal techniques such as rating and labeling contracts might increase readership. Meanwhile, Ayres and Schwartz (2014) have proposed a system where market sellers engage in the process of "term substantiation". This process would require terms that meet consumer expectations and, thus, can be enforceable. However, terms that are unfavorable could be disclosed in the "warning box" that has a government standard border. This way the "no reading problem" in consumer contracts could be solved. Arbel and Becher (2022) have recently suggested that technology might be useful for solving the problem of non-readership and, therefore, suggest employing the usage of a smart reader.

Smart Readers: A Conspectus

The term "smart reader" refers to an application which is built on a machine learning model (Arbel & Becher, 2022). For the sake of simplicity, Arbel and Becher coin the term and use it to refer to mobile applications that have LMs, such as GPT-3. In their recent paper, they use the GPT-3 technology to produce certain examples to show that the readability of contracts can increase among consumers. If GPT-3 as a model is indeed applied to a "smart reader" application, the functioning of it would be as follows (Figure 5.1).

Once a consumer opens the application and the text is scanned, the consumer can select any of the following options: (1) "explain", providing a succinct summary of the clause; (2) "example", providing hypothetical examples for understanding the clause; (3) "translation", translating contractual clauses from English to other languages; (4) "simplification", providing direct and easily understandable sentences for a lucid explanation of the clauses; (5) "meaning", asking queries/follow-up questions regarding the meaning of the text; (6) "benchmarking", allowing for comparisons with the terms of standard form contracts of the competitors in the market.

By explaining the terms and simplifying them for the layperson, there is a resultant reduction in the cost of processing information, leading to an uptake in the usage of smart reader technology. Additionally, Arbel and Becher (2022) have argued that smart readers may be useful in realizing the informed minority theory. This theory argues that as long as there is a substantial minority of consumers that read and negotiate contracts, the terms

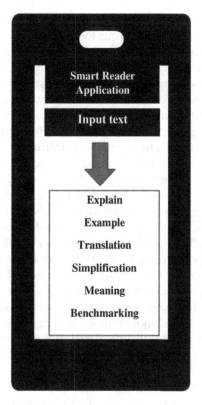

FIGURE 5.1
A sample "smart reader" application made by the author. The same is based on the idea of "smart readers" by Arbel and Becher (2022).

will reflect consumer preferences. More importantly, by reducing search costs and term optimism, smart readers might prove to be a valuable nudging tool to improve decision making. Nevertheless, smart readers are prone to mistakes, inevitably raising the cost that consumers may have to bear in some cases.

Problems with Smart Readers

Arbel and Becher (2022) have identified certain problems related to smart readers and, thereby, also with GPT-3. The first problem identified by them entails the errors in interpretation caused by smart readers. Their argument to address this problem is that smart readers may tend to make less mistakes than humans. Correspondingly, they also mention that even if the accuracy of the current technology does not match the human level, over time, the technology's accuracy will grow rapidly. However, this is quite a simplistic view. It is true that emotions and biases do cloud the judgment of humans, which might affect the interpretation of contractual provisions (Zacks, 2015). Nevertheless, the process of contract interpretation is extremely complex and does not necessarily rely on a literal understanding of the provision (Mitchell, 2018). In fact, in most cases, contractual interpretation requires the use of reasonable person standards to ascertain the meaning of specific clauses (Arnold v. Britton, 2015; Bank of India v. K. Mohandas, 2009). This can be done by understanding the circumstances, context, and relevant background. The simplistic approach of the smart reader technology would be unable to give the contractual term its proper contextual meaning, thereby leading to adversity for the consumer. Contractual interpretation may hinge on a particular term that is ambiguous and, therefore, as of this moment, even an application run by GPT-3 may not be equipped to provide higher accuracy in interpretation as compared to humans. Catterwell (2020) notes that contractual interpretation cannot be automated in its entirety. Nonetheless, logical design can be employed to automate different aspects of the process. Two important interpretation challenges that Catterwell rightly points out are, firstly, that there are no unifying theories, and secondly, oftentimes interpretation hinges on intuition. Thus, if a consumer gets an incorrect explanation of the contractual term, apart from the cost which will be incurred, he may forego using the technology altogether. The second challenge is that correct interpretation is specific to the parties and dispute. Therefore, even a large legal corpus trained GPT-3 model that can peruse through data may be unable to predict outcomes. Moreover, since contract clauses cannot be read in isolation, it would be necessary to show a contract excerpt to the smart reader (Kolt, 2022). Thus, for better results, a consumer, if interested in a question/follow up, may be required to provide an input of more than the clause. As noted by Kolt (2022), GPT-3 performed considerably better when the contract excerpt was shown, essentially showcasing that the technology uses the contract text to answer questions and not merely input data gained during pre-training. Inevitably, this increases the search cost of the consumers, and most may even forego using such technology (Seiler, 2012). Moving forward with this argument, Arbel and Becher (2022) provide an example of the lack of understanding among the consumers in reading privacy policies. The sellers may be able to take advantage of this by providing one-sided terms in the contract (Marotta-Wurgler, 2007). In the context of privacy, empirical evidence shows that consumers often ignore privacy policies (Obar & Oeldorf-Hirsch, 2020). Arbel and Becher (2022) may be right in saying that smart readers, with the GPT-3 technology, might give consumers the opportunity to understand intricate terms. However, Kolt's (2022) research shows that when GPT-3 was asked a simple question such as "Can Instructure (company) back up my data without asking me?", despite the obvious answer being "no", the model's output was a "yes" with high confidence. Considering this, if the consumer complies with the smart reader's answer, there might be devastating consequences, inevitably affecting consumers' right to access justice.

The reason for GPT-3's incorrect answer may probably be the anti-consumer bias that the model has (Kolt, 2022). This is far more insidious than a mere interpretation problem. Arbel and Becher (2022) argue that it is possible to have correlated errors where the smart reader would misinterpret or misread certain terms in specific ways. However, they argue that such models rely on statistical patterns and, therefore, the errors are rare. If one considers online contracts and website terms and conditions, they would be counted as atypical standard form contracts. Considering the large corpus GPT-3 has been trained on, it is possible that the data has a certain number of contracts. For the sake of simplicity, Kolt (2022) has divided these contracts, specifically contractual provisions, into three parts: (1) pro-company provisions, which exempt companies from liability, (2) pro-consumer provisions, which protect consumer rights, and (3) neutral provisions, which do not favor either side. Interestingly, when questions were posed to GPT-3, Kolt found that GPT-3 had an accuracy of 83.64% when answering pro-company questions, whereas GPT-3's performance was dismal, with an accuracy of 60%, when answering pro-consumer questions. A shocking 24 percentage point difference was noticed. The disparity in performance could be for three identifiable reasons. Firstly, there could be a bias in the model. Since GPT-3 is unsupervised and trained on a large corpus of web data, it is possible that the millions of website terms were pro-company, essentially due to the drafting of these contracts by company counsels in favor of the company. Secondly, there could be a bias in contracts. GPT-3's performance may have been poor due to the contract excerpt being pro-company, even though the contract provision may be pro-consumer. Thirdly, there could be a bias in the question formulated. If there is a pro-company bias in the model, the smart reader may not be able to provide apt guidance to the consumers. This may result in erosion of trust in the technology itself. Additionally, by relying on the interpretation/answers of the smart readers, the consumers may take risks, which may result in higher costs allocated to them. Therefore, Kolt (2022) elaborates that users of LMs must detect harmful biases and develop strategies. This could be done through policymakers laying down guidelines on the functioning of LMs. Additionally, techniques such as filtering of machine-generated data for future training models and the creation of sub-fields like prompt engineering, which aims at optimizing prompts, might also be beneficial in curbing the complications of the model (Zhao et al., 2021).

However, a discussion on the implications of smart readers on contractual interpretation would be incomplete without considering consumers' "duty to read".

Duty to Read Contracts

Duty to read (DTR) contracts, as the name suggests, states that the contracting parties would be responsible if they gave their assent to a written contract and, therefore, there is an inherent duty to read. The "duty to read" has long been a part of common law contractual jurisprudence, although codified statutes do not impose a general duty of readability of contracts (Calamari, 1974). Proponents of the DTR theory have justified it on primarily three grounds: economic efficiency (Busching v. Griffin, 1985), fairness (Knapp, 2015), and estoppel (Brubaker v. Barrett, 2011). The efficiency argument is grounded, in part, on the assumption that it eases administration. Ben-Shahar (2009) argues that the underlying economic reason for the DTR is to shift the burden of information acquisition to the passive party, generally, the consumers. A consumer would often use cost-benefit analysis in the reading of the contract. Thus, from an economic point of view, the cost of a precautionary step would increase when the consumer must invest more resources. All the causes of non-readership of contracts mentioned previously would increase the cost of readership and, thereby, reduce the incentive to navigate through contracts. However, the smart readers here may have the possibility of expanding the DTR (Arbel & Becher 2022). The courts have previously balanced the DTR doctrine with certain other doctrines in specific cases. For instance, under the doctrine of reasonable expectations, the courts analyze clauses from a reasonable man's point of view (Keeton, 1970). Under the doctrine of *contra proferentem*, the courts prefer an interpretation of an ambiguous contractual provision that is more favorable to the party who has not drafted the agreement (Boardman, 2006). From a fairness point of view, if a consumer with the ability and the opportunity to read the contract decided not to do so, it would be unfair if he was allowed to void the contract (Knapp, 2015). Lastly, the DTR doctrine has also been justified through the doctrine of estoppel by arguing that the one who affirms the contract induces the other party to perform some part of the agreement, resulting in a change of position for the performing party. The signer, thereafter, should be estopped from claiming that he is not bound by the agreement as he did not read it (Brubaker v. Barrett, 2011). The DTR doctrine has been criticized on several grounds. The most prominent criticism of the DTR doctrine is that there is a duty to read the contract, but there is no burden on the sellers to make the contract readable (Benoliel & Becher, 2019). The argument is withstood by the fact that, rather ironically, consumer law professors do not always read contracts or disclosures. Sovern's research points out that 57% of the law professors noted that they rarely read contracts and 48% said that they never read the required disclosures (Sovern, 2018; Sovern, 2008; Sovern, 2010). Thus, it is rather fallacious to assume that a person having no knowledge or understanding of the law has a duty to read.

Smart readers raise interesting questions regarding the DTR doctrine. Arbel and Becher (2022) rightly point out that "If smart readers are cheap and accessible, courts may find it natural to expect consumers to use them". In this situation, Arbel and Becher emphasize the words "cheap and accessible". If one were to assume that the technology is accessible and relatively cheap, the DTR doctrine, inevitably, must be widened. Courts have previously applied the DTR doctrine in cases even when the parties have argued that they had no formal education, were unable to speak or read in English, and were not represented by a legal counsel in their dealings. Therefore, it is very plausible that the courts may make the DTR doctrine wider after the adoption of the technology. As of now, there are several applications and companies that provide text summaries, flag risky contracts, and perform due diligence; most of these are applied by law firms in research and other commercial/litigation practices (Yu & Alì, 2019). Nevertheless, it would be a long time before any "smart reader" might hit the market. Even if it does, the problems associated with the technology may plague it from being adopted widely. Arbel and Becher (2022) are right in noting that "the technology is already sufficiently impressive to mislead an inexperienced person into believing that it is more effective than it actually is". Considering that the legal profession is often categorized as slow to adopt legal technology, the wide adoption of NL models or smart reader applications cannot be said to be a stone's throw away.

Conclusion

At first blush, it would appear that the use of LMs, particularly GPT-3, in smart readers may bolster the consumer understanding and readability of standard form contracts. As such, a case can be made for the use of GPT-3 technology in smart readers to bridge the gap between complex legalese and laypersons' understanding of contracts. However, the foregoing discussion reveals that as an emergent technology, GPT-3 is saddled with systemic issues, some of which are particularly alarming in this context. The anti-consumer bias, which has been evidenced through Kolt's (2022) empirical analysis of GPT-3, goes on to highlight the pitfalls at the heart of the very argument in favor of smart readers, as consumers may unwittingly rely on technologies that are skewed against them. To that end, the technology needs to be refined to achieve its truly pathbreaking potential. It is undeniable that GPT-3 has pushed the envelope insofar as LM technology is concerned. Nevertheless, reliance on the same should be placed with caution as far as contractual interpretation is concerned. Lastly, LMs, as with any technology, represent a minefield of opportunities that must be treaded upon with wonder and caution in equal measure.

References

Arbel, Y., & Becher, S. (2022). Contracts in the age of Smart Readers. *George Washington Law Review*. 90, 83.

Arbel, Y., & Shapira, R. (2020). Theory of Nudnik: The future of consumer activism and what can we do to stop it. *Vanderbilt Law Review*. 73(4), 929.

Arnold v. Britton [2015] UKSC 36.

Ayres, I., & Schwartz, A. (2014). The no-reading problem in consumer contract law. *Stanford Law Review*. 66(3), 545–610.

Bank of India v. K. Mohandas (2009) 5 SCC 313.

Becher, S. (2007). Behavioral standards and consumer standard form contracts. *Louisiana Law Review*. 68(1), 117–179.

Benoliel, U., & Becher, S. (2019). The duty to read the unreadable. *Boston College Law Review*. 60(8), 2255–2296.

Ben-Shahar, O. (2009). The myth of the 'opportunity to read' in contract law. *European Review of Contract Law*. 5(1), 1–28. https://doi.org/10.1515/ERCL.2009.1.

Boardman, M. (2006). Contra Proferentem: The allure of boilerplate. *Michigan Law Review*. 104, 1105–1128.

Brown, T., et al. (2020). *Language Models are Few-Shot Learners*. ARXIV. https://arxiv.org/abs/2005.14165.

Brubaker v. Barrett, 801 F. Supp. 2d 743, 751 (E.D. Tenn. 2011).

Burton, S. (2018). The case for plain language contracts. *Harvard Business Review*. https://hbr.org/2018/01/the-case-for-plain-language-contracts.

Busching v. Griffin, 465 So. 2d 1037 (1985).

Calamari, J. (1974). Duty to read- A changing concept. *Fordham Law Review*. 43(3), 341–362.

Catterwell, R. (2020). Automation in contract interpretation. *Law, Innovation and Technology*. 12(1), 81–112. https://doi.org/10.1080/17579961.2020.1727068.

Eisenberg, M. (1995). The limits of cognition and the limits of contract. *Stanford Law Review*. 47(2), 211–259.

Elman, J. (1990). Finding structure in time. *Cognitive Science*. 14(2), 179–211. https://doi.org/10.1016/0364-0213(90)90002-E.

Elwany, E., et al. (2019). *BERT Goes to Law School: Quantifying the Competitive Advantage of Access to Large Legal Corpora in Contract Understanding*. ARXIV. https://arxiv.org/abs/1911.00473.

Gehring, J., et al. (2017). *Convolutional Sequence to Sequence Learning*. ARXIV. https://arxiv.org/abs/1705.03122.

Gerhardt v. Continental Insurance Co. (48 NJ. 291, 225 A.2d 328 (1966)).

Goodfellow, I., et al. (2016). *Deep Learning (Adaptive Computation and Machine Learning Series)*. MIT Press.

Gross, D. (2020, June 15). I fed federal tax law to OpenAI's model and asked for a summary of a section. Input on the left, output on the right. *Pretty Nuts* [Tweet]. Twitter. https://twitter.com/danielgross/status/1272238098710097920.

Hendrycks, D., et al. (2021a). *CUAD: An Expert Annotated NLP Dataset for Legal Contract Review*. ARXIV. https://arxiv.org/abs/2103.06268.

Hendrycks, D., et al. (2021b). Measuring massive multitask language understanding. In *Proceedings of the 9th International Conference on Learning Representations*. https://doi.org/10.48550/arXiv.2009.03300.

Hillman, R., & Rachlinski, J. (2002). Standard-form contracting in the electronic age. *NYU Law Review*. 77(2), 429–495.

Hochreiter, S., & Schmidhuber, J. (1997). Long short-term memory. *Neural Computation*. 9(8), 1735–1780. https://doi.org/10.1162/neco.1997.9.8.1735.

https://doi.org/10.1017/S1472669619000021.

Jervis, F. (2020, June 15). GPT-3 performance on "write this like an attorney" is insane. It even includes relevant statutes if you mention a jurisdiction. This will put a low of lawyers out of work. Only the first 2 prompts [Tweet]. *Twitter*. https://twitter.com/f_j_j_/status/1283349995144359937.

Johnson, K. (2021, June 06). The efforts to make text-based AI less racist and terrible. *Wired*. https://www.wired.com/story/efforts-make-text-ai-less-racist-terrible/.

Johri, P., et al. (2021). Natural language processing: History, evolution, application, and future work. In *Proceedings of 3rd International Conference on Computing Informatics and Networks*. Springer. https://doi.org/10.1007/978-981-15-9712-1_31.

Jozefowicz, R., et al. (2016). *Exploring the Limits of Language Modelling*. ARXIV. https://arxiv.org/pdf/1602.02410.pdf.

Jurafsky, D., & Martin, J. (2008). *Speech and Language Processing*. Pearson.

Kauffman, M., & Soares, M. (2020). AI in legal services: New trends in AI-enabled legal services. *Service Oriented Computing and Applications*. 14, 223–226. https://doi.org/10.1007/s11761-020-00305-x.

Keeton, R. (1970). Insurance law rights at variance with policy provisions. *Harvard Law Review*. 83, 961.

Knapp, C. (2015). Is there are "duty to read"? *Hastings Law Journal*. 66, 1083–1112.

Kobayashi, G., et al. (2020). Attention is not only weight: Analyzing transformers with vector norms. In *Proceedings of the 2020 Conference on Empirical Methods in Natural Language Processing (EMNLP)*. http://dx.doi.org/10.18653/v1/2020.emnlp-main.574.

Kolt, N. (2022). Predicting consumer contracts. *Berkeley Technology Law Journal*. 37, Forthcoming.

Li, J., et al. (2016). *Visualizing and Understanding Neural Models in NLP*. ARXIV. https://arxiv.org/pdf/1506.01066.pdf.

Marotta-Wurgler, F. (2007). What's in a standard form contract? An empirical analysis of software license agreements. *Journal of Empirical Legal Studies*. 4(4), 677–713. https://doi.org/10.1111/j.1740-1461.2007.00104.x.

Martin, N. (2019, February 19). New AI development so advanced, it's too dangerous to release, says scientists. *Forbes*. https://www.forbes.com/sites/nicolemartin1/2019/02/19/new-ai-development-so-advanced-its-too-dangerous-to-release-says-scientists/?sh=2e84aefb4a80.

Mitchell, K. (2018). *Interpretation of Contracts*. 2nd edn. Routledge-Cavendish.

Niu, Z., et al. (2021). A review of the attention mechanism of deep learning. *Neurocomputing*. 452, 48–62. https://doi.org/10.1016/j.neucom.2021.03.091.

Obar, J., & Oeldorf-Hirsch, A. (2020). The biggest lie on the internet: Ignoring the privacy policies and terms of service policies of social networking services. *Information, Communication & Society*. 23(1), 128–147. https://doi.org/10.1080/1369118X.2018.1486870.

Phi, M., (2018, September 18). Illustrated guide to recurrent neural networks: Understanding the intuition. *Medium*. https://towardsdatascience.com/illustrated-guide-to-recurrent-neural-networks-79e5eb8049c9.

Plaut, V., & Bartlett, R. (2012). Blind consent? A social psychological investigation of non- readership of click-through agreements. *Law and Human Behavior*. 36(4), 293–311. https://doi.org/10.1037/h0093969.

Qiu, X., et al. (2020). Pre-trained models for natural language processing: A survey. *Science China Technological Sciences*. 63, 1872–1897. https://doi.org/10.1007/s11431-020-1647-3.

Radford, A., et al. (2018). Improving language understanding by generative pre-training. *Open AI Working Paper*. https://s3-us-west-2.amazonaws.com/openai-assets/research-covers/language-unsupervised/language_understanding_paper.pdf.

Radford A., et al. (2019). Language models are unsupervised multitask learners. *Open AI Working Paper*. https://d4mucfpksywv.cloudfront.net/better-language-models/language_models_are_unsupervised_multitask_learners.pdf.

Radin, M. (2013). *Boilerplate: The Fine Print, Vanishing Rights, and the Rule of Law*. Princeton University Press.

Sak, H., et al. (2014). *Long Short-Term Memory Based Recurrent Neural Network Architectures for Large Vocabulary Speech Recognition*. ARXIV. https://arxiv.org/abs/1402.1128.

Sarker, I. (2021). Machine learning: Algorithms, real-world applications and research directions. *SN Computer Science*. 2, 160. https://doi.org/10.1007/s42979-021-00592-x.

Seiler, S. (2012). The impact of search costs on consumer behavior: A dynamic approach. *Quantitative Marketing and Economics*. 11(2). http://dx.doi.org/10.1007/s11129-012-9126-7.

Sovern, J. (2008). The content of consumer law classes. *Journal of Consumer & Commercial Law*. 12(1), 48–51.

Sovern, J. (2010). The content of consumer law classes II. *Journal of Consumer & Commercial Law*. 14(1), 16–22.

Sovern, J. (2018). The content of consumer law classes III. *Journal of Consumer & Commercial Law*. 22(1) 2–10.

Tefula, M. (2020, July 21). Just taught GPT-3 how to turn legalese into simple plain English. All I gave it were 2 examples. Might build a term sheet and investment document interpreter out of this [Tweet]. *Twitter*. https://twitter.com/michaeltefula/status/1285505897108832257.

Vaswani, A., et al. (2017). Attention is all you need. In *Proceedings of the 30th International Conference of Neural Information Processing Systems*. https://arxiv.org/abs/1706.03762.

Wilkinson-Ryan, T. (2014). A psychological account of consent to fine print. *Iowa Law Review*. 99, 1745–1784.

Yu, R., & Alì, G. (2019). What's inside the black box? AI challenges for lawyers and researchers. *Legal Information Management*. 19(1), 2–13.

Zacks, E. (2015). Contract review: Cognitive bias, moral hazard, and situational pressure. *The Ohio State Entrepreneurial Business Law Journal*. 9(2), 379–427.

Zhao, T., et al. (2021). Calibrate before use: Improving few-shot performance of language models. *Proceedings of the International Conference on Machine Learning*. https://arxiv.org/abs/2102.09690.

6

The Impact of New Technologies on Taking Evidence in Civil Court Procedures: Is There a Need for a Conceptual Change?

Tjaša Ivanc and Cocou Marius Mensah

CONTENTS

The coronavirus epidemic has caused stagnation in the economy and some restrictions on access to justice. However, to maintain legal predictability and security in the new circumstances and to ensure the same level of citizens' rights in court proceedings, it is necessary to provide a certain legal framework that will allow the courts to function and communicate between courts and citizens. In the context of Covid-19, electronic evidence has become increasingly important and particularly relevant for national courts since it is necessary to use special tools to process and manage electronic evidence in an appropriate and secure manner.

While the law of evidence continues to develop and, at present, documents and data are likely to be transformed into electronic form, the use of IT should not diminish the procedural safeguards. Moreover, it should not deprive the parties of their rights to an adversarial hearing, to the production of original evidence, to have witnesses or experts, and to present and submit valuable materials. The national procedural legislations about evidence taking mainly deal with physical evidence, and the development of electronic evidence has not yet been taken into account in the majority of national rules of civil procedure. Electronic evidence must, as any other form of evidence, meet specific requirements in terms of the integrity of the electronic document and the identification of its author. A variety of issues may be considered within this context, such as the legal standards ensuring the trustworthiness of the documents (a legal requirement), the authenticity of electronic evidence, or digital archiving. A significant question that needs to be discussed is how digital data is or could

DOI: 10.1201/9781003215998-6

be assessed by the court and whether there is a need to establish new procedural rules for obtaining evidence in this regard. For example, digital evidentiary materials include e-mails, computer data, electronic records, audio or video recordings of party or witness statements, and personal information on social networks. This evaluation is closely linked to court culture, the ways of communication between the parties and the court (electronic filing and service), and the use of videoconferencing to hold the hearings.

However, it remains necessary to determine the place of information and communication technology (ICT) next to the oral and written character of procedural acts. It could be further argued that with new ICT, the principle of orality has gained new content. Above all, when using new technologies, the civil procedure rules have to overcome the obstacle of breaking the direct contact between the parties and the court in the procedure. This depersonalization will influence the principle of directness or immediacy (principle of presence) and the contact with the source of evidence. A procedure held by videoconference respects the orality and the publicity but not the presence of the parties; thus, the new technologies reveal the risk of losing the physical link that is essential for efficiency in litigation.

Due to recent developments in the use of ICT in different fields of everyday life and in the field of justice, the concept of e-justice used in the frameworks of both European procedures and national justice systems should be further defined. In the area of e-justice, the EU has already undertaken several initiatives. One of the most important ones is the e-Justice initiative, which aims to integrate all civil, criminal, and administrative law areas. The two most visible parts of European e-justice are the European e-Justice Portal[1] and e-CODEX (e-Justice Communication via Online Data Exchange).[2] The development of e-Justice[3] and e-courts is a strong European trend. Recent ongoing reforms in fields, such as electronic registers, databases for judicial decisions, electronic court files, electronic signatures, or case management systems, are a reality.[4]

Digital evidence must, as any other form of evidence, meet certain requirements in terms of the integrity of the electronic document and the identification of its author. Digital evidence deals with the definition of a document and whether the carrier of the relevant information is of significance. For example, are sounds, video recordings, or digital images to be regarded as documents? Or are they within the scope of an inspection as a means of evidence? In addition, how does the principle of the free evaluation of evidence relate to the evaluation of the credibility and authenticity of electronic documents and information produced and stored by electronic means? Moreover, problems with the applicability of ICT additionally arise in cross-border cases, as the different procedural structures in national jurisdictions and the tendency to use videoconferencing (VCF) are encouraged by EU civil judicial cooperation instruments. Specifically, obtaining evidence is more complicated when it has to be obtained from another country; barriers can be created by the physical distance between the court and the person to be examined. In this context, the chapter will analyze the challenges of the cross-border taking of evidence on the basis of the new EU evidence regulation. Furthermore, the Hague Conference on Private International Law (HCCH) recently published a guide to good practice on the use of video links under the Convention of 18 March 1970 on the Taking of Evidence Abroad in Civil or Commercial Matters (the Hague Evidence Convention).

This article will discuss the significant question of how digital information is or should be evaluated by the court. We will analyze whether there is a need to create new procedural rules of evidence taking, in this respect, digital evidentiary materials, including e-mails, computer data, electronic records, audio or video recordings of party or witness statements, and personal information on social networks. This evaluation is closely linked to court culture, the ways of communication between the parties and the court (electronic filing and service), and the use of videoconferencing to hold court hearings. The evidentiary stage is the core of civil litigation, and it is directly linked to the special principles of evidence taking. The main result of the introduction of electronic proceedings is certainly a shift from strictly fixed proceedings, recorded on paper, to dematerialized, more flexible proceedings (de Resende Chaves Junior, 2012, p.116).

The Interface of the Law and ICT

Theory (Lupo & Bailey, 2014, p.356) emphasizes that the relationship between law and the development of new technological solutions for implementation into civil procedures could be characterized as potentially conflicting.[5]

This conflict arises from the objectives of both subject areas. While the objective of civil procedure is to normatively ensure basic, regular, and general methods for the protection of subjective rights of the parties, technology serves as a functional tool and is "outcome-oriented" (Mohr & Contini, 2011), serving new commodities for all subjects involved in judicial procedures. On the one hand, the law is used to regulate technology; but on the other hand, it also employs ICT to pursue its own goals. Technologies thus may operate as an auxiliary tool for better achieving a legal aim (Palmerini, 2013).

The placement of ICT in regulated court environments requires sets of formal rules, and a range of activities must be given legal or valid effect. Formal regulation is needed in filing claims via online systems, hearings held via videoconferences, and electronic service of court documents. It means that the validity of such procedural actions is judged from a legal perspective and not only from a technological perspective, which enables the functionality and efficiency of individual e-justice services.[6]

Therefore, ICT may change the contents of protected legal interests (for example, the right to privacy). Still, the law must provide rules arising from technologies shaped by the features characterizing ICT use to provide judicial protection of legal interest. In some cases, the use of ICT implies the need to re-frame traditional[7] concepts or draw on new court proceedings concepts.

Scholars (Székely et al., 2011) further argue that there is no necessity to introduce new principles because of the rigidity and inertia of technology that follows legal instruments. This is driven by the fact that by the time the specific characteristic of new technologies and their impact on legal relationships are understood to adapt the law to these circumstances, the specific technology has already resulted in circumstances that cannot be modified *post factum* (obsolete). For this reason, principles and legal rules should be defined and created from the experience or at least expectation of using certain technologies (Székely et al., pp.182–183).

The main characteristic of ICT use in civil proceedings that must be emphasized is the shift from strictly fixed proceedings, recorded on paper, to dematerialized, more flexible proceedings.

Electronic Evidence and Its Admissibility

In its traditional meaning, evidence represents information by which facts tend to be proved (Keane & McKeown, 2014, p.2). The parties need to indicate the evidence necessary to prove the facts provided to the court. However, new technical possibilities mean that jurists and legislators are faced with entirely new challenges. E-mails, digital photographs, websites, audio and video recordings, and social media are only some of the new electronic communications that need to be analyzed when taking into consideration the position of ICT and evidence taking in civil procedures.

Electronic evidence is increasingly recognized as an essential source of evidence. However, central to this increasing use of electronic evidence are certain fundamental legal questions related to the definition of electronic evidence, types of electronic evidence, admissibility, assessment, the authenticity of the evidence, and the circumstances under which it can be obtained and then admitted into court. The awareness of the importance of electronic evidence and its place in civil procedures has become significant, which means that the differences between electronic evidence and traditional evidence must be examined, and the possible question of tailoring rules for electronic evidence needs to be considered. Comparatively, different legal systems reacted differently to this modern means of proof in civil proceedings.

In practice, different terminologies are used for electronic evidence; namely, the most commonly used terms are "digital" or "electronic" evidence. The term "digital" is used in computing and electronics and represents "information converted to binary numeric form as in digital audio and digital photography" (Mason, 2010, p.22).

The term "digital" is too broad, and the binary form is restrictive since it includes only one form of data (Mason, 2010). However, "electronic" evidence has "analog" and "digital" evidence, and it should be used as a "generative" term.

More broadly, electronic evidence is defined as "any information created or stored in a digital form whenever the computer is used to accomplish a task" (Chung & Byer, 1998, p.8).

As most Member States apply the analogy of traditional rules of evidence to electronic evidence, electronic evidence needs to fit into the conventional concept. If there is no separate category of evidence that refers to electronic evidence, then electronic evidence needs to correspond to the types of evidence recognized by national regulations. For now, it appears that there is only one procedure for all kinds of evidence; thus, the traditional procedural principles and rules apply to electronic evidence. As we will see below, the examined national regulations apply rules for documents when electronic evidence is in question.

A "technological-neutral" approach should be adopted so that there would be no sufficient differentiation between traditional documentary evidence and electronic evidence, and there would be "no evidential preference applied to any particular technology or mechanical device in adducing documentary evidence" (Law Reform Commission, 2009, p. 1).

However, with an analogous approach, similarities between traditional and electronic evidence need to be identified (Mason & Schafer, 2010).

Legal literature stresses that where the analogous approach is applied, three types of equivalence appear, namely:

- Equivalence between paper and electronic documents.
- Equivalence of electronic signature with the handwritten signature.
- Equivalence of the electronic mail with the postal mail (Insa & Lazaro, 2008).

According to the same research, legal practice considers electronic evidence equal to traditional evidence, arguing that the electronic evidence may be printed out and presented as a paper document (Insa & Lazaro, 2008).

Despite all of the difficulties defining and regulating electronic evidence presents, there are advantages that electronic evidence represents. Electronic evidence possesses certain important characteristics, which relate to reliability since this evidence is more objective than traditional evidence.

The legal practice has highlighted that electronic evidence is exact, complete, clear, precise, true, objective, and represents "neutral information" until the use of ICT was impossible to obtain (Insa & Lazaro, 2008). However, the difficulty that is most commonly expressed in practice relates to the evidential value and the lack of methods for the authenticity verification of the evidence, considering that there is no suitable legal basis for this area of electronic evidence in national regulations.

Furthermore, because of insufficient legislation, the preservation of electronic evidence presents a problem for legal practice (Insa & Lazaro, 2008). Another issue also relates to the necessity to use expert witnesses in the procedure, who will be able to determine the authenticity of electronic evidence with their special technical knowledge (Kodek, 2012).

The importance of how the evidence is collected and then presented in court relates to the means and format by which electronic evidence comes before the court. Electronic evidence usually refers to a sufficient volume of electronic documents, which may be gathered and then introduced to the court. The project "Admissibility of Electronic Evidence" obtained the results from the analysis of the examined legislation in EU Member States relating to the collection and presentation of electronic evidence in court. The findings show that in the examined jurisdictions, there are no special rules applicable solely to the collection and presentation of electronic evidence (Insa & Lazaro, 2008).

Most of the EU jurisdictions have no preservation procedures for electronic evidence compared to the US duty to preserve ESI. According to civil procedure legislation, the parties have no general legal obligation to preserve evidence regarding an anticipated proceeding or civil proceedings. However, legal jurisdictions in Member States provide special proceedings that can start even before the actual trial. If there is a concern that evidence will disappear or be destroyed, or otherwise be lost, a special proposal to preserve evidence can be filed to the court before a formal claim is filed. There are no specific procedures for preserving electronic evidence, so the procedures for preserving traditional evidence might be considered for the analogous application.

The German ZPO provides procedures to preserve evidence that can be launched before the actual trial commences:

> Where there is a concern that evidence might be lost, or that later in the procedure it will become difficult to use it, the party may file a petition to preserve evidence upon which the court may direct in the course of litigation or outside of the proceedings evidence by inspection, what witnesses are to be examined, or that an expert witness prepares a report.

("Beweissicherungsverfahren", section 485)

This provision covers only explicitly listed means of evidence and does not cover handing over the documents by the requested party.

A similar provision as in the German ZPO is included in the Austrian ZPO, which enables the preservation ("Beweissicherung") of evidence before or during the main hearing. The Austrian ZPO enables the preservation of certain means of proof, namely evidence by inspection, witnesses, and experts (Nunner-Krautgasser & Anzenberger, 2009).

Contrary to the Austrian and German jurisdictions, the Croatian (Article 287 CPA) and the Slovenian Civil Procedure Act (Article 264 CPA) do not limit the preservation of evidence to specific means of evidence.

In Estonia, pre-trial discovery is possible to provide securing of evidence or establishment of facts. The Estonian Civil Procedure Code does not specify grounds for securing evidence at the pre-trial stage. It is at the court's discretion to decide whether there is enough justification as to why the evidence has to be secured before the trial (Poola, 2015). If the reason for pre-trial proceedings was securing evidence, the applicant must commence the procedure (Poola).

Parties have no obligation to disclose (harmful) evidence to the other party in civil law jurisdictions. However, the court may order the opposite party to disclose certain evidence at the other party's request. Suppose the opposing party fails to disclose because the evidence has been destroyed on purpose or is not willing to hand over the evidence; in that case, the courts in most EU jurisdictions will consider this when freely assessing evidence (e.g., Austria (section 381 ZPO), Croatia (188 Article 233, paragraph 5 CCPA), Finland (Chapter 17 section 5 Code of Judicial Procedure), Slovenia (paragraph 5 Article 227 CPA), Spain (Article 329 LEC)).

The English Practice direction 31 paragraph 7 deals with the preservation of evidence and provides as follows:

> As soon as litigation is contemplated, the parties' legal representatives must notify their clients of the need to preserve disclosable documents. The documents to be preserved include Electronic Documents that would otherwise be deleted according to a document retention policy or otherwise deleted in the ordinary course of business. After the commencement of proceedings, the duty to preserve evidence begins. If the party fails to preserve evidence, adverse inferences may be imposed for not preserving the evidence.
>
> **(Mason, 2010 p.112)**

The duty to preserve documents has become a leading part of US litigation.[8] This duty arises "when the party has noticed that the evidence is relevant to the litigation or when a party should have known that the evidence may be relevant to a future litigation" (Braman et al., 2009, p.200).

The destruction or failure to preserve evidence after the party knew, or should have known, that the evidence may be relevant to litigation is called "spoliation". If a court deems a party's lack of compliance with a preservation order to be intentional or negligent destruction of evidence, a variety of sanctions can be imposed: from monetary penalties (such as attorney fees, costs, and/or pay-for-proof sanctions) to the exclusion of evidence, delay of the start of the trial, mistrial, or adverse inference jury instructions. In an extreme case, there can be a dismissal or judgment on the merits (Brownstone, 2006).

Requirements for the Admissibility of Electronic Evidence

The lack of formal rules for taking and adducing electronic evidence in most legal systems of the Member States also influences the question of requirements for the admissibility of electronic evidence. According to the findings of the project "Dimensions of evidence in the EU",[9] the European countries (Austria, Bulgaria, Croatia, Denmark, Finland, Germany, Greece, Hungary, Latvia, Lithuania, the Netherlands, Poland, Portugal, Romania, Slovakia, Slovenia, Sweden) have in common the free evaluation of evidence principle. This generally means that there are no legal rules for assessing evidence and that upon this principle, the judge is free to assess the admissibility of electronic evidence.

However, in Ireland, Section 22 of the Electronic Commerce Act 2000, electronic communication, an electronic form of a document, an electronic contract, writing in electronic form, or electronic signatures cannot be held inadmissible on account of its electronic form. The admissibility of electronic evidence depends on whether the admission comes from within common law or statutory exception to the hearsay rule (Cannon, 2007). Courts accept electronic evidence as admissible if the evidence is obtained as a result of a "mechanical process, without the intervention of the human mind". In practice, photographs, tapes, and video recordings have been held admissible (Cannon).

If the free assessment principle applies to electronic evidence, the courts need to be familiar with the unique characteristics that electronic evidence represents. However, related to its characteristics, electronic evidence can be manipulated, but the possibility of seeing it destroyed is less likely.[10] It is necessary to highlight that the authenticity, which relates to the question of whether the documents or data are genuine or authentic, is one of the main features when comparing written and electronic formats of documents.

With a paper document, there is a clear understanding of the original and what represents its copy. Contrary to the electronic form, legal literature (Mason & Schafer, 2010) emphasizes that the important distinction which is problematic is that when we deal with an electronic document, originals and copies are indistinguishable and that the sole act of a person working on the electronic form of the document will "create numerous copies" on the computer. In this respect, the experience from the American evidence gathering should be considered. Namely, legal practice is suggesting that "a new way of looking at how digital information should be considered by the courts" and new rules on electronic evidence, which will not be "based on tradition but on how digital information actually comes into creation" (Paul, 2008, as cited in Marcus, 2012, p.40). A significant question that needs to be discussed here is how digital information is or should be evaluated by the court and whether there is a need for the creation of new procedural rules of evidence considering examples of electronic evidentiary materials such as e-mails, computer data, electronic records, audio or video recordings of the party, witness or expert witness statements, personal information on social networks, and other new types of possible evidence.

Analysis of national regulations showed no special provisions on electronic evidence included in the civil procedure, and the analogy to traditional rules on evidence is applied whenever electronic evidence is being considered. This further poses the question of what conventional evidence category should different electronic evidence be assigned to. This aforementioned point is problematic and is considered in the following discussion.

According to Slovenian law, witnesses give statements and are examined orally (Article 238 CPA). Parties may submit written statements of the witnesses if the court orders them to do so or at the request of the parties (Article 236(a) CPA). Furthermore, experts may give their opinions orally or in written form, depending on the court's decision (Article 253 CPA). The interesting question in practice deals with the admissibility of pre-recorded witness statements submitted by a party as evidence.

Could such recordings be regarded as documents or should other rules apply, for example, the rules on inspection of evidence? If the rules on inspection are applied, the video recording would need to be authenticated, which means that a witness familiar with the video content (e.g., the person who captured the video images) would probably have to be examined. Another approach would be to verify whether the images have been altered in any improper way by using the opinion of an expert witness with special expert knowledge. In contrast, the reliability of written witness statements could be examined by inviting the relevant witnesses to give an oral testimony about the content of their written depositions.

In Austria and Germany, video or audio recordings are not considered to be a document and they generally fall under the rules for evidence by inspection (Articles 368 to 370 ZPO). The same solution may be found in the Croatian doctrine, which is based on the proposition that video, audio, and computer records can be inspected following the general rules of evidence inspection.

The judges have the discretion on how they classify electronic evidence. Nevertheless, the identification of under which traditional category certain electronic evidence falls is certainly difficult. It is necessary to consider whether broadening legal categories of evidence in the civil procedure would be more appropriate than applying traditional rules of evidence.

Paper versus Electronic Form

With the ever-increasing amount of electronic evidence admitted to courts, it is particularly important to provide a precise definition of what constitutes a document. The reason lies in the fact that there is no exact definition of an electronic document and the national regulations mainly apply an analogy of the rules for paper documents to electronic documents. This means that there is no fundamental difference when applying the rules of law of evidence between paper-based documentary evidence and electronic evidence. In all jurisdictions, particularly important evidence is documentary evidence.

As previously mentioned, electronic documents have special characteristics that may be judged differently when authenticity and securing are in question.

There is no legal definition of a written document in the Austrian, Croatian, Danish, German, Czech Republic, Slovenian, Spanish, Swedish, and Slovakian civil procedure law. The most widespread concept of a document for the purpose of civil procedure law is defined by legal theories as the written embodiment of thoughts (e.g., Austria, Croatia, Germany, and Slovenia). If thoughts do not represent a written embodiment, then there is no document, but they could fall, for example, within the rules of evidence by inspection.

However, the rest of the investigated Member States have definitions provided by their civil procedure rules, like Cyprus, Finland, Estonia, Lithuania, Latvia, Romania, and France. In France, Article 1316 of the Civil Code states that "documentary evidence, or evidence in writing, results from a sequence of letters, characters, figures or of any other signs or symbols having an intelligible meaning, whatever their medium and the ways and means of their transmission may be".

In England, according to rule 31.4 of the CPR, a document is defined as "anything in which information of any description is recorded". Andrews (2013) explains that this definition does not include information stored in human memory nor does it not cover no-documentary "things".

Unlike the common law discovery system, most civil code countries have no formal discovery process, and regulations only provide the basis for disclosing evidence in civil proceedings.

The use of electronic documents raises different important questions for legal practice, namely, is the use of electronic documents allowed under the regulations, and in which category of evidence does electronic evidence fall? (Rusmann, 2012).

The rules of the authenticity of electronic documents are essential to ensure the integrity of electronic information or enable persons to demonstrate their entitlement to a certain document.

According to the examined legal regulations, most Member States do not provide any definition for electronic documents but still refer to electronic evidence in other provisions that regulate written documents (e.g., Austria (sec. 292 ZPO, Slovenia Article 224 CPA)). Analogous application of written documents to electronic evidence is applied in the Czech Republic, Greece, Estonia, Lithuania, and Portugal.

Starting from the definition that a document represents the written embodiment of thought, in theory (Rusmann, 2012), there is a standpoint that electronic declarations of intent have no embodiment essential for the written expression of thoughts and that electronic documents are not covered by rules on documentary evidence, rather rules on inspection are applicable. Further arguments state that computer-stored documentation does not represent an embodied document. The visualization on the screen and the computer printout is only the image of the stored document. Electronic documents are, therefore, visual objects. However, they are subject to the free judicial evaluation of evidence.

Under the German ZPO, electronic documents belong to the rules of inspection as a means of evidence (section 371 (a) ZPO) (Ahrens, 2015, p.506), which states that the evidence obtained by visual inspection is offered by designating the object to be verified, visually and by referring to the facts in respect of which evidence is to be provided. However, printouts of electronic documents are equivalent to a document if they fulfill the condition that the document has been certified by the issuing public authority (section 416.a ZPO).

Evidentiary Value of Electronic Documents

In jurisdictions (Austria, Czech Republic, Estonia, France, Hungary, Poland, Slovakia, and Slovenia) where an electronic document is considered to be equal to a written document, the electronic document possesses the same probative value as a physical document if the electronic version fulfills certain conditions (the same as that applies for public documents, electronic signature).

According to German legislation, where electronic documents fall within inspection rules, the law refers to the evidentiary value of documents (public – paras. 415, 417, 418 ZPO and private documents – para 416 ZPO).

Regarding the evidential value, the judge is authorized to decide whether the presented electronic documents reflect the truth.

Each form of electronic documentation has its distinctive technical characteristics, which may give rise to particular procedural issues also connected to the question of reliability and authenticity. With e-mail, electronic signatures can be used to prove authorship, and other technical data may give evidence of the source of certain files. With video and audio recordings, possible witnesses with knowledge of the circumstances under which the evidence was produced, including the author of the recording, may give evidence. Furthermore, expert witnesses with technical knowledge may also be needed to give opinions about the authenticity of the recording in question. Generally speaking, procedural rules offer little guidance on how electronic evidence needs to be assessed when its authenticity is in question. The ultimate question for documentary evidence, electronic or physical, is whether it constitutes reliable proof. In case of doubt as to whether the electronic evidence is authentic, the guarantee of authenticity is needed.

In the common law system, an electronic document is regarded as authentic if the proponent produces evidence "sufficient to support a finding that the matter in question is what its proponent claims" (Esler & Schwerha, 2010, p.704). Federal Rules of Evidence 901 provides a list of non-exclusive means of proving authenticity, which includes: witness with knowledge, non-expert opinion, expert opinion, distinctive characteristics, witness identification, etc. (Esler & Schwerha).

In most civil law jurisdictions, in connection with documents, there is a hierarchy of documentary evidence for proving the content. Documents issued by public authorities have higher validity than private documents. There is also an important link between the document and signature. Authenticity relates to the issuer of the document and to the question of whether the document was truly issued by the person who is named as the issuer.

The Influence of the Videoconferencing Tools on the Depersonalized Character of Evidence Taking

In the study conducted within the EU project "Dimensions of evidence in the EU", the analysis has shown that a number of countries have designed and implemented ICT systems in order to support simplified procedures, e.g., payment orders and small claims. However, current new European trends show that there is a need for a secure and reliable decentralized IT system constituting interconnected and technically interoperable framework linking EU information systems. E-CODEX as an information technology tool is one of the solutions with the purpose of the EU Commission to ensure effective means of communication between parties and courts and between authorities in different Member States.[11] As a result, the Regulation on taking evidence (Regulation No. 2020/1783) facilitating effective procedure in civil matters has been amended.

The Recast Regulation No. 2020/1783 aims to improve the effectiveness and speed of judicial proceedings by simplifying and streamlining the mechanisms for cooperation in the taking of evidence in cross-border proceedings, at the same time reducing delays and costs for individuals and businesses. In doing so, it focuses extensively on encouraging courts to use all suitable modern communication technologies and on establishing secure and reliable decentralized IT systems for the communication and exchange of documents. Regulation No. 2020/1783 gives the requesting court the following options: (a) taking of evidence by the requested court (active court help); (b) direct taking of evidence by hearing witnesses, parties, and experts present in another Member State directly by videoconference or other remote means of communication (passive court help).

The European Union thus seems intent on responding to the need to simplify access to justice, integrating new technologies into the process at various levels, and promoting effective cooperation in one of the most sensitive parts of the process, namely the taking of oral evidence. Where evidence is to be taken by examining a witness, a party to the proceedings, or an expert present in another Member State, the Regulation states that the requesting court should take that evidence directly by means of videoconferencing or other remote communication technology, where that technology is available to the court and the court considers it is appropriate to use it taking into account specific circumstances of the case and the fair conduct of the proceedings. The Regulation further states that the court seized on the proceedings should instruct the parties and their legal representatives concerning the procedure for presenting documents or other material when the examinations are conducted using videoconferencing or other appropriate distance communication technology.

Videoconferencing (VCF), the telephone, and sound recordings could be considered oral elements of civil proceedings. It could be argued that with new ICT, the principle of orality has gained new content. If the new technologies enable renewed orality, they promote orality as well, but an orality that is "depersonalized". With the use of VCF, depersonalization can occur due to the lack of personal contact between subjects (court and parties) acting in the proceedings.

VCF represents a combination of both written and oral elements in the court's procedural activities. The recordings of the main hearing may be considered written elements. Such recordings offer numerous advantages since they can be heard again for the assessment of the evidence if there are doubts as to what was said, or even if there are doubts as to the manner in which it was said. Orality is also guaranteed. Since VCF can be used in geographically distant places, the main characteristic is that physical contact with the parties, witnesses, or experts is absent. Proceedings held by VCF respect orality and publicity, but not the actual presence of the parties, the witnesses, or the experts.

According to Slovenian law, witnesses give statements and are examined orally (Article 238 CPA). Parties may submit written statements of the witnesses if the court orders them to do so or at the request of the parties (Article 236(a) CPA). Furthermore, experts may give their opinions orally or in written form, depending on the court's decision (Article 253 CPA). In practice, the interesting question regards the admissibility of witness or expert statements recorded on video and submitted by a party as a written statement. Could such recordings be regarded as documents, or should other rules apply, e.g., the rules on inspection of evidence by the court?

If the rules on inspection are applied, the video recording would need to be authenticated, which means that a witness familiar with the video content (e.g., the person who captured the video images) would probably have to be examined. Another approach would be to verify whether the images have not been altered in any improper way by using the opinion of an expert. In contrast, the reliability of written witness statements could be examined by inviting the relevant witnesses to give oral testimony about the content of their written depositions.

The Slovenian Civil Procedure Act regulates the use of VCF in civil proceedings (Article 114(a)) and the possibility of direct service (including electronic service) between the counsel of the parties in civil proceedings (Article 139(a) CPA). The law provides only one rule on the use of VCF for main hearings, and this provision also relates to the examination of parties, witnesses, and experts.

With the consent of the parties, the court has the right to allow the parties and their counsel to be at another place at the time of the hearing and to perform procedural acts there, under the condition that audio and video transmissions have been provided for between the site of the hearing and the place, or places, where the party (parties) and their counsel are located, and vice versa. When the court decides that a hearing will be made via a VCF system, it issues a decision against which there is no appeal. This is an option rather than an obligation for the courts. VCF eliminates the cost of traveling long distances and consequently reduces the time needed for justice to be effectively administered. VCF is certainly a new form of orality by which a hearing is held at a distance.

The Regulation on evidence taking provides that the requesting court may ask the requested court to use communication technology for the taking of evidence, in particular, VCF and teleconferencing (sound transmission). Suppose the court in Slovenia directly hears a witness situated in another Member State by means of a video link – in that case, the use of VCF is allowed only if the conditions stated in Article 17 of the Regulation are met. This means that the procedure for the direct taking of evidence from the requesting court would apply (Galič, 2010). If the requested court executes the hearing of a witness, then the taking of evidence could be realized through the representatives of the requesting court and the parties via VCF (Article 12). The requesting court may also request the participation of court representatives according to the conditions set by the requested court (Article 12(3) and (4)). Since the Slovenian CPA allows the taking of evidence via VCF, the only impediment to implementing such a request may be the lack of technical equipment at the courts.

In Slovenia, there are at present no rules on how to use VCF in civil litigation. Awareness of the lack of regulation is important because there are some significant differences between face-to-face civil litigation in the courtrooms and the main hearing held via VCF. One good example of a better practice is the English Videoconferencing Guidance in Annex 3 of Practice Direction 32 (Kawano, 2012), which covers the use of VCF in both ways (a) in a courtroom, whether through VCF equipment permanently placed therein, or through a mobile unit, and (b) in a separate studio or conference room. The guidance is applicable to cases in which VCF is used for the taking of evidence as well as for other parts of legal proceedings (e.g., interim applications, case management conferences, pretrial reviews). The guidance also contains some important remarks for courts using VCF. It states that although the use of VCF may be an ideal means of saving time and money, it is, however, "inevitably not as ideal as having the witness physically present in court". There is also guidance for the judge when deciding about the use of VCF in a particular case. The judge needs to consider whether the use of VCF will enable an overall cost saving, i.e., "whether its use will be likely to be beneficial to the efficient, fair and economic disposal of the litigation". There are considerably greater limitations in the degree of control that the court may exercise when having a witness physically present than a witness giving a statement at a remote site (Videoconferencing Guidance in Annex 3 of Practice Direction, p. 32).

When we recall the traditional principle of orality, we can see that new technologies may influence the basic elements of this principle. It seems that the traditional need for physical attendance of the parties, witnesses, and experts is eclipsed. Even a court inspection via VCF could become a reality. The principle of the traditional written form has already been influenced by ICT, at least to a limited extent, in some procedures in the Slovenian judicial system. The e-Justice Portal allows electronic enforcement, land register entries, and insolvency proceedings; in some cases, the use of electronically prepared forms is obligatory. Furthermore, the CPA introduced the (promising) possibility of audio and video recordings of hearings (e-hearings). Thus, Article 125(a) of the Slovenian CCP determines that the presiding judge may order the audio or video recording of the hearing. The judge will inform the parties and the other participants at the hearing.

Currently, the common practice of Slovenian courts is to make an audio recording of the hearing of witnesses, the parties, or, in particular, court experts. In contrast, the entire main hearing is not recorded. Following the rules mentioned above, clerks transcribe the audio recordings and transcriptions are then served to the parties.

Certainly, it may also be stated that audio recordings disburden the judge in taking away the need to dictate the party, witness, and expert statements for the record at the main hearing. Furthermore, the parties, their representatives, witnesses, and experts are not interrupted when stating their thoughts for the court record.

Proceedings held by videoconference respect the orality and the publicity, but not the presence of the parties or witnesses and experts. Recorded testimony lacks the immediacy of direct personal perception of the judge of the recorded or broadcasted statement via the video link given by the party, witness, or expert in person. If these new technologies enable renewed orality, they further provoke an orality that is more "depersonalized". According to Soraya (2011), the new forms of expression that ICT offers influence the traditional orality so the term "secondary orality" should be introduced, reflecting the hybrid form between the written and oral.

This lack of personal presence may endanger efficiency and fairness in the civil procedure if the physical presence is necessary for the judge to get a personal impression in order to evaluate evidence. The aforementioned problem points to the importance of the decision of what position should the ICT get in the civil procedure.

The use of videoconferencing enables the oral and written forms of acts. The written form principle is manifested in the video recordings of the main hearings. Recordings offer advantages in that evidence can be heard again during the evaluation of the evidence and if there is any doubt about what was said or how it was said. However, recorded testimony lacks the immediacy of live testimony at the hearing (Marcus, 2012).

With the use of ICT, these principles may gain new and broader content. Namely, the main hearing could be considered as an oral manifestation in the civil procedure. The Estonian amendments to the Code of Civil Procedure are a good example of how the legislator has considered the specific characteristic of new technologies. In the explanatory memorandum to the CCP, it has been explained that the traditional principle of orality is impractical and ineffective with regard to the developments in the communications systems (Poola, 2015).

When using VCF, some special procedural guarantees need to be considered, for example, the administration of the oath in some jurisdictions, assuring that the witness understands who is at the local site and what are the different roles of all participants and the use of electronic documents at the hearing. National regulations generally lack special guidance in the procedural and technical sense for courts and parties when using VCF in civil procedure.

The Hague Conference on Private International Law (HCCH) lately published a guide to good practice on the use of video links under the Convention of 18 March 1970 on the Taking of Evidence Abroad in Civil or Commercial Matters (the Hague Evidence Convention). The Guide provides an overview of the latest developments on the use of video links in the taking of evidence under the Hague Evidence Convention and good practices, which can be followed by its Member States. It was drafted as a result of an extensive survey HCCH conducted, for which all Member States presented and shared their most frequent practice (the HCCH Survey). The Guide points out that the use of video links may not be appropriate in all circumstances where a person is to testify before a court. It is important that the use of VCT continues to be considered complementary to traditional methods of obtaining evidence, which means that the primary method is still personal attendance in the courtroom. In the Guide, a study executed across a variety of appellate courts of one Contracting Party (the United States) is analyzed. The study revealed that some judges believed they asked fewer questions when examining a witness by a video link and were less likely to interrupt an argument (Guide to good practice under the 1970 Evidence Convention).

Videoconferences lower the cost of traveling long distances, and the speed enables access to justice to be effectively guaranteed. Kodek warns of two obvious disadvantages when using VCF at the main hearing (Kodek, 2012). First, as it was already emphasized, the judge is deprived of a personal impression of the parties and witnesses, which is important when determining the credibility and admissibility of the evidence. The second disadvantage relates to the position of witnesses who will feel less involved in hearing evidence and the importance of the role as an examined witness. If a witness is a party who gives testimony, there is an even greater risk in terms of establishing his credibility by subjecting himself to VCF. Furthermore, there is a danger of manipulation in terms of assisting the witnesses with signs and facial expressions, which is not possible when a witness testifies in a courtroom (Kodek).

In McGlinn v Waltham Contractors Limited (2007), the court considered two factors for allowing the use of video links: the weight of the witness's evidence can be of sufficient importance or purely ancillary. If it was ancillary, it would be less important that the witness appears in person and, therefore, the court would be more likely to allow VCF. Second, the court also asked whether there was a real, as opposed to fanciful reason, why VCF evidence is being sought. If it is only fanciful, the court would definitely be less inclined to grant the application for VCF (Hwang & Nicholls, 2012). From US experience, the immediate advantages of videoconferencing are lowered costs and increased efficiency (Roth, 2000). Included in Rule 43 of the Federal Rules of Civil Procedure, remote witness testimony is accepted through the use of videoconferencing in the courtroom. The lawyers and judges have concerns that the difficulties with determining the credibility and competencies of the witness are greater with remote testimonies than with witnesses appearing in the courtroom.

In some countries, like the US and England, ICT in civil procedure has gained considerable importance. In England, most disclosure is now being processed electronically (Hollander, 2015, p.155). According to Andrews, disclosure enables four functions: ensuring equality of access to information; enabling settlement of a dispute; avoiding "trial by ambush", the situation when a party is unable to respond to surprise arguments at the final hearing; and assisting the court when reaching a determination of facts, when deciding on the judgment on the merits (Andrews, 2013).

In 2004, there was a need for a review of electronic disclosure in England. The outcome was the new e-disclosure Practice Direction and e-disclosure questionnaire, which considers developments in the US (Hollander, 2015). Parties are now required to discuss the use of technology in managing electronic documents and the conduct of procedures before the first case management conference.[12] The purpose of the e-disclosure questionnaire is for parties to provide information regarding the scope, extent, and most appropriate format for disclosure of electronic documents (Hollander).

Summary

New technologies are modern methods of expression in court procedures. They appear to bring speed and quality to the procedure. The first virtue reflects speeding up communications and efficiency in the exchanges between parties and the court. The new technologies could also contribute to a fair and equitable procedure with respect to adversarial principles. The adversarial system in electronic proceedings will certainly be more intense and authentic if the electronic filing of claims, defenses, and evidence (witness and party statements, expert opinions) could be made available and accessed through an e-system.

However, it remains necessary to determine the place of ICT next to the oral and written character of procedural acts. It could be further argued that with new ICT, the principle of orality has gained new content. Above all, when using new technologies, the civil procedure rules have to overcome the obstacle of breaking the direct contact between the parties and the court in the procedure. This depersonalization will influence the principle of directness or immediacy (principle of presence) and the contact with the source of evidence. A procedure held by VCF respects the orality and the publicity, but not the presence of the parties; thus, the new technologies reveal the risk of losing the physical link that is essential for efficiency in litigation. As far as the use of videoconferences is concerned, these would be useful in cases where the distance was such that the cost of travel would be disproportionate. In other words, the judges should appreciate the specific opportunity of being able to have a direct, physical appearance of the party in question.

The important aspect is to examine to what extent the principles and evidentiary rules should apply to traditional paper-based documents and electronic documentary evidence. One can argue that the ICT-neutral approach should be adopted so that the term documentary evidence should apply to traditional paper-based documents and electronic documents. This would mean that there would be no fundamental differences applying the law of evidence between traditional paper-based documentary evidence and electronic evidence. For admitting documentary evidence, there is a need for a clear definition of a document. The law of evidence, of course, must continue to accommodate the traditional notion of a document since such documents will continue to be relevant to court proceedings. However, the concept of "document" must also be updated to accommodate electronically generated or stored information in electronic form. Generally, countries do not include a specific definition of electronic evidence in their legal codes, and consequently, electronic evidence is considered equivalent to traditional evidence.

For the suggested changes, we can speak in favor of the adaptation of national justice systems towards those changes that would add specific regulations for the distinct definition of electronic evidence. On the other hand, there should be rules of minimum standards at the international level. There is certainly a need to develop certain European rules that would guarantee at least a minimum unified approach in the treatment of electronic evidence in civil procedures.

Notes

1. https://e-justice.europa.eu/home.do.
2. e-CODEX, the digital platform for cross-border legal data exchange within the EU. See the official website: https://www.e-codex.eu/.
3. The use of "e-justice" not only requires modifications in court organization but also serves as a ground for the traditional principles of the civil procedure to be re-examined.
4. Richard Susskind launched a website at the outset of the corona crisis, in order to create a platform to share experiences of "remote" alternatives to traditional court hearings. The website provides an overview of interesting developments on a global level and helps justice

workers including judges, lawyers, officials, and court technologies share to best practices on the use of remote alternatives to traditional court hearings. https://remotecourts.org/news .htm.

5. There is an extensive body of literature discussing relationship between law and technology: Hildebrandt, 2009; Cockfield, 2004; Cockfield & Pridmore, 2007; Tranter, 2011.

6. The entanglements between formal rules and technology is analyzed in three case studies illustrating different approaches using ICT, namely Money Claim Online, handling small claims in England and Wales; Civil Trial Online, project developed in Italy to digitize the entire civil procedure, and speed cameras used in Australia. See: R. Mohr and F. Contini, 2011: 994–1019.

7. Harsági argues that the use of modern technologies may lead to "relative reconstruction of some traditional principles" and that the impact is more intense in litigious proceedings than in non-litigious proceedings (Harsági, 2012, p. 125–133).

8. According to the US practice, the failure to preserve evidence or the intentional destruction of evidence may lead to severe consequences. Sanctions proportioned to the degree of wrong-doing may be imposed by a court. In the case *United States v. Philip Morris USA, Inc.*, 2004, U.S. Dist. LEXIS 13580 (D.D.C., 21 Jul. 2004), the court issued an order on the preservation of evidence, which required for the defendant to preserve potentially relevant evidence and documents. The defendant disregarded the court preservation order and deleted e-mails after 60 days. Consequently, the court, imposed Philip Morris monetary sanctions of USD 2.75 million, which was USD 250,000 per each of eleven managers and officers, who failed to follow an e-mail "print and retain" policy.

9. The project was carried out by 11 foreign universities based on 28 national reports under the successful management of Prof. Dr. Vesna Rijavec, University of Maribor, Faculty of Law. This European research cooperation is a key scientific research network, which joined the most influential experts in the field of European civil procedure law. They are working together, exchanging their expertise in order to contribute to a better understanding of national and unified requirements in this area.
   The project aimed to explore whether a common core of European Law of Evidence exists (and taking evidence in particular). By providing a framework of common core principles, the project will serve as a starting point for further harmonization or unification processes in this field. One of the major objectives was the research of the functioning of Regulation (EC) No 1206/2001 and cross-border taking of evidence within the EU that contributed to the rise of mutual trust, elimination of obstacles in cross-border civil proceedings as well as provision of information to individuals and businesses in order to improve their access to justice, and trainings for legal practitioners with emphasis on communication technologies. Official website: http://www.acj.si/en/presentation-evidence.

10. Schafer, B. and Mason, S. (2010). The Characteristics of Electronic Evidence. In S. Mason, P. Argy, and D. Begg (Eds.), Electronic Evidence: Disclosure, Discovery and Admissibility (2nd ed.). Butterworth.

11. The Commission considers e-CODEX as the main tool and the gold standard for establishing an interoperable, secure, and decentralized communication network between national IT systems in cross-border civil and criminal proceedings. Proposal for a Regulation of the European Parliament and the Council on a computerized system for communication in cross-border civil and criminal proceedings (e-CODEX system), and amending Regulation (EU) 2018/1726.

12. Case management conference is a hearing conducted in order to determine further steps in the case. Parties need to inform the court of those issues in relation to disclosure they agree on and those matters on which they disagree. Where the parties cannot agree, the court will give written directions or even consider making and order that the parties must complete the Electronic Documents Questionnaire (Mason, 2015).

Bibliography

Ahrens, H.J. (2015). *Der Beweis im Zivilprozess*. Otoschmidt.

Amrani-Mekki Soraya, S. (2011). L'impact des nouvelles technologies sur l'ecrit et l'oral en procedure civile. *La parole, l'écrit et l'image de la justice*, Quelle procedure pour le XXI siecle, Les entretiens d'Aguesseau, PULIM, 157, from http://www.uv.es/coloquio/coloquio/ponencias/6tecmek2 .pdf.

Andrews, N. (2013). *Andrews on civil proceedings, vol. 1: Court proceedings*. Intresentia, 263–270.

Braman, R.G., Daley, M.J., & Withers, K. J. (2009). International overview of discovery, data privacy & disclosure requirements. A project of the Sedona Working Group on International Electronic Information Management, Discovery and Disclosure (The Sedona Conference, 2009).

Brownstone, R.D. (2006). Preserve of perish; destroy or drown – E-Discovery morphs into electronic information management. *North Carolina of Law & Technology*, 8(1), 15–20.

Cannon, R. (2007). Ireland. In Mason, S. (Ed.), *Electronic evidence: Disclosure, discovery and admissibility* (p. 335). LexisNexis Butterworths.

Cockfield, A.J. (2004). Towards a law and technology theory. *Manitoba Law Journal*, 30, 383–415.

Cockfield, A.J., & Pridmore, A. (2007). Synthetic theory of law and technology. *Minnesota Journal of Law, Science and Technology*, 8(2), 475–513.

de Resende Chaves Junior, J.E. (2012). Proceedings on the web. In Kengyel, M., & Nemessanyi, Z. (Eds.), *Electronic technology and civil procedure* (pp. 101–124). Springer.

Esler, B., & Schwerha, J.J. (2010). United States of America, Chapter 18. In Mason, S. (Ed.), *Electronic evidence* (p. 704). LexisNexis Butterworths.

Evidence Convention, The use of Video-Link (2020). The Hague Conference on Private International Law – HCCH Permanent Bureau. https://assets.hcch.net/docs/569cfb46-9bb2-45e0-b240 -ec02645ac20d.pdf.

Galič, A. (2010). Komentar 114.a člena ZPP. In Ude, L. & Galič, A. (Eds.), *Pravni postopek: Zakon s komentarjem*, book 4 (pp. 91–94). Uradni list Republike Slovenije, GV založba.

Guide to good practice under the 1970 Evidence Convention, The use of Video-Link (2020). The Hague Conference on Private International Law – HCCH Permanent Bureau. https://assets. hcch.net/docs/569cfb46-9bb2-45e0-b240-ec02645ac20d.pdf.

Harsági, V. (2012). Digital technology and the character of civil procedure. In Kengyel, M., & Nemessanyi, Z. (Eds.), *Electronic technology and civil procedure* (pp. 125–133). Springer.

Hildebrandt, M. (2009). Techonology and the end of law. In Claes, E., Devroe, W., & Keirsbilck, B. (Eds.), *Facing the limits of law* (pp. 443–464). Springer – Verlag.

Hollander, C. (2015). *Documentary evidence*. London: Sweet and Maxwell.

Hwang, M., & Cheah Nicholls, A. (2012). When should video conferencing evidence be allowed? *Singapore Law Gazette*, LexisNexis, from http://www.lawgazette.com.sg/2012-09/.

Hwang, M.S.C., & Cheah Nicholls, A. (2015). When should video conferencing evidence be allowed? *Singapore Law Gazette*. http://www.lawgazette.com.sg/2012-09/.

Insa, F., & Lazaro, C. (2008). Admissibility of electronic evidence in court: A European project. In Mason, S. (Ed.), *International electronic evidence* (p. 5). British Institute of International and Comparative Law.

Kawano, M. (2012). Electronic technology and civil procedure – Applicability of electronic technology. In Kengyel, M., & Nemessanyi, Z. (Eds.), *Electronic technology and civil procedure* (pp. 3–27). Springer.

Keane, A., & McKeown, P. (2014). *The modern law of evidence*, 10th ed. Oxford University Press.

Kodek, G.E. (2012). Modern communications and information technology and the taking of evidence. In Kengyel, M., & Nemessanyi, Z. (Eds.), *Electronic technology and civil procedure* (pp. 271–274). Springer.

Law Reform Commission. (2009). *Consultation paper, documentary and electronic evidence (LRC CP 57–2009)*.

Lupo, G., & Bailey, J. (2014). Designing and implementing e-Justice systems: Some lessons learned from EU and Canadian examples. *Laws*, 3(2), 353–387. https://doi.org/10.3390/laws3020353.

Marcus, R.L. (2012). The impact of digital information on American evidence – Gathering and trial– The straw that breaks the Camel's back? In Kengyel, M., & Nemessanyi, Z. (Eds.), *Electronic technology and civil procedure* (p. 46). Springer.

Mason, S. (2010). The characteristics of electronic evidence in digital format. In Mason, S. (Ed.), *Electronic evidence*, 2nd ed. (pp. 21–24). LexisNexis Butterworths.

Mason, S. (2015). *Electronic disclosure: A casebook for civil and criminal practitioners*. PP Publishing, p. 47.

Mason, S., & Schafer, B. (2010). The characteristics of electronic evidence in digital format. In Mason, S. (Ed.), *Electronic evidence*, 2nd ed. (p. 21). LexisNexis.

McGlinn v. Waltham Contractors Limited 2007, Westlaw citation: 2007 WL 763659.

Mohr, R., & Contini, F. (2011). Reassembling the legal: The wonders of modern science in court-related proceedings. *Griffith Law Review*, 20(4), pp. 994–1019.

Nunner-Krautgasser, B., & Anzenberger, P. (2009). Evidence in civil law – Austria. In Braman, R.G., Daley M.J., & Withers K.J. (Eds.), *International overview of discovery, data privacy & disclosure requirements. A project of the sedona working group on international electronic information management, discovery and disclosure*. The Sedona Conference.

Palmerini, E. (2013). The interplay between law and technology, or the RoboLaw project in context. In E. Palmerini & E. Stradella (Eds.), *Law and technology: The challenge of regulating technological development*. Pisa University Press. http://www.robolaw.eu/RoboLaw_files/documents/Palmerini_Intro.pdf.

Paul, G. (2008). *Foundation of digital evidence*. Chicago, IL: American Bar Association.

Poola, M. (2015). *Evidence in civil law – Estonia*. Institute for Local Self-Government and Public Procurement Maribor.

Roth, M.D. (2000). Laissez_Faire Videoconferencing: Remote witness testimony and adversarial truth. *UCLA Law Review*, 1(1), 191.

Rusmann, H. (2012). Electronic documents, security and authenticity. In Kengyel, M., & Nemessanyi, Z. (Eds.), *Electronic technology and civil procedure* (p. 248). Springer.

Sgarlata Chung, C., & Byer, D.J. (1998). The electronic paper trail: Evidentiary obstacles to discovery and admission of electronic evidence. *Boston University Journal & Technology Law*, 4, 8.

Székely, I., Szabó M.D., & Vissy, B. (2011). Regulating the future? Law, ethics, and emerging technologies. *Journal of Information, Communication & Ethics in Society*, 9(3), 180–194.

Tranter, K. (2011). The laws of technology and the technology of law. *Griffith Law Review*, 20(4), 754–762.

7

Legal Analytics on the Role of Intellectual Property and Digital Health System in Overcoming Public Health Emergency

Ranti Fauza Mayana, Ahmad M. Ramli, and Tisni Santika

CONTENTS

Introduction

The chaos and panic started in December 2019; hospitals in Wuhan, Hubei Province, China, began to report patients with unidentified novel pneumonia cases, and after, researchers rapidly analyzed this novel virus known as a coronavirus (Covid-19) (Zhu et al., 2020). World Health Organization (WHO) has declared Covid-19 as a pandemic (WHO, 2020). In a relatively short time, the world faced a multidimensional crisis (Welsch, 2021) and a fundamental dilemma. In the several years before the pandemic, there was a massive concern about poverty and poor health (United Nations Department of Economics and Social Affairs, 2020), but this pandemic brought a deterioration in health status, as well as crashed the economy, particularly in developing countries. This is collateral damage, for the evidence that links poverty with high disease or health problems is convincing because poverty directly decreases purchasing power (Leisinger, 2012). However, although the economic factor is a determinant factor of public health, the quest for comprehensive policies regarding public health emergency can make a large difference to the performance and success rate of public health measures even in low-income countries (The Conversation, 2020).

Covid-19 has shown potential for long haul pandemic with global effects, a high risk of fatality, and insufficient health capacity (Rajan et al., 2021), and to rely on vaccines alone is high risk and insufficient to handle the outbreak (Moore et al., 2021); thus, it's critical to comprehensively conduct approaches including infection prevention such as contact tracing, quarantine, case isolation, physical distancing, and the implementation of safety and health protocol. The approach of existing health services needs some transformation and adaptation under the consideration of risk and urgency during the pandemic (Wanat et al., 2021). In an attempt to reduce the risk of virus transmission, health facilities are trying to minimize direct encounters by providing alternative services that rely on alternative devices, for example, phone applications, telemedicine, and other digital platforms to support and promote distance healthcare and administration (Ricciardi et al., 2019).

In the term of the pandemic caused by respiratory viruses, there are underlying causes of spread: high levels of contact, population mobility (Auliya & Wulandari, 2021), and interaction across the globe; thus, every country faces the risk of potential outbreaks and spread of this infectious disease (Saunders-Hastings & Krewski, 2016) until it causes the pandemic in a relatively short time that affects any country regardless of location and socio-economic status (Ross AGP, Crowe SM, 2020).

Concerning the nature of the infectious disease, it's highly critical to understand the routes and timings of transmissions (Saunders-Hastings & Krewski, 2016), and the digital health system becomes a reliable tool in providing the data (Golinelli et al., 2020). The study concerning digital intervention for tracking infectious disease was previously conducted by Eysenbach on flu-related disease in 2006, and the result shows that now it's possible to measure what was formerly immeasurable by utilizing the internet and user-generated data technology on social media (Eysenbach, 2006).

Many countries have shifted their healthcare method toward digital health systems, considering the efficiency of utilizing technology to replace conventional manual contact tracing as the part of new normal during the long haul of the Covid-19 pandemic (Huang et al., 2020). The massive utilization of digital health devices and applications could change the nature of healthcare (Ferguson, 2012) by using digital technology to increase patient engagement, improve health quality, transform healthcare/treatment, speed up the response and treatment, reduce healthcare costs and minimize human error.

Indonesia launched "Peduli-Lindungi/Care-Protect" as an online surveillance-mapping tool to conduct early detection of the infectious risk of Covid-19. Through the decree of Ministry of Communication and Informatics Number 171/2000, the "Peduli-Lindungi/Care-Protect" application has been declared as the official tool in the implementation of health surveillance to combat Corona Virus Disease 2019 (Covid-19) using some main strategies: tracing, tracking, warning, and fencing and further encouraging patients to seek online services on health consultations rather than direct meeting services (Norton Rose Fulbright, 2020; SGPP Indonesia Student, 2020).

A similar approach was also implemented in the US, where, after the pandemic erupted, the number of online health consultations has massively increased up to ten times compared to the recorded data in pre-pandemic times (Webster, 2020). Similar situations occurred in Australia with the COVIDSafe.app as part of the digital health system that is based on voluntary uptakes and initiatives, protected by legislation (Goggin, 2020). Further, electronic prescriptions and healthcare services access are now being fast-tracked under Government's National Health Plan in Australia to provide vital healthcare services while keeping the protocol of social distancing and isolation to support this; more than 90% of pharmacies and 94% of public hospital beds are registered and connected (Agency, 2020).

The adequacy of national health systems is the determinant aspect of public health preparedness in mitigating the pandemic. This adequacy builds by comprehensive policies, basic infrastructure, funding, and capable human resources with adequate skills (Revere et al., 2011), which leads to a functioning healthcare delivery system. However, the nature of the pandemic forces to minimize physical contact to prevent the spread of the virus, and thus, conventional delivery of the healthcare system faces several challenges, especially the limitation of direct contact between patients and healthcare providers shifting to the digital tendency of the public health system in the future (Budd et al., 2020). The pandemic brought both challenges and opportunities for a rapid shift towards digital health in terms of disease prevention, consultation, and health treatment (Blandford et al., 2020), and in this condition, it's important to search for creative, effective, and proactive solutions by optimizing human creativity as endless resources to provide common interest, ensuring the right to health, and promoting well-being for all (Rosenthal, 2009).

This chapter will summarize several existing shreds of evidence about public health urgencies and importance with the focus on those that particularly affect the people with limited access to healthcare in the pandemic situation and further analyze the legal challenges in the development of digital health systems as the pillar of public health and how the digital health system potentially provides an effective and promising solution.

Literature Review

Public Health System and Public Health Emergency

The function of public health covers a wide scope of elements and a wide range of activities and measures. Public health aims to increase the health of the whole people and community by emphasizing disease prevention, health protection, and the well-being of society (Binns & Low, 2015). Emergency preparedness is a core public health function; unfortunately, public health emergency preparedness activities – even in developed countries – operate largely in the background until an event of concern or health emergency occurs (Khan et al., 2017). Based on the findings on 182 countries facing the Covid-19 outbreak and crisis, many countries were at a low level of public health preparation and readiness even before the pandemic (Ramanathan et al., 2020).

The system of public health holds a critical, integral, and important role in preparing responses and strategies to mitigate the risks and to support recovery from health crises and emergencies (Public Health Ontario, 2020). The state, local community, and each individual must be prepared to cooperate, coordinate, and collaborate between cross-sector partners and organizations in times of emergency at every level, scale, type, and severity. The public health system is expected to articulate and execute the strategic arrangements and mechanism for the coordination as the protection and response to public health emergency conditions and preparedness (Nelson et al., 2007).

National health emergency plans should develop a national stockpile of healthcare, essential medicines, vaccines, and health product supplies to meet emergency needs including public health law that authorizes the government to take such actions as reasonably important and necessary to prevent and mitigate public health crisis (WHO, 2013). Public health emergency preparedness and response capabilities include three major planning models: assessing the current stage in terms of organizational roles and responsibilities, resource elements, and performance, followed by determining the strategies and activities in short-term and long-term goals and then developing plans on capacity building and organizational initiatives (Centers for Disease Control and Prevention, 2018).

During an emergency like a pandemic, healthcare providers are relied on to prevent excess deaths, facilitate surveillance, and mitigate the escalation of infected patients, and to perform these tasks they need to be aware of public health threats and emergency conditions (Baker & Porter, 2005); therefore, public health emergency preparedness (PHEP) also requires legal preparedness concerning broad legal policy concerning public health emergency, legal guidance, law designation, interpretation of current existing law and concern about the liability of each party and ethical concerns on infringement of privacy (Botoseneanu et al., 2011). Moreover, PHEP also requires public awareness and competency to make an effective legal response to all public emergencies (Gebbie et al., 2008).

Public awareness and education through risk communication shall be the part of health disaster risk reduction (World Health Organization, 2017), and public competency to conduct the right response to a public emergency requires comprehensive and transparent information on a certain level, conflict of norms concerning confidentiality and privacy arises between the right to privacy as a fundamental right and the public interest in general (Ventrella, 2020). One of the most venerable obligations of medical ethics is the principle of medical confidentiality (Eysenbach, 2009). However, the principle of confidentiality is not absolute and may be overridden by public interest as a response to the high importance of information availability and accessibility for public health emergency preparedness (PHEP). In short, there is a common belief and recognition that public interest may on several conditions (including health emergencies) justify a breach of the principle of confidentiality (Papadodima et al., 2008).

Health literacy is a valuable and powerful tool, because the better and more well informed the public, health professionals, and policymakers, the bigger it affects the success rate in affecting awareness, knowledge, or even clinical outcomes (Okan et al., 2020), and digital media platforms provide the wide reach and fastest way of information spread. Taking Vietnam as an example, the government commits to broadcasting and announcing the seriousness of the threats caused by Covid-19. The Ministry of Health develops an online portal as an actual, factual, and reliable source of official news and updates concerning Covid-19 that posts each new case in detail including the type of infection, location, and what kind of action is taken (Le & Nguyen, 2020).

The accurate and actual information is the key to improved problem solving, decision making, and care delivery (Dalrymple, 2011) to set out the public health response to the health crisis. Some adjustments need to be taken in particular conditions of health emergencies. However, there are five interrelated components that functioned as the pillar of public health in fulfilling the right to health:

1. **Availability**: Availability means the medicines and healthcare treatments should be available in sufficient quantities to maintain the health of the community (Sorato & Taddele, 2019). It's critical to guarantee an adequate supply, for example, by fostering effective procurement strategies for essential medicines, or in case the adequate supply is unavailable, the government may interfere with the funding or subsidies for scaling up production and distribution (Raja et al., 2006).

2. **Accessibility** means that the medicine and health treatment are accessible; this required multifaceted aspects that start from price to financial and human resources, and a certain level of political commitment to provide general infrastructure (Scarpetta et al., 2021). The existence of a vaccine, drugs, medicine, or other products can only benefit society when the patient can access and make use of it, thus ensuring access to the vaccine is in the interest of us all, for without this, no one is safe (United Nations Human Rights, 2020).

3. **Acceptability**: It's important that public health measures are both acceptable and appropriate. To achieve this, health system research plays an important part in the effort to invent the health product and treatment that serves both efficacy

and accessibility, another important determinant element is the level of trust in information from the government (Lazarus et al., 2021), for example, in Muslim-majority countries like Indonesia, Halal certification is an important element of medicine and/or vaccines to be accepted (Rahmah & Barizah, 2020).

4. **Effectiveness**: To achieve the effectiveness of both preventive and curative measures, an appropriate testing standard is required to conduct a clinical test of the health products and to conduct clinical trials of the users. Vaccine effectiveness is often confused with vaccine efficacy, but these two have a different context. Efficacy is proved by clinical research, while vaccine effectiveness is about how a vaccine reduces disease proved by real case data (Weinberg & Szilagyi, 2010).

5. **Affordability**: The cost of medicine, healthcare, and treatment is an important issue and often becomes the most common obstacle in the implementation of the right to health (Guzman et al., 2021). The development of new healthcare products will be of no value if they cannot be made available and accessible to those who need them; therefore, the affordability of Covid-19 vaccines is gaining increasing attention (Karim, 2020). The effort to provide effective and good quality medicine and healthcare at affordable prices would greatly be benefited from the proportional involvement and intervention of the public sector to assure the availability, acceptability, effectiveness, and affordability of medicine and healthcare for poor populations (Wouters et al., 2021). Financing and subsidies also can help the scale-up of medicine and healthcare products to ensure global availability and accessibility, particularly for a vulnerable and poorer group of society (Hotez & Bottazzi, 2020).

Public health emergency that affects a wide range of society in many countries demands a different approach from pharmaceutical companies in considering the production or manufacture of medicine and healthcare products (U.S Department of Health and Human Services et al., 2020). The market approach of pharmaceutical companies was mostly used as the consideration of product decisions, therefore the many pharmaceutical companies focused, the invention and innovation of health products targeting those with higher purchasing power. The damaging effect on public health (Civaner, 2012) raises as the result of commercial consideration that later determines the production portion of medicines (Chiu, 2005). It's economically logical that pharmaceutical companies focus on the product with strong demands and potential profit to cover their investment in innovation, but there are certain public health costs of relying on profit-driven policy for the pandemic response, for example the inadequate access and adequate portion of medicines and health products for relatively poorer society in developed and underdeveloped countries, therefore, government intervention with public health sensitive approach is needed (Heled et al., 2020).

Results and Discussions

Legal Analytic on the Role of Intellectual Property in Providing a Digital Health System, Essential Medicines, and Vaccines for All

Public health surveillance is a highly important undertaking to conduct disease navigation and control by serving as an early warning system and monitoring the epidemiology of a condition to identify public health emergencies to set priorities in designing public health policy and strategy. Along with technological advancement, many public health surveillances conducted with the support of electronic and digital devices lead to the formulation and development of digital health (Vokinger et al., 2020).

The digital health as a technology strategically depends on digital tools, technologies, and services to transform healthcare delivery (Snowdon, 2020) is urgently needed because the conventional contact tracing method for infectious disease has been proven to be inadequate during the Ebola outbreak for causing incompletion in contact identification and detain in contact tracing pace, for example, in alerting of involved contact or suspected case that required isolation (Danquah et al., 2019). The utilization of digital health has already been utilized for Ebola, SARS, and other infectious diseases; however, numerous governments and health systems have been steady in adopting and implementing these digital technologies (Alwashmi, 2020).

Since the beginning of the Covid-19 outbreaks, many smartphone apps have been developed and utilized in the context of contact tracing as health surveillance measures by public authorities (Adeniyi et al., 2020). The aim of contact tracing is for public health authorities to warn and quickly identify as many contacts as possible with the confirmed case of Covid-19 (Hernández-quevedo et al., 2020), suggest them to self-quarantine in case of no symptoms or light symptoms, prioritizing rapid tests, and isolating them if they developed symptoms. Although the implementation of contact tracing is challenging, former epidemic infectious diseases have been beneficially controlled with the help of contact tracing and case isolation procedures (Tom-Aba et al., 2018); thus, online surveillance-mapping tools provide the potential to conduct early warning and detection of a disease outbreak compared to conventional epidemic tools (Ahn et al., 2021).

In this matter, digital tools serve extra huge contribution to contact tracing efforts (Akarturk, 2020) by improving the efficiency, speeding up the overall process substantially and making it far less time consuming than contact tracing carried out manually by public health authorities, maintaining the accuracy of data management and automated processes based on the seriousness of their condition, and enabling early access to healthcare and reducing the risk of transmission (Eckmanns et al., 2019).

Digital tools also prevent the burden of the data collection process and management conducted by the human resource by allowing electronic self-reporting by cases, locations, and contacts (Freifeld et al., 2008); this is considered to be a great help concerning the nature of the Covid-19 pandemic where the virus spreads through the interaction or contact between people, mainly when an infected person is in close contact with another person, current studies and evidence suggest one of the main ways the virus spreads is by respiratory droplets (Galbadage et al., 2020).

Numerous countries developed online contact tracing for the surveillance of Covid-19 (Kamel Boulos & Geraghty, 2020), for example, the Surveillance and Outbreak Response Management and Analysis System (SORMAS) that has been widely used in Nigeria and Ghana, aligned with the Africa-wide Integrated Disease Surveillance and Response (IDSR) and digital health platforms, and as a result, the utilization of these digital platforms keeps the countries in a good position to keep pace with the emergence of new Covid-19 cases and enhances national health system preparedness and response capabilities (Grainger, 2020). SORMAS is also widely used in Nepal and Fiji island (Tom-Aba et al., 2020).

From Boston, the free website "healthmap.org" (Freifeld et al., 2008) and the mobile application named "Outbreaks Near Me" provide real-time intelligence information for a diverse audience including local health departments, libraries, governments, and international travelers concerning emerging infectious disease. Through an automated process, "healthmap.org", "Outbreaks Near Me" updating 24/7/365, the system monitors, organizes, integrates, filters, visualizes, and disseminates online information about emerging outbreaks, diseases, and provides early detection of a health emergency (Bhatia et al., 2021).

In Indonesia, there are several digital health initiatives conducted both by the public and private sector, for example, the 24-hour-homecare/telemedicine and teleradiology introduced in Makassar by its previous city municipal to increase the accessibility to healthcare services for the people living in the underserved areas by providing health transportation facility named *"Dottoro'ta"* to take the patients to the healthcare services and vice versa. Each *"Dottoro'ta"* is equipped with a stock of medicines, oxygen, medical devices, and health monitoring tools. Other examples are mobile services for the national health insurance that covers the registration, general information, invoicing service, selective health screening, diagnosing, therapy and laboratory services, patient information, and digital claim verification.

Private sectors in Indonesia have more initiatives regarding the development of digital health services as a consequence of the user's preference for using digital healthcare applications because of their practicality and convenience. There are several examples like *AloDokter*: a mobile application platform that focuses on providing high-quality medical information that is easily understood by users, *Go-Med* provided services that enable users to order medicine via an online application. *HaloDoc*: a mobile application that connects patients, doctors, insurance, and pharmacies into one simple healthcare application, providing doctor consultation services through chat, voice call, and video call with verified doctors that registered by the Indonesian Doctors' Association and Indonesian Medical Councils. *HaloDoc* also provides *telepharmacy* services in collaboration with more than 1,000 trusted pharmacies in Indonesia to deliver medicine to the customer within an hour.

The current Covid-19 pandemic is undoubtedly challenging conventional health systems and services to their core (Cummins & Schuller, 2020), and digital health systems play huge roles in developing adequate public health systems (Nordic Innovation, 2018) and are going to play more important roles in the future; thus, it's important to break down the challenges and develop potential solutions. The **first** and the most basic challenge is access to technology. Digital health systems required digital technology, and without having sufficient access to this technology, the digital health system would not optimally benefit the patients and healthcare providers. There are significant proportions of the population who are unable to access technology or connectivity, especially in low- and middle-income countries (Mitchell & Kan, 2019). Access to technology and connectivity is of paramount importance in developing digital health systems, a study conducted by researchers at Oxford University Big Data Institute concludes that a minimum of 60% of a country's population would need to be involved for the digital health surveillance approach to be functionally effective (Ferretti et al., 2020), and to reach the highest percentage possible of digital network connectivity, government interventions are needed. These interventions should not only be in providing connectivity and developing tracing devices, but the government also needs to work comprehensively with the stakeholders and experts to establish an accessible, transparent, and reliable digital health system (Farizi & Harmawan, 2020).

Second, the gap in digital health literacy is the ability to seek, find, understand and appraise health information from electronic sources and apply the knowledge gained to solve a health problem. The gap in digital health literacy is influenced by various factors such as age, educational background, digital skill, geography and motivation for seeking information, the availability of adequate infrastructure, and the lack of political commitment (Ortiz, 2017).

Third, regulatory factors in terms of the uncertainties surrounding digital health policies and regulations, especially related to privacy, data protection, and other ethical concerns (Cummins & Schuller, 2020); the global digital strategy emphasizes that health-related data are to be classified as sensitive personal data, or personally identifiable information, that require high safety and security standards; therefore, it stresses the need for a strong legal and regulatory base to protect privacy, confidentiality, integrity, and availability of data and the processing of personal health data and to deal with cyber security, trust-building, accountability, and governance ethics, equity, capacity building, and literacy, ensuring that good quality data are collected and subsequently shared to support planning, commissioning, and transformation of services; also, it's critical to maintain transparency and effectively communicate about the data security strategies (World Health Organization, 2020). Each country shall adopt its own digital health action plan built on the strategy, within its own national context in a way that is sustainable, respects its sovereignty, and best suits its culture and values, national health policy, vision, goals, and available resources.

Fourth, lack of accountability of commercial sectors that provide digital infrastructures and conduct health-related data collection. Digital technology and digital data collection still struggle with proper data protection, the utilization of data, the level of secrecy and access, and the accountability of the provider that are mostly commercial companies in case of data violation. The digital health system, especially in developing countries, has a certain degree of dependency on the business sector; therefore, it's critical to formulate comprehensive regulations and business ethics related to data and technology utilization. Digital health has become a part of daily life as well as massive growth business; thus, it's important for government, business, and other stakeholders to collaborate in preparing proper digital infrastructure completed with proper regulation and technical support to protect the community as the users and business entities in the industry (Deloitte Indonesia, 2019).

Last: Technological risk. The massive amount of data stored in digital health systems requires advanced technology for data storage, security, privacy, and interoperability of health data. These technological risks lead to a low level of community acceptance in several regions due to several risks like data loss, incorrect data input, and breaching of medical ethics. Therefore, substantial efforts needed to be made to normalize the use and inclusivity of digital health at a societal level while applying risk management in the digital health system operation. The providers shall ensure and commit their best effort and are accountable in case they do not.

Considering the number of potential challenges, government intervention especially in formulating an adequate legal framework is needed, authorities must investigate the proper development and utilization of the digital health system considering healthcare quality, patient safety, data privacy, legal and ethical matters (Garell et al., 2016) by focusing to streamline the regulatory processes and to promote innovation and to educate the designers of digital health services concerning relevant regulation and legislation and how to relate and act on such regulations; a proper legal framework would simplify and speed up the development of digital health services, and for monitoring process, the government can formulate a constructive legal framework to investigate, assess, and verify digital health services according to existing regulations and legislations(Garell et al., 2016).

Acknowledge that institutionalization of digital health in the national health system requires a decision and commitment by countries, recognize that successful digital health initiatives require an integrated strategy and promote the appropriate use of digital health and global collaboration and advance the transfer of knowledge on digital health, strengthen governance for digital health at global, regional, and national levels, advocate people-centered health systems that are enabled by digital health and the continuous and sustained implementation of the strategy and action plan completed with monitoring and evaluation (World Health Organization, 2020).

Conclusion

Many countries have shifted their healthcare methods towards digital health systems considering the efficiency of utilizing technology to replace conventional manual contact tracing as the part of new normal during the long haul of the Covid-19 pandemic; however, there are several challenges in the implementation: *first*: the access to technology, *second*: low levels of digital and health literacy, especially in aging population, *third*: the uncertainties surrounding digital health policies and regulations especially related to privacy, data protection, and other ethical concerns, *fourth*: lack of accountability of commercial sectors that provide digital infrastructures, and *last*: technological risk. Therefore, substantial efforts need to be made to normalize the use and inclusivity of digital health at a societal level and to optimally ensure the availability, accessibility, acceptability, effectiveness, and affordability of health services, essential medical services, and vaccines; government intervention especially in legal preparedness and developing broad legal policy concerning public health emergency, legal guidance, law designation, interpretation of current existing law, and concern about the liability of each party and ethical concerns is needed.

References

Adeniyi, E. A., Awotunde, J. B., Ogundokun, R. O., Kolawole, P. O., Abiodun, M. K., & Adeniyi, A. A. (2020). Mobile health application and Covid-19: Opportunities and challenges. *Journal of Critical Reviews, 7*(15), 3481–3488. http://www.epistemonikos.org/documents/750daedffca4123e799dbbf72a2655c76a438e58.

Ahn, E., Liu, N., Parekh, T., Patel, R., Baldacchino, T., Mullavey, T., Robinson, A., & Kim, J. (2021). A mobile app and dashboard for early detection of infectious disease outbreaks: Development study. *JMIR Public Health and Surveillance, 7*(3), 1–16. https://doi.org/10.2196/14837.

Akarturk, B. (2020). The role and challenges of using digital tools for COVID-19 contact tracing. *The European Journal of Social & Behavioural Sciences, 29*(3), 3241–3248. https://doi.org/10.15405/ejsbs.283.

Alwashmi, M. F. (2020). The use of digital health in the detection and management of COVID-19. *International Journal of Environmental Research and Public Health, 17*(8). https://doi.org/10.3390/ijerph17082906.

Auliya, S. F., & Wulandari, N. (2021). The impact of mobility patterns on the spread of the COVID-19 in Indonesia. *Journal of Information Systems Engineering and Business Intelligence, 7*(1), 31–41. http://e-journal.unair.ac.id/index.php/JISEBI.

Australian Digital Health Agency. (2020). *Media Release: Technology Delivers Social Distancing for Healthcare in Fight against Covid-19.* Australian Digital Health Agency.

Baker, E. L., & Porter, J. (2005). The health alert network: Partnerships, politics, and preparedness. *Journal of Public Health Management and Practice, 11*(6), 574–576. https://doi.org/10.1097/00124784-200511000-00017.

Bhatia, S., Lassmann, B., Cohn, E., Desai, A. N., Carrion, M., Kraemer, M. U. G., Herringer, M., Brownstein, J., Madoff, L., Cori, A., & Nouvellet, P. (2021). Using digital surveillance tools for near real-time mapping of the risk of infectious disease spread. *Npj Digital Medicine*, 4(1), 1–10. https://doi.org/10.1038/s41746-021-00442-3.

Binns, C., & Low, W. Y. (2015). What is public health? *Asia-Pacific Journal of Public Health*, 27(1), 5–6. https://doi.org/10.1177/1010539514565740.

Blandford, A., Wesson, J., Amalberti, R., Alhazme, R., & Allwihan, R. (2020). Opportunities and challenges for telehealth within and beyond, a pandemic. *The Lancet Global Health*, 8(11), E13164–E1365. https://doi.org/10.1016/S2214-109X(20)30362-4.

Botoseneanu, A., Wu, H., Wasserman, J., & Jacobson, P. D. (2011). Achieving public health legal preparedness: How dissonant views on public health law threaten emergency preparedness and response. *Journal of Public Health*, 33(3), 361–368. https://doi.org/10.1093/pubmed/fdq092.

Budd, J., Miller, B. S., Manning, E. M., Lampos, V., Zhuang, M., Edelstein, M., Rees, G., Emery, V. C., Stevens, M. M., Keegan, N., Short, M. J., Pillay, D., Manley, E., Cox, I. J., Heymann, D., Johnson, A. M., & McKendry, R. A. (2020). Digital technologies in the public-health response to COVID-19. *Nature Medicine*, 26(8), 1183–1192. https://doi.org/10.1038/s41591-020-1011-4.

Centers for Disease Control and Prevention. (2018). *Public Health Emergency Preparedness and Response Capabilities*. U.S Department of Health and Human Services. https://www.cdc.gov/cpr/readiness/00_docs/CDC_PreparednesResponseCapabilities_October2018_Final_508.pdf.

Chiu, H. (2005). Selling drugs: Marketing strategies in the pharmaceutical industry and their effect on healthcare and research. *Explorations: The Undergraduate Research Journal*. http://undergraduatestudies.ucdavis.edu/explorations/2005/explorations.pdf#page=89.

Civaner, M. (2012). Sale strategies of pharmaceutical companies in a "pharmerging" country: The problems will not improve if the gaps remain. *Health Policy*, 106(3), 225–232. https://doi.org/10.1016/j.healthpol.2012.05.006.

Cummins, N., & Schuller, B. W. (2020). Five crucial challenges in digital health. *Frontiers in Digital Health*, 2(December), 1–5. https://doi.org/10.3389/fdgth.2020.536203.

Dalrymple, P. W. (2011). Data, information, knowledge: The emerging field of health informatics. *Bulletin of the American Society for Information Science and Technology*, 37(5), 41–44. https://doi.org/10.1002/bult.2011.1720370512.

Danquah, L. O., Hasham, N., MacFarlane, M., Conteh, F. E., Momoh, F., Tedesco, A. A., Jambai, A., Ross, D. A., & Weiss, H. A. (2019). Use of a mobile application for Ebola contact tracing and monitoring in northern Sierra Leone: A proof-of-concept study. *BMC Infectious Diseases*, 19(1), 1–12. https://doi.org/10.1186/s12879-019-4354-z.

Deloitte Indonesia. (2019). *21st Century Health Care Challenges: A Connected Health Approach*. Deloitte Konsultan Indonesia, 1–114.

Eckmanns, T., Füller, H., & Roberts, S. L. (2019). Digital epidemiology and global health security; an interdisciplinary conversation. *Life Sciences, Society and Policy*, 15(1). https://doi.org/10.1186/s40504-019-0091-8.

Eysenbach, G. (2006). Infodemiology: Tracking flu-related searches on the web for syndromic surveillance. *AMIA. Annual Symposium Proceedings / AMIA Symposium. AMIA Symposium*, 244–248.

Eysenbach, G. (2009). Infodemiology and infoveillance: Framework for an emerging set of public health informatics methods to analyze search, communication and publication behavior on the internet. *Journal of Medical Internet Research*, 11(1), 1–10. https://doi.org/10.2196/jmir.1157.

Farizi, S. Al, & Harmawan, B. N. (2020). Data Transparency and Information Sharing: Coronavirus Prevention Problems in Indonesia. *Jurnal Administrasi Kesehatan Indonesia*, 8(2), 35. https://doi.org/10.20473/jaki.v8i2.2020.35-50.

Ferguson, B. (2012). The Emergence of Games for Health. *Games for Health Journal*, 1(1). https://doi.org/10.1089/g4h.2012.1010.

Ferretti, L., Wymant, C., Kendall, M., Zhao, L., Nurtay, A., Abeler-Dörner, L., Parker, M., Bonsall, D., & Fraser, C. (2020). Quantifying SARS-CoV-2 transmission suggests epidemic control with digital contact tracing. *Science*, 368(6491), 0–8. https://doi.org/10.1126/science.abb6936.

Freifeld, C., Mandl, K. D., Reis, B. Y., & Brownstein, J. S. (2008). HealthMap : Global infectious disease monitoring through automated classification and visualization of internet media reports.

Journal of the American Medical Informatic Association, 15(2), 150–157. https://doi.org/10.1197/jamia.M2544.

Galbadage, T., Peterson, B. M., & Gunasekera, R. S. (2020). Does COVID-19 spread through droplets alone? *Frontiers in Public Health, 8*(April), 1–4. https://doi.org/10.3389/fpubh.2020.00163.

Garell, C., Svedberg, P., & Nygren, J. M. (2016). A legal framework to support development and assessment of digital health services. *JMIR Medical Informatics, 4*(2). https://doi.org/10.2196/medinform.5401.

Gebbie, K. M., Hodge, J. G., Meier, B. M., Barrett, D. H., Keith, P., Koo, D., Sweeney, P. M., & Winget, P. (2008). Improving competencies for public health emergency legal preparedness. *Journal of Law, Medicine and Ethics, 36*(SUPPL. 1), 52–56. https://doi.org/10.1111/j.1748-720X.2008.00261.x.

Goggin, G. (2020). COVID-19 apps in Singapore and Australia: Reimagining healthy nations with digital technology. *Media International Australia, 177*(1), 61–75. https://doi.org/10.1177/1329878X20949770.

Golinelli, D., Boetto, E., Carullo, G., Nuzzolese, A. G., Landini, M. P., & Fantini, M. P. (2020). Adoption of digital technologies in health care during the COVID-19 pandemic: Systematic review of early scientific literature. *Journal of Medical Internet Research, 22*(11). https://doi.org/10.2196/22280.

Grainger, C. (2020). *A Software for Disease Surveillance and Outbreak Response.* Deutsche Gesellschaft für Internationale Zusammenarbeit (GIZ) GmbH. https://health.bmz.de/wp-content/uploads/studies/GHPC_SORMAS_full_version_final.pdf.

Guzman, J., Hafner, T., Maiga, L. A., & Giedion, U. (2021). COVID-19 vaccines pricing policy options for low-income and middle-income countries. *BMJ Global Health, 6*(3), 10–13. https://doi.org/10.1136/bmjgh-2021-005347.

Heled, Y., Rutschman, A. S., & Vertinsky, L. (2020). The problem with relying on profit-driven models to produce pandemic drugs. *Journal of Law and the Biosciences, 7*(1), 1–23. https://doi.org/10.1093/jlb/lsaa060.

Hernández-quevedo, B. C., Scarpetti, G., Webb, E., Shuftan, N., Williams, G. A., Birk, H. O., Jervelund, S. S., Krasnik, A., Vrangbæk, K., & Hernández-quevedo, C. (2020). Effective contact tracing and the role of apps: Lessons from Europe. *Eurohealth, 26*(2), 40–44.

Hotez, P. J., & Bottazzi, M. E. (2020). Developing a low-cost and accessible covid-19 vaccine for global health. *PLoS Neglected Tropical Diseases, 14*(7), 1–6. https://doi.org/10.1371/journal.pntd.0008548.

Huang, Z., Guo, H., Lee, Y. M., Ho, E. C., Ang, H., & Chow, A. (2020). Performance of digital contact tracing tools for COVID-19 response in Singapore: Cross-sectional study. *JMIR MHealth and UHealth, 8*(10). https://doi.org/10.2196/23148.

Kamel Boulos, M. N., & Geraghty, E. M. (2020). Geographical tracking and mapping of coronavirus disease COVID-19/severe acute respiratory syndrome coronavirus 2 (SARS-CoV-2) epidemic and associated events around the world: How 21st century GIS technologies are supporting the global fight against outbr. *International Journal of Health Geographics, 19*(1), 1–12. https://doi.org/10.1186/s12942-020-00202-8.

Karim, S. A. (2020). COVID-19 vaccine affordability and accessibility. *The Lancet, 396*(10246), 238. https://www.thelancet.com/journals/lancet/article/PIIS0140-6736(20)31540-3/fulltext.

Khan, Y., O'Sullivan, T., Gibson, J., Brown, A., Henry, B., Genereux, M., Nayani, S., Tracey, S., & Schwartz, B. (2017). Conceptualizing the essential elements of public health emergency preparedness in Canada. *Prehospital and Disaster Medicine, 32*(S1), S197. https://doi.org/10.1017/s1049023x17005167.

Lazarus, J. V., Ratzan, S. C., Palayew, A., Gostin, L. O., Larson, H. J., Rabin, K., Kimball, S., & El-Mohandes, A. (2021). A global survey of potential acceptance of a COVID-19 vaccine. *Nature Medicine, 27*(2), 225–228. https://doi.org/10.1038/s41591-020-1124-9.

Le, T. V., & Nguyen, H. Q. (2020). How Vietnam learned from China's coronavirus mistakes. *The Diplomat.* https://thediplomat.com/2020/03/how-vietnam-learned-from-chinas-coronavirus-mistakes/.

Leisinger, K. M. (2012). Poverty, disease, and medicines in low- and middle-income countries. *Business and Professional Ethics Journal, 31*(1), 135–185. https://doi.org/10.5840/bpej20123116.

Mitchell, M., & Kan, L. (2019). Digital technology and the future of health systems. *Health Systems and Reform*, 5(2), 113–120. https://doi.org/10.1080/23288604.2019.1583040.

Moore, S., Hill, E. M., Tildesley, M. J., Dyson, L., & Keeling, M. J. (2021). Vaccination and non-pharmaceutical interventions for COVID-19: A mathematical modelling study. *The Lancet Infectious Diseases*, 21(6), 793–802. https://doi.org/10.1016/S1473-3099(21)00143-2.

Nelson, C., Lurie, N., Wasserman, J., & Zakowski, S. (2007). Conceptualizing and defining public health emergency preparedness. *American Journal of Public Health*, 97 Suppl 1, 9–11. https://doi.org/10.2105/AJPH.2007.114496.

Nordic Innovation. (2018). A nordic story about smart digital health. *Nordic Innovation*, 1–25.

Norton Rose Fulbright. (2020). Contact tracing apps in Indonesia. Council of Europe Portal, 2020(April), 5–7. https://www.coe.int/en/web/data-protection/contact-tracing-apps.

Okan, O., Sørensen, K., & Messer, M. (2020). COVID-19: A guide to good practice on keeping people well informed. *The Conversation*, 19 March 2020.

Ortiz, D. N. (2017). Digital health literacy. In *Digital Health Literacy*. https://doi.org/10.4018/ijmd-wtfe.20210101.oa4.

Papadodima, S. A., Spiliopoulou, C. A., & Sakelliadis, E. I. (2008). Medical confidentiality: Legal and ethical aspects in Greece. *Bioethics*, 22(7), 397–405. https://doi.org/10.1111/j.1467-8519.2008.00654.x.

Public Health Ontario. (2020). *Public Health Emergency Preparedness Framework and Indicators: A Workbook to Support Public Health Practice* (Issue May). https://www.publichealthontario.ca/-/media/documents/w/2020/workbook-emergency-preparedness.pdf?la=en.

Rahmah, M., & Barizah, N. (2020). Halal certification of patented medicines in Indonesia in digital age: A panacea for the pain? *Systematic Reviews in Pharmacy*, 11(12), 210–217. https://doi.org/10.31838/srp.2020.12.34.

Raja, R., Mellon, P., & Sarley, D. (2006). *Procurement Strategies for Health Commodities : An Examination of Options and Mechanisms within the Commodity Security Context* (Issue July). U.S. Agency for International Development. https://pdf.usaid.gov/pdf_docs/PNADH233.pdf.

Rajan, S., Khunti, K., Alwan, N., Steves, C., Greenhalgh, T., Macdermott, N., Sagan, A., & Mckee, M. (2021). In the wake of the pandemic preparing for long COVID. http://www.euro.who.int/en/about-us/partners/.

Ramanathan, K., Antognini, D., Combes, A., Paden, M., Zakhary, B., Ogino, M., Maclaren, G., & Brodie, D. (2020). Health security capacities in the context of COVID-19 outbreak: An analysis of international health regulations annual report data from 182 countries. *The Lancet*, 395(January), 19–21.

Revere, D., Nelson, K., Thiede, H., Duchin, J., Stergachis, A., & Baseman, J. (2011). Public health emergency preparedness and response communications with health care providers: A literature review. *BMC Public Health*, 11(1), 337. https://doi.org/10.1186/1471-2458-11-337.

Ricciardi, W., Pita Barros, P., Bourek, A., Brouwer, W., Kelsey, T., Lehtonen, L., Anastasy, C., Barry, M., De Maeseneer, J., Kringos, D., McKee, M., Murauskiene, L., Nuti, S., Siciliani, L., & Wild, C. (2019). How to govern the digital transformation of health services. *European Journal of Public Health*, 29, 7–12. https://doi.org/10.1093/eurpub/ckz165.

Rosenthal, G. (2009). Economic and social council. *The Oxford Handbook on the United Nations*, 03048(March). https://doi.org/10.1093/oxfordhb/9780199560103.003.0007.

Ross, A. G. P., & Crowe, S. M. (2020). Planning for the next global pandemic. *International Journal of Infectious Diseases*, 38, 89–94. https://www.ncbi.nlm.nih.gov/pmc/articles/PMC7128994/pdf/main.pdf.

Saunders-Hastings, P. R., & Krewski, D. (2016). Reviewing the history of pandemic influenza: Understanding patterns of emergence and transmission. *Pathogens*, 5(4). https://doi.org/10.3390/pathogens5040066.

Scarpetta, S., Pearson, M., Colombo, F., Lopert, R., Dedet, G., & Wenzel, M. (2021). Access to COVID-19 vaccines: Global approaches in a global crisis. *OECD Policy Responses to Coronavirus*, March, 1–30.

SGPP Indonesia Student. (2020). Can a Covid-19 tracing app save Indonesia. *Sr.Sgpp.Ac.Id*. http://sr.sgpp.ac.id/post/can-a-covid-19-tracing-app-save-indonesia.

Snowdon, A. (2020). HIMSS defines digital health for the global healthcare industry. https://www.himss.org/news/himss-defines-digital-health-global-healthcare-industry.

Sorato, M. M., & Taddele, B. W. (2019). Availability and affordability of essential medicines and patient satisfaction on pharmacy services: Case of two public hospitals in Gamo Zone, Southern Ethiopia. *CPQ Medicine*, 7(4), 1–19.

The Conversation. (2020). Coronavirus response: Why Cuba is such an interesting case. *The Conversation*. https://theconversation.com/coronavirus-response-why-cuba-is-such-an-interesting-case-135749.

Tom-Aba, D., Nguku, P. M., Arinze, C. C., & Krause, G. (2018). Assessing the concepts and designs of 58 mobile apps for the management of the 2014–2015 west Africa ebola outbreak: Systematic review. *JMIR Public Health and Surveillance*, 4(4), 1–10. https://doi.org/10.2196/publichealth.9015.

Tom-Aba, D., Silenou, B. C., Doerrbecker, J., Fourie, C., Leitner, C., Wahnschaffe, M., Strysewske, M., Arinze, C. C., & Krause, G. (2020). The surveillance outbreak response management and analysis system (SORMAS): Digital health global goods maturity assessment. *JMIR Public Health and Surveillance*, 6(2). https://doi.org/10.2196/15860.

U.S Department of Health and Human Services, Food and Drug Administration, Center for Drug Evaluation and Research, Center for Biologics Evaluation and Research, & Center for Veterinary Medicine. (2020). Good manufacturing practice considerations for responding to COVID-19 infection in employees in drug and biological products manufacturing guidance for industry (Issue June). https://www.fda.gov/regulatory-.

United Nations Department of Economics and Social Affairs. (2020). The long-term impact of COVID-19 on poverty. *Policy Brief, 86*, 1–5.

United Nations Human Rights. (2020). *Human Rights and Access to Covid-19 Vaccines*. United Nations. https://doi.org/10.5771/9783845273945.

Ventrella, E. (2020). Privacy in emergency circumstances: Data protection and the COVID-19 pandemic. *ERA Forum*, 21(3), 379–393. https://doi.org/10.1007/s12027-020-00629-3.

Vokinger, K. N., Nittas, V., Witt, C. M., Fabrikant, S. I., & Von Wyl, V. (2020). Digital health and the COVID-19 epidemic: An assessment framework for apps from an epidemiological and legal perspective. *Swiss Medical Weekly*, 150(19–20), 1–9. https://doi.org/10.4414/smw.2020.20282.

Wanat, M., Hoste, M., Gobat, N., Anastasaki, M., Boehmer, F., Chlabicz, S., Colliers, A., Farrell, K., Karkana, M.-N., Kinsman, J., Lionis, C., Marcinowicz, L., Reinhardt, K., Skoglund, I., Sundvall, P.-D., Vellinga, A., Verheij, T., Goossens, H., Butler, C., … Tonkin-Crine, S. (2021). Transformation of primary care during the COVID-19 pandemic: Experiences of healthcare professionals in eight European countries. *British Journal of General Practice*, BJGP.2020.1112. https://doi.org/10.3399/bjgp.2020.1112.

Webster, P. (2020). Virtual health care in the era of COVID-19. *Lancet*, 395(10231), 1180–1181. https://www.ncbi.nlm.nih.gov/pmc/articles/PMC7146660/pdf/main.pdf.

Weinberg, G. A., & Szilagyi, P. G. (2010). Vaccine epidemiology: Efficacy, effectiveness, and the translational research roadmap. *Journal of Infectious Diseases*, 201(11), 1607–1610. https://doi.org/10.1086/652404.

Welsch, F. (2021). The COVID-19 pandemic: A multidimensional crisis. *Gaceta Medica de Caracas*, 128(December), S137–S148. https://doi.org/10.47307/GMC.2020.128.S2.2.

World Health Organization (WHO). (2013). Chapter 11: Public health emergencies. *Advancing the Right to Health: The Vital Role of Law*, 165–180. http://www.who.int/healthsystems/topics/health-law/chapter11.pdf.

World Health Organization (WHO). (2017). Communicating risk in public health emergencies. In *A WHO Guideline for Emergency Risk Communication (ERC) Policy and Practice*. https://apps.who.int/iris/bitstream/handle/10665/259807/9789241550208-eng.pdf?sequence=2.

World Health Organization (WHO). (2020). *WHO Announces COVID-19 Outbreak a Pandemic*. WHO. https://www.euro.who.int/en/health-topics/health-emergencies/coronavirus-covid-19/news/news/2020/3/who-announces-covid-19-outbreak-a-pandemic.

Wouters, O. J., Shadlen, K. C., Salcher-Konrad, M., Pollard, A. J., Larson, H. J., Teerawattananon, Y., & Jit, M. (2021). Challenges in ensuring global access to COVID-19 vaccines: Production, affordability, allocation and deployment. *The Lancet Global Health*, *397*(March), 1023–1034. https://www.ncbi.nlm.nih.gov/pmc/articles/PMC7906643/pdf/main.pdf.

Zhu, N., Zhang, D., Wang, W., Li, X., Yang, B., Song, J., Zhao, X., Huang, B., Shi, W., Lu, R., Niu, P., Zhan, F., Ma, X., Wang, D., Xu, W., Wu, G., Gao, G. F., & Tan, W. (2020). A novel coronavirus from patients with pneumonia in China, 2019. *New England Journal of Medicine*, *382*(8), 727–733. https://doi.org/10.1056/nejmoa2001017.

8

Automation in Labor Sector and Its Effects and Challenges on Job Creation: An Analytical Study

Jagdish Khobragade and Simran Bais

CONTENTS

Introduction

The concept of automation is not a new phenomenon altogether, and pertinent issues have arisen regarding its promises and effects. Going back to more than half a century ago, under the guidance of US President Johnson, a National Commission was established, which had the objective to critically determine the effect of technology on the state of the economy and labor employment, declaring that the jobs are not wrecked by automation; rather it "can be the ally of our prosperity if we will just look ahead".[1] The impact of automation is primarily universal. It has its prospects to transform the circadian work activities of everyone ranging from a gardener to a commercial bank, and everything will witness a tremendous transformation.[2]

Impact of Automation on Labor Market

To examine the possible impact of automation technologies, a research program was conducted. The key comments are enlisted and discussed in detail.[3] The new age of automation is marked in such a way that the automation is not only able to conduct a range

DOI: 10.1201/9781003215998-8

of physical activities, but they are also capable of undertaking activities that essentially involve the application of cognitive capabilities like driving a vehicle, sensing emotions that were earlier considered difficult to automate.[4] The entire economic sphere is included as the automation of activities leads to productivity growth and other benefits, which are also inclusive of individual processes and businesses. In such cases, productivity acceleration is required, particularly in the nations where the working-age population declines. Analyzing it from the lens of microeconomic levels, the business expansion attains a competitive advantage from automation technologies, which is attributable to not just the labor law reductions but also increasing the efficiency in output in lower periods. To understand the potentiality of automation, the focus was on individual activities rather than the entire occupations. However, it can be noted that every work is being equipped with partial automation. Therefore, in this line of thought, about half of all the listed activities people are paid to do in the world's workforce could potentially be automated through the adaptation of technologies. It is to be noted that across various activities, occupations, and wage and skill levels, the extent of automation and its impact tends to vary. As multiple activities are trapped in automation, workers can be engaged in noetic work alongside the machines. Some of the physical activities which are prevalent in the sphere of manufacturing and retail trade, collection, and processing of data are the activities that are borne to be automated. There can be skill-biased forms of automation, tending to raise the productivity and efficiency of high-skilled workers even as they reduce the demand for lower-skilled workers involving routine-intensive occupations.[5] The middle-skilled workers are disproportionately affected by other automation.[6] The activities of both low-skill and high-skill workers, due to technological development, are more prone to automation. These effects of polarization could be minimized. Across various geographies and sectors, automation does have wide-ranging effects. The effects of potential get impacted in the industrial sector predominately in manufacturing and agriculture sectors, which include certain physical activities that demand high technical aspects to be automated; however, developing countries possessing lower wage rates could restrict the adoption of automation to some extent. The process of automation cannot happen in a day or so. The key factors which should be taken into consideration are *technical feasibility*; the need of the hour is to invent technology, facilitate integration, and adapt solutions that play an essential role in the automation of specific activities. This adoption of automation certainly has an impact on the business case; therefore, the *cost of developing and enforcing the solutions* should be taken into consideration for the phase of automation. Since the impact of automation directly lies in the *labor supply*, *demand*, and *economy*, these factors and the economic benefits derived from *increased quality and labor cost savings* are other essential ingredients to be considered for the recipe of automation.

In conclusion, the *regulatory and social acceptance* forms a primordial part of determining the rate of adoption of automation. After getting into the analysis of the factors as formulated above, it is presumed that it will take decades for the integration of automation's effect on the incumbent work activities to get operationalized on a total basis. The impact on a macroeconomic level can be slow in terms of the scale of economies; at the microeconomic level, the effect could be expedited in terms of an individual worker whose activities can be automated on a large scale, thus enhancing efficiency, or a company whose industry is deranged by the competitors using automation. The issue of mass unemployment has taken a nuclear point for the debate concerning automation; however, the point is to bypass the socio-demographic aging patterns in both the developed and developing economies. The demand of the world economy expects working of every division of human labor and robots in collaboration. It is imperative to note that unless automation is deployed in the broader sense, the marginal surplus of human work is probably less likely to occur than a deficit of human labor. However, the age of automation would transform the *nature, form,* and *content of work* tremendously. Through the process of automation, people will be required to perform work that will run complementary to the tasks the machine performs. This will trigger competitive practices across various businesses. It is undeniable that for businesses, the performance benefits of automation are inevitable and clear, but the pertinent issue arises for policymakers. They should, in a way, embrace the opportunity of economic growth from their productivity incentive and put in place the policies to encourage investments to perpetuate progress with innovation. However, there must be an evolution of innovative policies that would aid the institutions as well as the workers to get themselves adapt to the impact on the employment ecosystem. This can be attained through emphasizing acquiring new skills, which are ancillary to the technological age, and the mindset of the people should be developed in such a way that integration of machines and labor becomes a day-to-day affair and not an exception.

International Collective Labor Agreements

Internationally, at the human level, the goal is to establish a uniform set of guidelines concerning working hours and advanced training options to create a level playing field. The idea seems simple, but its implementation is a very tedious task. Though there is a prevalence of binding agreements under international law on fundamental aspects such as the prohibition of child labor, minimum wages, and occupational safety, the standards in these aspects are addressed on an international front in the name of the International Labour Organization Standards.[7]

At the international or European level, there is no harmonized collective approach to labor law, and this is particularly due to different national labor law systems. It has been observed that in some countries, specific topics were essentially unregulated by labor law. Because of the corresponding statutory regulations in different nations, the collective agreement would be rendered invalid. To eliminate this degree of uncertainty for companies, Transnational Collective Agreements ("**TCAs**") came to light for mitigating and regulating group-wide issues.[8] Through the means of TCAs, only the minimum required standards were specified, and there was no binding nature created by TCAs, the proper implementation of which depends on the respective national law.

Protecting the Interests of the Vulnerable Workers from Shifting Gears of Technology

By revolutionizing progressive tax systems, apportioning subsidies, or offering a reward as a bypass route to unconditional primary income government could introduce higher minimum social standards. It was contended by the US economist Richard Freeman that the age of automation demands employees' prudent investment in the development of new IT or IT-related systems and another point expressed by the economist was that the employees should buy the robots that replace their jobs.[9] The benefit acquired through this mechanism over the unconditional basic income is that working is not subjected as unattractive. As a result of substitution, the state could reduce the maximum working hours by bringing in necessary amendments to law so that the incumbent jobs can be allocated to various employees which could also come as a relief to curbing the menace of unemployment in the economy.[10] Governments should implement upskilling and reskilling programs for adult workers on a large scale, based on talent planning strategies. Also, governments need to refocus education systems to develop logical thinking, reasoning, curiosity, openness to new ideas, collaboration, creativity, leadership, and systems thinking and the same can be achieved by collaborating with the private sector and academia.

The employment crisis can be further decelerated by permitting the working-age population to decline.[11] The evidence on the subject matter suggests that a rise in unemployment is prevented by reducing the working hours substantially; however, its contribution to the reduction in existing unemployment is almost *nil*.[12] As depicted in Figure 8.1, the pictorial representation can be referred to to understand the effects of automation on the workforce and to analyze whether the robots own humans or *vice versa*. The duty can be cast upon the lawmaker of the respective states to determine which jobs or category of jobs it wants to be executed by humans exclusively. Therefore, to facilitate this vision further, the state can come up with a type of *"human quota"* in any sector and could decide what social work it is ready to support (*because full automation cannot happen overnight*) and whether it wants to bring "made by humans" label or a tax which can be levied for the use of machines.[13]

The arguments proposed in this chapter focus on the point that automation can be observed in the changing nature of work and not jobs. The question from the economic context arises, *how can we get better at sharing the wealth created by technology equally?* One of the routes to solve the problem posed can be the government's support in the form of providing subsidies to innovative start-ups, particularly in the inception stage. The welcoming of new ideas into the market could result in the modernization of the economy. New jobs can be created if these ideas are not tainted due to state bureaucratic or political interventions because it is only then that the economy can realize its most tremendous potential. Therefore, all these ideals must be tested on the touchstone of their economic feasibility and socially fair manner, and then only they can be implemented in society. However, the lawmakers should look into all these proposals so that the interests of the laborers are also protected along with the advancement in technology.

FIGURE 8.1
Effects of automation on the economy (Pic Credit: MIT Technology Review, 2015).

Comprehending the Issue from the Lens of International Labor Collective Agreements in the Age of Automation

A cover of the New York magazine depicted a picture of humanoid robots walking on the street of the city and giving handouts to a human beggar; this pictorial representation sparked an intensive debate by the mainstream on the reduction in the number of jobs due to automation.[14] The same issue was discussed on the subject matter of job automation and its consequences of introducing modern automated work in the incumbent workplaces; also an imperative factor which the article took into consideration was the extensive analysis of the relationship between humans and machine labor and how the workers interact with the advanced machinery at the industrial level. The policy and academic debates were sparked only on the number of workers replaced due to automation following the quantitative approach only, and relatively less attention was paid to the quality of the jobs.[15] It has been observed that these jobs of the future are being taken for granted because they will require high technical skills to get oneself acquainted with the machinery and algorithms complemented by AI, which in a way attempts to displace or absorb the routine dangerous tasks that the fortunate workers who remain employed had access to such gratifying jobs. Therefore, instead of fearing and losing hope, the regulators should ensure that courses on technical skills are provided and encourage the workforce to acquire modern skills which are in line with the age of automation.

Electronic Personality and Dehumanization of Workers: Discerning the New Future

The age of automation characterizes a fundamental question that centers itself on *whether robots can be made to have rights and duties as humans possess?*

To tackle such issues, a draft report was formulated by the EU Parliament, which scrutinized the question of the possibility of conferring the legal status for robots basically, which would serve as a medium of the electronic personality. It was contended in the draft report that the necessity of bringing in the concept of electronic personality is mainly for the robots who are involved in sophisticated tasks or where the robots have taken smart autonomous decisions or have interacted with third parties in an independent fashion. Therefore, in these circumstances, conferring robots with specified rights and obligations will act as a measure of security.[16]

The draft report further explains that conferment of legal rights and obligations to non-humans cannot be considered a neutral process. It can prove to be beneficial in some contexts. However, it can also put the other party in jeopardy and can abuse its conferment. The process of providing electronic personality to the robots will ensure that the owner is liable towards the act of the robots, then the commercial partners, creditors who interact with the robots would have no meaningful redress in case of damages.

In furtherance, a conflation between these two entities will be created by assigning a personality to the non-natural physical being, especially in a sphere where both the humans and machines share the same physical space in a work environment as it will hamper human dignity, because apart from the machines, the humans are already under the control and supervision exerted by their subjects and therefore, working side-by-side with robots can increase the feeling of alienation.[17] The EU Parliament Report expressed the concerns and risks integrated with dehumanization and the emergence of intelligent workers, stating that "human contact is one of the fundamental aspects of human care and that replacing the human factor with robots would inherently dehumanize caring practices". However, it was observed that no concerns were reiterated concerning the work of caregivers or work involving human touch.[18]

Need of Necessary Skillsets for Employees

The multidisciplinary support endorsed by AI and machine learning is bound to change the requirement for a future employee. The content of the work, which is mainly repetitive, will be eventually absorbed by the technology, and hence, the need for employees engaged in such work will be eliminated drastically. It has been observed that the strength of the factory workers is reduced tremendously, and humans are becoming the controller of the machines; an example to understand this proposition is the automotive industry, where many production points are automated on a total basis. If the demand for such workers gets lower, then the companies' demand will escalate for recruiting highly qualified and technical employees. Therefore, it can be reiterated that as per the common belief, better education and training will help.[19]

The labor market in the age of automation aspires employees to be prepared with creative solutions and skilled and experts in mathematics and science.[20] It would be required that the employee must necessarily possess a firm grip on the analytical and technical concepts and examine the software critically; the employee should integrate work with the machine and must be acquainted with the basic structures of the algorithm. All the skills enlisted above raise the bar of demand for the employees with strategic working. The important point of consideration is not just overseeing or supervising the machines but striking a string of coordination between them. The interface which is prevalent between humans and robots and the field of accountability among humans must be synchronized. This paves the way for the future executive staff to be well versed with social as well as interdisciplinary competence.

Task-Based Conceptual Approach

Few researchers propose that the central unit of production should follow a task-based approach (*as elucidated in* Figure 8.2), and in this particular respect, some of the tasks are carried out through laborers and few other tasks are continued either by labor or by capital. This conceptual framework of a task-based approach facilitates the resolution of dichotomies where automation is conceptualized as a replacement of labor. A direct displacement effect reducing labor demand is observed in this replacement process. If the other economic forces involved don't counterbalance this displacement effect, the result will lower labor demand, wages, and employment. The counterbalancing force that should be taken into account is that the cost of production gets reduced by automation, thus increasing productivity. Also, there exists a comparative advantage in labor and capital concerning distinct tasks, which means that the relative productivity of labor varies across different tasks.[21] The process of automation is effectively characterized as an expansion in a lot of tasks that can be assembled with the aid of capital, and if subsequently, it is found that the capital is sufficiently cheap, then the AI would result in a barter of money for labor; this is the displacement effect as depicted in Figure 8.2.

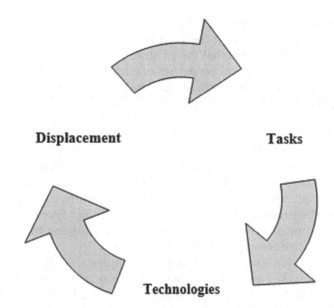

FIGURE 8.2
The diagram depicts the cycle of automation and its impact on labor markets.

The displacement effect could reverberate in reduced demand for labor and the equilibrium wage rate. Another point of significance is the fact that due to technological improvements, the wages of the workers get reduced, which often gets sidelined. From the lens of economics, it is contended that considering the labor market imperfections, inelastic labor supply, or quasi-labor supply, a depression in the demand for labor also leads to lower employment. As opposed to the standard approach dependent on the factors enlarging the technological changes, a task-based approach guides the ulterior route of productivity through the magnification of technological developments by reducing wages and employment.[22]

A vital facet should be on the generation of new tasks counterbalancing the inherent deleterious effects of automation on labor share, and the demand for labor ignores a budding constraint that is the latent incongruency existing between the requirements of new technologies and the skills involved in the workforce. This is simply because new tasks require the inception of modern skills which should be acquired by force, and therefore, this particular adjustment takes place over a relatively long period. The standard of living can be enhanced by the induction of AI. The short-term displacement effect could be substantial; however, the potential and frequently cited solution to this particular puzzle lies in the concept of universal basic income, which the government should provide to all the members of society by way of cash transfers.

Robotization and Its Implications

For the automatic systems and their certifications, high standards are set. Firstly, the system must be able to learn independently. That is, it should be able to optimize its skills. This can be achieved by the human programming individual production steps or demonstrating the same to the system and the IT system emancipating experience during its work and independently imbibing the suggestions for improvement or even learning by itself. This, therefore, mandates that the programmer of the automation system must include and understand the physical properties of employees and their cognitive abilities in the pretext of relevant tasks and accordingly make use of this information while the program is being generated. The life-long interactive learning processes from the human partner and responding to human needs are the mitochondria of AI and a functioning production IT system.[23] In furtherance, the robot should function so that it must pull down highly complex plans as needed by the customer base and produce them autonomously. The IT system must develop comprehensive collective intelligence and interact with other devices and humans at large. A production robot is significantly designed so that the robot possesses all the human capabilities like perception, adaptability, and cognition; to achieve this, the robot must be programmed dynamically and rigidly. Therefore, in this manner, the operating human must synchronize the system's functioning to cater to individual needs even if the system does not recognize them itself.

As illustrated in the below graph (in Figure 8.3), an empirical study from the Netherlands points to the fact that spikes are witnessed when the firm invests in technological upgradation. Therefore, the first curve represents that the firms which inculcated the automation events are steadily experiencing growth, and in contradistinction, the firm without automation events are not experiencing steady growth.

Escalating Demand for the New Skilled Jobs

Through the deployment of AI systems, it has been observed that the advancement of technology will result in the development and creation of new jobs. The machine learning systems also identified the emergence of three different types of jobs because of AI.[24] The description of these particular jobs is enlisted in the following manner:

1. *Trainers* who will train AI systems, for example, reducing the error of a language translator, tagging data in a training dataset, adapting to the chatbot system to mimic the behavior of a particular kind or match the culture of a company.
2. *Explainers* who will interpret the outputs which are generated by AI systems to improve transparency and accountability, for instance, explaining how an AI

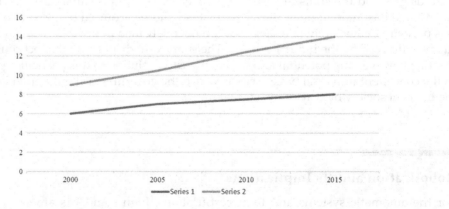

FIGURE 8.3
Employment at firms making major automation investments and not in the Netherlands (Bessen et al. 2019).

system arrived at a particular decision facilitating the decision makers about appropriate uses of AI throughout an organization.

3. *Sustainers* will monitor the work of a system to ensure that they are working in the manner which is intended, for instance, installing content filters in machine learning for a chatbot.

Conclusion

The contribution of artificial intelligence to the labor market has a significant impact on the economy. It is the technological development which cannot be avoided, and it has both positive and negative impacts. Although it is important to take note of the fact that the technological transformation will not take place suddenly, the way we perceive it in its full sense, however, it can cast its spell modestly in certain sectors with noteworthy changes. In economics, human capital is one of the important factors which cannot be replaced by automation. Therefore, in this context, a commitment from the side of the government is warranted to evolve some deep technical training and education for skill development and serious efforts in lowering the hardships and struggles of the people in the workplace. If this commitment is followed by the nation, then the future full of machines can be made tolerable to a large extent. Hence, it is imperative to note that the sophisticated analytical models need to be reinvigorated and integrated into the workforce strategy so that economic and social frictions are minimized in relation to the mismatch between demand and supply.

References

1. US Dept of Education, Health and Welfare (1966). *Report of the National Commission on Technology, Automation, and Economic Progress.*
2. Acemoglu, Daron, and Pascual Restrepo (2017). *The Race between Machine and Man: Implications of Technology for Growth, Factor Shares, and Employment.* Working paper, MIT.
3. McKinsey Global Institute (2017). *A Future That Works: Automation, Employment, and Productivity;* Research Insight impact.
4. McKinsey Global Institute (2016). *The age of analytics: Competing in a data-driven world;* Research Insight impact.
5. David H. Autor, Frank Levy, and Richard J. Murnane (2003). *The skill and content concerning the recent technological innovations,* Quarterly Journal of Economics.
6. David H. Autor (2015). *Still, there are so many Jobs? The History and future of Automation and Workforce,* Journal of Economic Perspectives 29 (3).
7. Chang Jae-Hee and Huynh Phu (2016). *The Future of Jobs at Risk of Automation.* International Labour Organization.
8. Stefano Valerio (2018). *Negotiating the algorithm: Automation, AI and labour protection,* Employment Policy Department Geneva Working Paper No. 246.
9. Rotman David (2015). *Who will own the Robots?* MIT Technology Review Technology Policy.
10. International Labour Organization (2021). *ILO Monitor: COVID-19 and the World of Work.* 7th edition.

11. The Federation of International Employers (2016). *The rise and rise of the shorter work week*, Latest News.
12. Scherf Wolfgang (2015). *From a Macroeconomic perspective, reductions in working hours cannot meet expectations* Economist. Die Presse.
13. Becker (2016). *The revolution of Industrial 4.0*, HR Performance p. 62.
14. Lohr Steve (2017). *Robots will take jobs, but not as fast as people fear*, *The New York Times*.
15. Frey and Osborn (2013). Dauth *et al.* 2017.
16. Mady Delvaux (2016). *Draft Report with recommendations to the Commission on Civil Law Rules on Robotics (2103(INL))*, Committee on Legal Affairs Rapporteur, European Parliament.
17. Publications Office of the European Union (2019). *Annual review of working life*, Luxembourg, Eurofound.
18. Moore, Phoebe, Akhtar, Pav, Upchurch, Martin (2018). *Digitalization of Work and Resistance* in Moore, Phoebe, Upchurch, Martin, Whittaker, Xanthe (eds.) *Humans and Machines at Work: Monitoring, Surveillance and Automation in Contemporary Capitalism*, Palgrave Macmillan.
19. Krischke and Schmidt (2015). *"Kollege Roboter"*, Focus Magazine 66.
20. Aaronson, D., Agarwal, S., Hotchkiss, J.L., and Kelley, T. (2019). *Job displacement and financial outcomes*, Economics Letters, 177, p.18–21.
21. Acemoglu, Daron (2009). *Introduction to Modern Economic Growth*. Princeton, NJ: Princeton University Press.
22. Acemoglu, Daron, and David Autor (2011). *"Skills, Tasks and Technologies: Implications for Employment and Earnings"*. Handbook of Labour Economics 4:1043–171.
23. W. W. Norton & Company. (2014). The second machine age *Work, progress, and prosperity in a time of brilliant technologies*.
24. PWC (2018). *The macroeconomic impact of artificial intelligence*.

9

Water Scarcity: A Global Threat to Access Human Right to Clean Water

Anuja Mishra and David W. Tushaus

CONTENTS

Introduction

"Access to Safe water is a fundamental human need therefore a basic Human Right".

Kofi Annan,

Former Secretary General of United Nations

Accessibility to safe drinking water is a basic necessity, and it is an indicator of human development. Demand for clean drinking water is increasing on a daily basis while the water resources are decreasing rapidly due to excessive uses of water resources for agriculture, industry, and fulfilling the drinking and domestic water demands of an increased population. Wise management of existing water resources and finding new technologies have become challenging to the governments of different nations, specifically in developing countries (Raju 2018). "Water is the essence of life and potable drinking water is the indispensable necessity of human health" (Van Derslice 2011). Water serves as a lubricant and forms the saliva and fluid in the body, regulates temperature for cooling and heating, regulates metabolism, and controls diseases in the body. Along with air, it is a substance without which no one can survive (Maddalena). Since water is such an important component of our physiology, it would make sense that water should be accessible equally to all people at the same time. Quality of water is just as important as quantity.

The right to water has increased in importance through national and international legal instruments and case laws in the last 20 years. As water and life are surprisingly

DOI: 10.1201/9781003215998-9

interlinked, so they should get a proper place as a legal right in any human rights instrument or bill of rights. It may or may not be included in the list of formally recognized rights. Nowadays, the right to clean water has a leading role, which is directly linked to the right-based approach (Gleick 1996).

The state is obliged to give access to clean water free from all contaminations. States own the responsibility to manage and make arrangements for public health, sanitation, agriculture, water supply, irrigation, drainage, and fisheries as these all are subjects of the state (Raju 2018). Since the "right to life" is a fundamental right which is extended toward the environment, which is clean and good health in the form of right, the right to water became part of these rights without which proper health and wellness cannot be ensured. We can assume the importance of safe drinking water with this thing that it is on the top of the essentials for human health and it is an essential ingredient of health protection policies. Various international policy forums have reflected this, which include the International Conference on Primary Health Care, held at the international level in Alma-Ata, Kazakhstan (the former Soviet Union), 1978, World Water Conference, Mar Del Plata Argentina, 1977, Millennium Development Goals (MDGs) 2000 adopted by the United Nations General Assembly and Johannesburg World Summit for sustainable development, 2002. Ten years from 2005 to 2015 had been declared as the international decade for action, "Water for life" by the UNGA[1] (WHO 2006). There are also other treaties which recognize the right to safe drinking water like CEDAW (UN 1979) and CRC (UN 1989), and the African Charter 1990 (Charter 1990). General Comment 15 is an instrument which is still a key document for understanding the **right to safe drinking water**. UNCESC[2] issued it in 2002 (Coomans 2011). General Comment 15 was the first document which authoritatively defined all the contents and scope of the human right to water, as well as corresponding state obligations. This document has given an affirmation to the right to clean water as a human right that:

> "Everyone is entitled to sufficient, safe, acceptable, physically accessible and affordable water for personal and domestic uses".

(Albuquerque 2012)

It is noticeable that a wide range of international documents recognize the right to clean water and reaffirm its importance while stating:

> "For leading a life with human dignity human right to water is indispensable. For realization of other human rights it is a prerequisite".

(Tully 2005)

The United Nations General Assembly considered the importance of the right to clean water and therefore adopted the United Nations Millennium Declaration (Resolution A/res/55/2) in 2000, which established the MDGs.[3] In these Millennium Development Goals, the "right to safe drinking water" is also one of the important objectives to be achieved. The global efforts to meet the Millennium Development Goals and Johannesburg Targets are of no importance to people without access to potable water.

Universal Declaration of Human Rights 1948 was the first document which recognized the right to clean water at an international level (Assembly 1948). On July 28, 2010, the United Nations General Assembly passed Resolution 64/292, which recognizes the right to clean water explicitly. This Resolution also acknowledged that the right to clean water is quintessential for the actual realization of all human rights (Assembly 2010). This Resolution was a call for all the international organizations, states, and nations to provide financial resources and help in capacity building.

At international and national levels also, different countries have given recognition to the right to water in different ways. In India, this right has been recognized but it is not a fundamental or a directly mentioned constitutional right; instead it is an expanded right which has evolved out of article 21 that is the right to life in the Indian Constitution (Cullet 2013). Expansion of the right to clean water is a result of continuous judicial activism (Mishra 2018). At the level of government, different policies have been made and adopted like water policy 1987, water policy 2002, and water action plan for India 2020, which focuses on a participatory approach and regulation of the use of water from various aspects. Still, the right to water is in danger as the availability and easy accessibility of clean water is not in a comfortable zone despite various efforts at the international, national, and judicial levels. Consequently, the right to clean water for common people is violated (Kathpalia and Kapoor 2002). Though this right has legal recognition in one or another way, various policies at the international and national level are also supporting it; still an additional effort is required for the actual realization of the right and protecting the violation of the right.

This chapter talks about the human right aspect of water, how water became scarce, though it is a natural resource, what have been the reasons behind the scarcity of water. Further, the chapter will discuss how scarcity of water is a problem of violation of innumerable human rights. Finally, a conclusion will be derived.

Water as a Scarce Natural Resource

The scarcity of water is demonstrated by more than 1.2 billion people who lacked access to clean drinking water even 20 years ago (WHO 2003). Approximately 70% of the Earth is surrounded by water. Available fresh water is only about 2.5% of the total water resources (WHO 2001); however, only 1% of the total water is safe for drinking without purifying (EU 2012). According to an estimated calculation by the United Nations, a population of 2.8 billion will be facing the problem of non-availability of freshwater by[4] 2025 (Agency 2008). It is estimated that sustainable water resources of Western Asia and Northern Africa have been already effectively used up (Secretary-General 2011). These numbers of deprived people just throw a light that the situation is painful while the reality is worst (WHO 2004). In rural areas, around 900 million people, who have an income of less than one dollar per day, face a lack of access to sufficient water for running their livelihoods (Prüss, Kay et al. 2002). Violations of some human rights occur from not obtaining water and sanitation facilities (Twas 2002). Rights concerning adequate housing are limited without sanitation systems. Here, the most important thing is the "right to life and health". Lack of fresh water does not only spread various diseases but also causes death and costs a large amount of economic burden on the society (Dinka 2018).

Sub-Sahara and South Asian countries are critically suffering from scarcity of fresh water. Almost 30% of the inhabitants of these countries are surviving under critical circumstances. Oceania is the leading country in this respect with an enormous lack of safe drinking water, where 53% of the total population is suffering badly (UN 2012). Even some developed countries in Europe are facing scarcity of fresh water, specifically in the Eastern European states. In some of these states, there is a scarcity of clean water on a large scale (Twas 2002). Throughout the world, many people are underprivileged to access fresh water sources; either it is very limited or non-existent. In many states, they must walk far to get fresh water for their drinking and domestic usage. The physical non-accessibility of water is an additional hurdle. In African countries, females are the worst victims of this situation; they have to cross an average distance of 6 kilometers to get a suitable water source (UNFPA 2002). The lack of safe drinking water triggers a chain reaction. First, it results in poverty and poverty means incapability to access necessities with dignity. This is perhaps the most pervasive human rights violation (Mahapatra (2005–2006)).

It is frequently questioned about the right to water, what this right really entitles, and whose responsibility it should be at a primary level to provide these facilities. For this reason, many issues regarding the right are taking place; for instance, privatization is an indication that states are becoming dependent on third actors and shifting over their responsibility to provide potable water to people to private companies. Another big hurdle here is poverty at the global level that may cause the inaccessibility of water. If companies start providing water and water services, then the directly affected masses will be those who are not so strong economically (Gleick 1998). They may not be able to afford the vital substance. Companies do not want to invest in areas with poor inhabitants because profit will be relatively low there (UNESC 2003). It is frequently argued that water should not be a human right, which should be dependent on market profits for companies, nor should it be available according to the economic or social development of individuals. In this present era, there are various conflicts and the right is being violated in many forms and for different reasons (Gleick 1998). But when we talk about the scarcity of water, then it is very much relevant to know what is the scarcity of water and what are the different kinds of scarcities of water, what are the different shapes of scarcity of water around the world, and what may be the various reasons which create the situation of scarcity of water. It is required to understand water scarcity for the formulation of policies and plans at the global, national, regional, and local levels (Liu, Yang et al. 2017).

When a person does not have enough water for his drinking, washing, bathing, other domestic purposes, and running his livelihood properly, then the person is water insecure. When a large number of people are facing a lack of water for a noticeable period of time, then that particular area, where the people are living, is called water scarce. Though there is no particular definition of water scarcity which is accepted commonly (Rijsberman 2006), it can be addressed in the way that crisis of water, water shortage and water stress, and water deficiency creates the situation of water scarcity (Seckler, Barker et al. 1999).

Water scarcity may be either physical (absolute) or economical. If the natural resources of water are out of the reach of a particular community, area, or region and cannot meet the requirement, then it is called "physical water scarcity". While, on the other hand, if the water scarcity was due to poor management of water resources and lack of technology and investment, then it is economic water scarcity (Anonymous).

Demonstrating the global accessibility to water and the most crucial areas, Figure 9.1 presents the picture of the present world full of obstacles in the form of a lack of safe drinking water, which excessively affects inhabitants worldwide enormously (Figure 9.2).

In the coming decades, the Earth will consist of the same level of water as it contains at present; however, the global population over the following decades is projected to grow, which will also increase additional demand for drinking water. At present, the global scheme of development is neither desirable nor sustainable. Considering the future. Indeed, the current situation regarding accessibility and availability of safe drinking

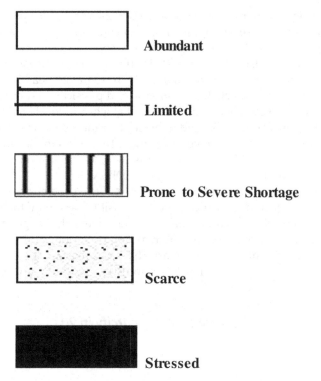

FIGURE 9.1
Global accessibility to water.

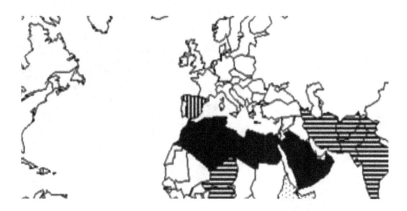

FIGURE 9.2
Expected water scarcity in 2025 in different ways. Source: Third World Academy of Sciences. (TWAS 2002).

water will prove a bigger obstacle in the coming years if serious and instant efforts are not taken soon (UNESC 2003).

A further serious obstacle that the world is tackling is common water resources. The need of human beings is causing water shortages. Pollution of the environment and water sources is leading to environmental collapse at a wider level in the common area. This creates conflicts between basic human needs and social development, the environment, its biodiversity, and economic growth, such as industrial development. Here, another significant issue is states sharing common water resources (UNESC 2003). Measures have to be taken towards the present issue by recycling, storing, and consuming water, and a more sustainable way of doing this has to be recognized. Besides tools for sanitation, access to water is essential to reduce the huge lack that currently is affecting a large amount of the global population negatively, in the form of health issues, which frequently lead to death (Twas 2002).

Figure 9.3 shows various impacts like economic and physical on the availability of fresh water, which will lead to scarcity of water globally at a grand level by 2025.

At the global level, law frameworks and referred documents of various international and national organizations worldwide have recognized the problematic situations occurring due to a shortage of fresh drinking water (Murthy 2013). Identification of water as a human right has set the right in focus internationally (UNDESA).

Mere identification is not sufficient. For assurance of the protection of the right to clean drinking water, it is required to sort out the responsible reasons which create the situation of scarcity of water. Water scarcity has become a great challenge in the 21st century. Rapid population growth, overuse of water, water pollution, and improper and non-implementation of governmental policies are the basic reasons for creating water scarcity. FAO reported that fisheries, agriculture, forestry, livestock, and crops are causing

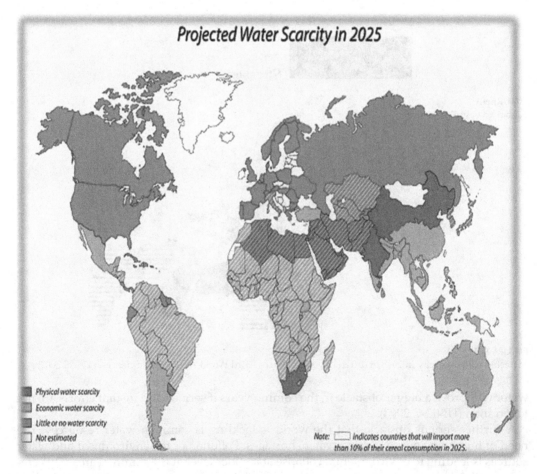

FIGURE 9.3
Projected water scarcity in 2025. Source: IWMI water scarcity map (IWMI 2000).

water scarcity and are a victim of water scarcity as well. Almost 70% of the world water is being consumed in these sectors while there is competition with other water sectors (Chakkaravarthy and Balakrishnan 2019). If a particular community or people at large are facing scarcity of water and a particular area is a water scarcity-prone area, then there have been some reasons behind it like economic development and shifting of food habits that have also added to the increase in the demand for water, consequently creating a burden on water resources (Liu, Yang et al. 2017).

Basic Reasons for Scarcity of Water

Following are the basic reasons which are responsible for creating a scarcity of water:

1- **Population growth:** As the population grows and incomes grow, demand for water also grows. Over-population is creating the over exploitation of water resources, which is a great reason for scarcity of water (LeRoy 1995). The present world population is 7.5 billion, which is going to increase 2.3 billion more by the year 2050, while the water resources are going to remain the same or may be less than the present time, then how is it possible for the planet to satisfy the thirst and water demand for other purposes of this increased population (Leah 2017). Growing incomes are also creating water scarcity problems because of an increase in the use of water-intensive products like meat and energy sources from fossil fuels, which wealthy people demand (Falkenmark 1990).

2- **Climate change:** It is expected that climate change can increase the global temperature at an average of 1.4 degrees Celsius to 5.8 degrees Celsius [34.52 Faharenheit to 42.44 Faharenheit] by 2100 and may bring changes in the precipitation level at various degrees across the globe. It will affect the availability and quality of water on a large scale, and lack of water will create severe health issues unless the nations which will be facing water-scarce situations are able to adapt (DeNicola, Aburizaiza et al. 2015). Climate change will bring droughts in different parts of the world like Somalia, shortage of water in Rome, water floods in Jakarta, and different other calamities in different parts of the world. Analysis by a specialist is not required for the realization of the growing water crisis across the world. Climate change causes deprivation of rainwater to most parts of sub-Saharan Africa, the Middle East, and Central America. The worrying situation is that climate change is increasing water-related problems in other parts also; specifically, people who live on the banks of rivers can lose a lot. Almost 21 million people are at risk of facing floods around the globe. Horribly, this number may increase from 21 to 54 million by the year 2030. The countries, which have exposure to rivers, are mostly developing or least developed countries, and this makes them more vulnerable to climate change (Vairavamoorthy, Gorantiwar et al. 2008). In Stockholm, water specialists, innovators of industry, and researchers of the related area gather for celebrating World Water Week to tackle the most burning issue of water crisis in the month of August, every year (Leah 2017).

3- **Over-exploitation and depletion of groundwater:** The underground water level contains 30% of the total fresh water available on the Earth. Water is extracted daily for agricultural, drinking, and other domestic and industrial purposes. In

India, it is at a higher rate than in any other country. Wells are a great source of groundwater in India. Almost 54% are decreasing, which shows that water is used at a more rapid rate than it is being refilled. In the coming 20 years, India's underground water will be under pressure. Water supplies are at risk and susceptible globally to the growing and hidden threat (Leah 2017).

4- **Water uses for agriculture:** Almost 70% of the available fresh water on the Earth is used for agricultural purposes, out of which 60% is wasted. It happens because of the spongy or soaking water irrigation system (Pereira, Oweis et al. 2002). Application of proper methods is not sufficient for the growing tradition of growing more and more crops. These kinds of uses are the basic reason for drying out of the rivers and underground water resources specifically in those countries which produce a large amount of food like China, India, Australia and the USA (Mancosu, Snyder et al. 2015). These countries are either close to reaching their limits of underground water resources or have reached up to that limit. However, water is the basic need for growing crops to meet the food demands of the growing population (Stott 2015).

5- **Poor water supply infrastructure:** Water infrastructure like treatment plants, supply systems, pipes, and sewer systems are built poorly or in a status of disrepair, which causes great loss of treated water. In the United States alone, 6 gallons of water is lost per day due to leaky water supply systems. Installment and repairing of built water infrastructure is expensive, which means that many localities most of the times ignore the issues of water infrastructure until there is a disastrous situation as happened in California in 2017 (Leah 2017).

6- **Ignorance of natural water infrastructures:** If we have a plentiful and healthy natural ecosystem, then it will work like a "natural infrastructure", which plays a vital role in healthy and clean water. It filters the pollutants in water; works like a buffer system, stops floods and storms, and regulates a healthy water supply. To recharge underground water supplies, plants and trees are the essential elements. Without the trees, rainfall is wasted (Pedro-Monzonís, Solera et al. 2015). Rainwater will slide across the dry land and plain areas instead of being absorbed into the soil. Deforestation, rapid urbanization, and various other causes are limiting the natural benefits which they provide. Watersheds which are covered by forests are at risk around the globe. In the last 14 years, 22% of the forest cover of watersheds has been lost (Stott 2015).

7- **Water pollution:** Water pollution is a problem, which creates scarcity of water especially in those areas which do not have a good sewage system. Water pollution may be in many forms like from oil, chemicals, and fecal matter. It makes a lot of issues for human beings (Goel 2006).

8- **Distance:** Throughout the world, water scarcity arises because of the distance from the water resources. In those Deserts and scheduled areas, where earlier people could get water effectively, but now they do not find any source of water, this situation also creates water scarcity in that particular area (Anonymous).

9- **Illegal dumping of garbage and other pollutants:** This is another significant reason for water scarcity. Frequent dumping of industrial garbage by the industries in the nearby rivers and lakes as it is a cheap method of disposing of garbage. It pollutes the river and lake water and creates a shortage of clean water, which is a great problem for those people who are dependent on these water bodies (Chowdhary, Bharagava et al. 2020).

Water scarcity may have disastrous effects. Water scarcity may cause starvation as there will be no water for agriculture, no production of food, and lack of education as children, mostly girls, will remain busy getting water for their families. It may cause various kinds of diseases among suffering people, which may result in a pandemic and loss of lives. Lack of water may create sanitation issues also. Water scarcity may create poverty and lead toward migration waves. It may cause biodiversity loss and destruction of habitats.

Though these situations may be avoided or overcome through education, improvement in the farming-related practices, lessening of use of chemicals in farming, improving sewage systems, improved water supply infrastructures, water recycling, and using advanced technologies related to water conservation and rainwater harvesting.

Water Scarcity and Violation of Various Human Rights

Water scarcity is a very common problem even in those countries which have plenty of water. Though there may be innumerable reasons responsible for this problem, it creates a disastrous situation. Due to this, not only the human right to clean water suffers but also uncountable human rights are being violated. Scarcity and unavailability of water is the basic reason for the lack of sanitation. People find themselves helpless and unable to perform their day-to-day hygiene activities like using toilets. This also creates hurdles in school healthcare facilities. Scarcity of water also creates failure of sewage systems, which again leads to a threat of contracting dangerous diseases like cholera surges. Scarce water is also expensive water (Chakkaravarthy and Balakrishnan 2019).

In water-scarce areas, women and children are the worst sufferers as they often take the responsibility to collect water, specifically women and girls. It takes more time to collect the water when it is in far regions, which results in walk outs from schools. In this way, it affects the education of innumerable girl children. To carry water from distant places is not only a physical burden which affects health very badly but also poses risks to child safety and sexual exploitation of women and girls, which includes a danger to life (Rijsberman 2006).

Suggestions to Overcome the Problem of Scarcity of Water

The problem of scarcity of water may be resolved by using some specific technologies, which may prove helpful in better accessibility of human right to safe and clean water. We may also focus on the following steps:

1- **Identification of new water resources:** Use of various modern technologies including remote sensing and geophysical surveys and field investigations may be helpful in assessing the new water resources.

2- **Efficiency of water resources:** Rehabilitation of urban water distribution networks and water treatment systems for reducing leakages of water and water contamination, promotion of reuse of wastewater for the purpose of agriculture are some effective steps required to be taken in this reference.

3- **Plans for urban water scarcity:** There should be proper planning for the needs of water in urban areas so that the risk of running out of water may be avoided.

4- **Expansion of technologies for ensuring climate flexibility:** The water resources flexible toward climate change should be developed. Solar water networks should be used for use of deeper underground water reserves. Small-scale water retentions should be encouraged. Underground water recharge and rainwater harvesting are very helpful and required things to be done nowadays.

5- **Changing behaviors:** Behavioral change in our day-to-day life is the quintessential thing. Some small camps, meetings, and courses should be organized to spread awareness among the people at school, college, colony and community, and national levels. Environmental clubs at school levels may be an innovative step.

6- **Making plans for water needs at the national level:** Water requirements at national, state, and local levels for drinking, health, sanitation, agriculture, and other domestic purposes should be understood and it should be ensured that the national policies reflect this (Noorani 2021).

Conclusion

Scarcity of water, water pollution, and damage to clean water resources and ecosystems are preventable problems. There must be an investment in sustainable water governance. An estimation by the Organization for Economic Co-operation and Development shows that an additional investment of 500 billion dollars is required to secure the world from scarcity of water. The advantages of clean and sufficient water, a healthy ecosystem for humanity at large, are uncountable. States must follow and implement a right-based governance program of seven steps like capacity building, public participation and public engagement monitoring, legal mapping and strengthening, development of right-based plans, implementation, and evaluation. For the better realization of human rights to clean water, the state should prevent or provide immediate solutions to water-related problems, water-related disasters, and pollution of water under or above the land.

Our present rate of growth and development can paralyze our food and energy supply chains as great threats have been posed to water resources and a situation of water scarcity is seen in most places around the globe. This will be a violation of many human rights, not only a violation of a right to clean water. **We are at the beginning of the entrance of an incredible change**, so let us take steps together toward a real change in the coming future.

Notes

1. United Nations General Assembly.
2. United Nations Committee on Economic, Social and Cultural Rights.
3. Millennium Development Goals.
4. When sources are not sufficient for satisfaction of long-term requirements, then scarcity of water occurs.

References

"The rights to water and sanitation: Meeting the millennium goals". Retrieved 30/06/2021, from http://www.righttowater.infoprogress-so-farmeeting-the-millennium-development-goalsthe-millennium-development-goals.

Agency, E. E. T. (2008). *Water Scarcity. Water Facts.*

Albuquerque, C. d. (2012). *On the Right Track, Good Practices in Realising the Rights to Water and Sanitation.* Lisbon. http://www.un.org.

Anonymous. "What is water scarcity? Causes, effects and solutions to water scarcity (water deficit)". Conserve Energy Future. Retrieved 28/06/2021, from https://www.conserve-energy-future.com/causes-effects-solutions-of-water-scarcity.php.

Assembly, U. G. (1948). "Universal declaration of human rights." *UN General Assembly* 302(2): 14–25.

Assembly, U. G. (2010). "Resolution 64/292: The human right to water and sanitation." In *64th Session.* http://www. un. org/es/comun/docs.

Chakkaravarthy, D. N. and T. Balakrishnan (2019). "Water scarcity-challenging the future." *International Journal of Agriculture, Environment and Biotechnology* 12(3): 187–193.

Charter, A. (1990). *African Charter on the Rights and Welfare of the Child.* Addis Ababa: OAU.

Chowdhary, P., et al. (2020). "Role of industries in water scarcity and its adverse effects on environment and human health." In *Environmental Concerns and Sustainable Development.* Springer, 235–256.

Coomans, F. (2011). "The extraterritorial scope of the international covenant on economic, social and cultural rights in the work of the United Nations Committee on economic, social and cultural rights." *Human Rights Law Review* 11(1): 1–35.

Cullet, P. (2013). "Right to water in India–plugging conceptual and practical gaps." *The International Journal of Human Rights* 17(1): 56–78.

DeNicola, E., et al. (2015). "Climate change and water scarcity: The case of Saudi Arabia." *Annals of Global Health* 81(3): 342–353.

Dinka, M. O. (2018). "Safe drinking water: Concepts, benefits, principles and standards." In *Water Challenges of an Urbanizing World.* London: IntechOpen, 163–181.

E.U. (2012). *Water Protection and Management.* European Union, Brussels.

Falkenmark, M. (1990). "Rapid population growth and water scarcity: The predicament of tomorrow's Africa." *Population and Development Review* 16: 81–94.

Gleick, P. H. (1996). "Basic water requirements for human activities: Meeting basic needs." *Water International* 21(2): 83–92.

Gleick, P. H. (1998). "The human right to water." *Water Policy* 1(5): 487–503.

Goel, P. (2006). *Water Pollution: Causes, Effects and Control.* New Age International.

Kathpalia, G. and R. Kapoor (2002). "Water policy and action plan for India 2020: An alternative." *Alternative Future* 1: 1–33.

Leah, S. (2017). *7 reasons we're facing a global water crisis.* Retrieved 24/07/2021, from https://www.wri.org/insights/.

LeRoy, P. (1995). "Troubled waters: Population and water scarcity." Colorado Journal of International Environmental Law and Policy 6: 299.

Liu, J., et al. (2017). "Water scarcity assessments in the past, present, and future." *Earth's Future* 5(6): 545–559.

Maddalena, D. F. "The importance of water on human health' health centric research."

Mahapatra, I. K. A. R. (2005–06). *Right to Water and Sanitation National Family Health Survey -2005-06.* Delhi: WaterAid India, 6.

Mancosu, N., et al. (2015). "Water scarcity and future challenges for food production." *Water* 7(3): 975–992.

Mishra, A. (2018). "Right to water as a human right and Indian constitution: An analysis of various judgments of Apex Court of India." *IOSR Journal of Humanities and Social Science* 23(4): 45–48.

Murthy, S. L. (2013). "The human right (s) to water and sanitation: History, meaning, and the controversy over-privatization." *Berkeley Journal of International Law* 31: 89.

Noorani (2021). *Water scarcity: Addressing the growing lack of available water to meet children's needs.*

Pedro-Monzonís, M., et al. (2015). "A review of water scarcity and drought indexes in water resources planning and management." *Journal of Hydrology* 527: 482–493.

Pereira, L. S., et al. (2002). "Irrigation management under water scarcity." *Agricultural Water Management* 57(3): 175–206.

Prüss, A., et al. (2002). "Estimating the burden of disease from water, sanitation, and hygiene at a global level." *Environmental Health Perspectives* 110(5): 537–542.

Raju, K. (2018). "Water technology landscape and right to clean drinking water—An Indian perspective." *The Journal of Engineering Research.* Retreived 20/08/2020 from http://docs.lib.purdue.edu.

Rijsberman, F. R. (2006). "Water scarcity: Fact or fiction?" *Agricultural Water Management* 80(1–3): 5–22.

Seckler, D., et al. (1999). "Water scarcity in the twenty-first century." *International Journal of Water Resources Development* 15(1–2): 29–42.

Secretary-General (2011). *The Millennium Development Goals Report 2011.* Geneva: United Nations, 52.

Stott, L. (2015). *Water Scarcity – The Main Causes.* Esuela de Organizacion Industrail. Springer, Cham: Brussels.

Tully, S. (2005). "A human right to access water? A critique of general comment no. 15." *Netherlands Quarterly of Human Rights* 23(1): 35–63.

Twas (2002). *Report of Safe Drinking Water: The Need, the Problem, Solutions and an Action Plan.* Italy: Third World Academy of Sciences, 8.

UN (1979). "Convention on the elimination of all forms of discrimination against women."

UN (1989). "Convention on the rights of the child."

UN (2012). *"Water for life."* Retrieved 25/07/2021, from https://www.un.org.

UNDESA (2014). "The right to water and sanitation." *International-Timeline.* Retrieved 25/07/2021, from https://www.un.org.

UNESC (2003). *General Document E/C.12/2002/1.* s. a. c. r. Committee on economic. Geneva: The United Nations.

UNFPA (2002). *"Water: A critical recourse."* Retrieved 27/07/2021, from https://www.unfpa.org.

Vairavamoorthy, K., et al. (2008). "Managing urban water supplies in developing countries–Climate change and water scarcity scenarios." *Physics and Chemistry of the Earth, Parts A/B/C* 33(5): 330–339.

VanDerslice, J. (2011). "Drinking water infrastructure and environmental disparities: Evidence and methodological considerations." *American Journal of Public Health* 101(S1): S109–S114.

WHO (2001). *"Health in water resources development."* Retrieved 25/07/2021, from https://www.who.int.

WHO (2003). *"Water sanitation and health."* Retrieved 27/07/2021, from http://www.who.int/water_sanitation_health/hygiene/en/.

WHO (2004). *Guidelines for Drinking-Water Quality.* World Health Organization. Geneva.

WHO (2006). *"Guidelines for drinking water quality."* 1. World Health Organization: Geneva.

10

The Development of the Transportation Sector in the Ages of Automation

Maria A. Bazhina

Transportation is known to be the engine of all spheres of economy. It is difficult to imagine trade or construction projects without transportation. It satisfies the needs of other economic branches. Therefore, transportation is a core sector of every country's economy. In this sense, we can admit such dependence: the more transportation is developed, the more well-developed the economy is.

Nowadays the speed of all processes is much faster than earlier. To be competitive, it is necessary to furnish with the tempo of life. The transportation sphere is not an exemption. The modern cargo turnover requires an increase in speed. On the one hand, the process of automation helps to cope with such goals by means of new technologies and approaches. On the other hand, the era of digitalization will completely change the transportation process. The ability of the transportation branch to satisfy the needs of other branches opens up future sustainable economic growth and societal well-being.

In this context, it is worth analyzing the prospects of the transportation sector in the age of automation. Therefore, it is necessary to highlight *several aspects*.

1. *The implementation of digital documents expands the horizons of the transportation process.*
2. *The appearance of automated kinds of transport requires the elaboration of new rules for their application.*
    I. Coming to the first aspect concerning *the implementation of digital documents*, it is worth mentioning that there is no single approach to this question among all countries. The reasons for such situations go deep into the peculiarities of legal regulation. There are two levels of regulation of transportation relationships: international and domestic levels, which differs from each other.

International conventions give rules only for international transportation for those countries that are members of these agreements. For instance, the Additional Protocol on electronic CMR made to Convention on the contract for the international carriage of goods by road (CMR, 1956) [1]. It was entered into force on June 5, 2011. Today, about 26 countries joined this agreement devoted to the introduction of e-CMR, such as Russia, the Czech Republic, Bulgaria, Denmark, Estonia, France, Spain, Switzerland, Sweden, Turkey, Slovenia, Spain, etc. The official implementation of e-CMR was in January 2017, when the first international carriage of goods between Spain and France was fulfilled.

DOI: 10.1201/9781003215998-10

e-CMR is proved to be an effective instrument that is used in international turnover. The electronic form of the document has its invaluable advantages. Firstly, the use of e-CMR reduces expenses. It is proved that processing costs were cut by 3–4 times. It is connected with the acceleration of administrative work, namely, the reduction of paperwork, the cancellation of the necessity to make copies and scans. Secondly, implementation of e-CMR increases the transparency because of the accuracy of data, control of the dispatch, and receiving of sending, the access to the information about the sending in real-time mode. These benefits support the improvement of logistics in the carriage of goods. Therefore, the competitiveness of all parties engaged in the carriage of goods with the use of e-CMR increases. Thirdly, using e-CMR helps to provide traffic security. e-CMR is bound with the eCall. In case of an emergency, a message is automatically sent to the emergency services.

In spite of the above-stated advantages, not all countries are using such electronic documents. Nevertheless, some countries give only their official support but do not ratify the protocol. According to the Regulation on electronic information about carriage of goods (eFTI) adopted by the Committee on transport and tourism of the European Parliament on May 4, 2020, all members of the EU have to move to the e-CMR not later than 2025. However, this demand does not mean that carriers are limited by the implementation of only electronic documents. The paper form of documents is not excluded. This results in the conflict of using different forms of documentation in international relations by various countries.

The above-given example shows that the effectiveness of introduction of electronic documents depends on the number of states involved in its use. Only global use of electronic transport documents would lead to the transparency, accuracy, and acceleration of document turnover.

Proceeding to the national level of transportation regulation, it is found that electronic documentation exchange in transportation for different countries has diverse stages of development. Let us consider some examples of transportation legislation systems.

In Russia, there is no single system of electronic transport document turnover. There is a complex chain of restraining reasons. The first obstacle is connected with the system of legal regulation of transport relationships itself. There is no one act devoted to transport regulation. Each kind of transport has its own codified act (codes or charters):

- Transportation by road is regulated by the Federal Law "Charter of road transport and urban land electric transport", dated November 8, 2007 No. 259-FZ.
- transportation by air is based on the Federal Law "Air Code of the Russian Federation", dated March 19, 1997 No. 60-FZ.
- Transportation by rail is governed by the Federal Law "Charter of Railway Transport of the Russian Federation", dated January 10, 2003 No. 18-FZ.
- Transportation by water has two acts: the Merchant Shipping Code of the Russian Federation, dated April 30, 1999 No. 81-FZ and Code of Inland Water Transport of the Russian Federation, dated March 7, 2001 No. 24-FZ.

The absence of the single act-regulated transportation relationships leads to the enlargement of juridical acts [2]. The same issues on different kinds of transport are governed by different acts, including subordinate legislation. The fragment regulation of transport relationships means that there are different forms of consignment depending on the kind of transport. Moreover, the number of documents and their titles are not the same for various kinds of transport. The same applies to electronic document turnover: legislation on each kind of transportation has its own rules.

Thus, road transportation is regulated by the Decree of the Government of the Russian Federation "On approval of the Rules for the carriage of goods by road and on amending paragraph 2.1.1 of the Road Traffic Rules of the Russian Federation", dated December 21, 2020, No. 2200. In air transport, the Order of the Ministry of Transport of Russia "On approval of the form of an electronic consignment note in civil aviation", dated August 10, 2018 No. 300, is applied. The possibility of using an electronic waybill on railway transport is established by clause 113 of the Rules on goods, empty freight wagon carriage by rail, containing the procedure for redirecting the transported goods, empty freight wagons with a change in the consignee and (or) railway station of destination, drawing up acts on the goods, empty freight wagon carriage by rail, drawing up a railway waybill, terms and procedure for storing goods, containers at the destination railway station, approved by Order of the Ministry of Transport of Russia, dated July 27, 2020, No. 256.

The above-mentioned examples show that differences in legal paperwork cause an overwhelming majority of obstacles to exist in transportation and logistics. Especially in multimodal transport, this isolation according to the kind of transport impedes the performance of transportation and logistics operations. The need to reissue documents when using various types of transport significantly complicates the turnover of goods.

Thus, it seems appropriate to develop a unified form for all types of transport. In the Russian legislation, this goal is declared in the document approved by the order of the Government of the RF No. 1734-r. (November 22, 2018), known as the Transport Strategy that is applied until 2030. The act emphasizes the need to introduce electronic document management according to the "single window" principle. This means that the documents should be available not only to the direct participants in the legal relationships but also to state bodies (for example, tax authorities) that exercise control over transport activities. Achievement of this goal would allow solving the following problems in the transport sector. First, the introduction of electronic document management would make the transportation process more transparent and accessible. Currently, in practice, situations arise when carriers (in particular, large companies) refuse to submit documents to court nitration, citing the fact that this is impossible due to the special nature of their maintenance and storage or for other reasons. Such circumstances may negatively affect the establishment of the circumstances of the case and, as a consequence, the determination of the responsible persons. Secondly, the introduction of such innovations would make it possible to increase the turnover of goods carried out through multimodal transport, in particular, intermodal transport, in which only a container with cargo is reloaded from one vehicle to another in specially equipped centers. At the same time, re-registration of documents is not required, which significantly reduces time costs. Thus, we are talking not only about changes related to the form of documents but also about a qualitatively new approach to the regulation of document flow. The foregoing substantiates the need for unification of documents drawn up for multimodal transport, as well as for transport by various modes of transport. However, the Russian Federation has not yet adopted a law regulating multimodal transport. Despite the fact that the Ministry of Transport of Russia has already prepared five draft laws of the Federal Law "On direct mixed (combined) transportation" (project ID 02/04 / 05-20 / 00102210): the draft law of January 26, 2009, the draft law of March 30, 2015, the draft law of February 21, 2019, the draft law of January 9, 2020, and the draft law of May 22, 2020, none of them was adopted. The reason for this is that none of the draft laws regulates fundamental issues among others the procedure and mechanism for the application of a single document during transportation. Thirdly, the introduction of electronic document management would create a basis for the unification of management of electronic documents.

Thus, the construction of a common (single) system of electronic document turnover in the transportation sphere is a chance for Russia to build a safe, reliable, and sophisticated transportation and logistics system.

According to the European legislation, the efficacy and effectuality of transport can be essentially increased by using communication technology including the integration of electronic documentation turnover in countries – members of the EU. The open access for all participants of transportation relationships to the transport-related information has to improve traffic management and simplify administrative procedures. Such provisions are stated in the Regulation "On Union guidelines for the development of the trans-European transport network and repealing Decision No. 661/2010/EU", adopted by the European Parliament and of the Council of EU, dated December 11, 2013, No. 1315/2013.

It is very clear from the observation that there are bi-leveled differences in transportation regulation. On the one hand, different kinds of transport have their own rules including electronic document turnover. On the other hand, the national systems of transportation legislation are not equal to the international legislation. These doubled differences are the obstacle to the development of the ecosystem of transportation document turnover.

Presented in the national and European Union strategies, this vision proposes measures that will streamline the creation of a unified system of electronic documentation. According to the analysis of international and national legislation, we tend to underline as the sample of successful documentation turnover the experience of the International Federation of Freight Forwarders Association (FIATA) [3]. FIATA has elaborated a document used by a freight forwarder within the framework of an international freight forwarding agreement. This document is called FBL (Negotiable FIATA Multimodal Transportation Bill of Landing – multimodal bill of landing). A distinctive feature of the document is the ability to be negotiable, that is, act as a security (like a bill of landing issued for the carriage of goods by sea). This bill is an example of a single document that can be used on different modes of transport. The unified form brings the benefit for the development of a new generation of transportation: free of paperwork and human mistakes, the time-consuming process of administrative work, etc. An electronic version of the FBL already exists. However, the problem remains in the development of a mechanism for monitoring, issuing, and checking such an electronic waybill to meet existing requirements. Following the tendencies of digitalization, FIATA is looking forward to facilitating the interoperability system of documentation that relies on an open and collaborative approach.

Thus, the creation of a single environment of electronic document turnover requires the unification of international and national legislation in the sphere of transport document exchange. This tendency of building an ecosystem of e-documentation has to correlate with the idea of creating a more interoperable transport system.

Along with the problem of doubled level regulation of transportation documentation turnover, it is necessary to underline that the enhancement of transport and logistic traffic depends on the standardization and harmonization of transportation and trade documentation. At first glance, trade and transport are two separate branches of economic activity. However, these two areas are very close to each other as carriage of goods is the kind of instrument to fulfill the obligation on goods supply. Therefore, transport operations are usually inside the contractual obligation of goods supply. Using metaphoric language, we can say that these two contracts are a kind of "Russian Matryoshka": relationships on supply contain relationships on carriage of goods inside. This explanation presents the necessity to enable the "partnership" between two systems of documentation. The integration of two systems of documentation may leverage its strength to make transportation faster, transparent, more clear, secure, etc. This will result in the intensification of territorial, economic, social cohesion.

From the results of the research of the first issue of the article, it is concluded that to build up a system of transport documentation that can support the development of transportation (so-called "ecosystem"), there is a need to step up action at multivarious levels.

II. Coming to the second issue of the article, it is obvious that the guideline for the development of the transportation sector is defined by *the appearance of automated kinds of transport in the transportation sphere*. Technological shoot forward is expected to change the traditional conceptualization of transport.

Before analyzing legal aspects of automated vehicle appearances, it is necessary to consider the main differences in the sense of the notion "automated kinds of transport". The term "automated vehicle" is widely spread today. However, this term is connected with such notions as "autonomous", "self-driving", "driverless". These terms must be differentiated, though not all scientists and legal sources do it in the same way.

First of all, let us define the difference between "automated" and "autonomous". There is an opinion that these two notions identify the various degree of human intervention. Therefore, an autonomous car is seemed to have a higher level of intelligence and independence from the driver than that an automated car has [4]. However, in some documents, these terms are used as synonyms. For instance, in the Framework document on automated/autonomous vehicles, adopted by the World Forum for Harmonization of Vehicle Regulations (WP.29) at its 178th session (see ECE/TRANS/WP.29/1147, para. 27), that has been endorsed by the Inland Transport Committee of UNECE at its February 2020 session (hereafter – the Framework document on automated/autonomous vehicles), the terms "automated" and "autonomous" are written with a slash. According to the precise study of the Framework document, it becomes obvious that these terms do not have the same meaning. The context is devoted either for automated vehicles or autonomous vehicles. That reason explains the manner of the term arrangement.

The deference between the term "self-driving" and the term "automated" is also a subject for proper research. To understand the full range of features belonging to each, term it is necessary to go back to the legal roots of these term appearances. According to the Annex B to Document No. 8: Automated Vehicles: Policy and Principles Discussion Document, adopted on the 75th session of the Global Forum for Road Traffic Safety by the Inland Transport Committee of the Economic Commission for Europe (UNECE) (hereafter – Policy on Automated Vehicles 2017) (Geneva, September 19–22, 2017), the term "automated vehicle" is primarily connected with the term "Self-Driving System" (abbrev. – SDS) that means the combination of hardware and software that are combined together to perform the entire dynamic driving task on a continuous basis, regardless of whether it is limited to a specific Operational Design Domain (ODD).

The term "self-driving system" was used in documents of UNECE. However, not all documents follow this terminology. In "A Framework for Automated Driving System Testable Cases and Scenarios", issued by the US Department of Transportation, National Highway

Traffic Safety Administration (NHTSA), September 2018, the term "automated driving system" is used instead of the term "self-driving system". This tendency was also taken by UNESE. Thus, one of the latest documents – the Framework document on automated/autonomous vehicles – contains only the term "automated driving system". Nevertheless, nowadays, the term "self-driving system" is used along with the term "automated driving system". For instance, the terminology that is used in the Safety Report of one of the famous companies – ARGO AI, a self-driving technology platform company. The report "Developing a Self-Driving System You Can Trust", dated April 2021, contains the term "self-driving system" [5].

Once used, the terms give birth to similar terms such as "self-driving innovations", "self-driving technologies", etc. The correlation of these terms is not given in any single document or research paper. For instance, pursuant to the official website of NHTSA, the terms "self-driving cars" and "fully automated cars" are used as synonyms [6]. Academic research studies have shown up the same result: the term "self-driving cars" is the same as the term "fully automated cars" [7].

As a result, there is a mixture of terms in the developing sphere of the high technology industry. To estimate the most appropriate vision of the notion's correlation, it is worth analyzing the levels of SAE – the gradation made by the Society of Automotive Engineers (SAE) in the J3016 Standard "Taxonomy and Definitions for Terms related to On-Road Motor Vehicle Automated Driving System". This document is the basic one taken by UNESE as the core to allocate types of automated vehicles. SAE adopted six levels of driving mode for on-road vehicles, including Levels 0–2 that require the human driver to monitor the driving environment (detecting, recognizing, and classifying objects and events and preparing to respond as needed) and Levels 3–5 that transfer the function of monitoring to the SDS or ADS. According to this classification of SAE, SDS (ADS) can be found in vehicles of Levels 3–5, namely, a conditionally automated vehicle, a highly automated vehicle, a fully automated vehicle.

Based on the above-mentioned classification, the definition of each kind of automated vehicle differs in the description of what the system and the driver are allowed to do. Let us consider them in detail.

1. The conditionally automated vehicle is an automobile, which has an SDS (ADS) that matches the third level (L3) of classification made by the Society of Automotive Engineers. The ADS recognizes time restrictions of the Operating Design Domain (ODD) and gives a transfer demand to the driver. The driver is a fallback-ready user while he or she does not need to uninterruptedly control the traffic environment. Nevertheless, they have to be ready and able to renew the dynamic driving task (DDT) in case of system request or system failure.

2. The highly automated vehicle is an automobile, which has an SDS (ADS) that matches the fourth level (L4) of classification made by the Society of Automotive Engineers. The system recognizes all driving tasks. It is capable of monitoring the driving environment. Under certain conditions, the system may issue a transfer demand to the chauffeur. The driver has to undertake the dynamic driving task to continue the trip without Operating Design Domain. However, the SDS (ADS) can execute the total dynamic driving task if the chauffeur is unable to satisfy the transfer demand. The driver does not need to uninterruptedly control the driving environment.

There are some uncertainties against the status of the driver. Some consider that all those who are present in the vehicle are considered to be passengers while the SDS (ADS) is involved, though they may still be executing the strategic driving task. Others consider the user still as a chauffeur while the SDS (ADS) is involved. Thus, it is important to continue further deliberation on this point as the determination of the driver's position influences the legal consequences (especially in determining the liable person in case of losses).

3. The fully automated vehicle is an automobile, which has an SDS (ADS) that matches the fifth level (L5) of classification made by the Society of Automotive Engineers. According to it, none of those who are in the vehicle is expected to execute any part of the dynamic driving task. In other words, a driver may not be present. When the SDS (ADS) is engaged, all occupants are viewed as passengers. It means that driver intervention is not needed. However, the driver may have the option to control the driving under the following conditions: (1) if the vehicle contains appropriate equipment and (2) if the driver has the required skills and qualifications (licenses). In this case, human occupants may choose to perform the DDT. On the whole, according to the details of the term "fully automated vehicle", it can be assumed that the term "driverless" is equivalent to the term "fully automated vehicle".

In summary, there is a mixture of terms applied to the same objects. Thus, the driving system can be defined with the term "self-driving" and the term "automated driving". However, the majority of legal documents use the term "automated driving system" to indicate the system installed in the vehicle. Speaking about "the car", such collaborations as "self-driving car", "automated car" are in use.

As a result, the term "self-driving" is close to the term "automated". The term "self-driving" refers to three modes of automation (Levels 3–5): conditionally, highly, and fully. In this sense, the terms "self-driving" and "automated" are terms that are more general. The term "driverless" is synonymous with one of the modes of automation, namely, "fully automated vehicle". To avoid the confusion of terms, it is necessary on the national and international levels to identify the meaning of the applied terms. As the terminology becomes an ever more central part of the newly appeared technologies, each notion should be adjusted by different countries and confirmed in the legal documents. The international harmonization of terminology has to be elaborated in accordance with the principles of consistency, entirety, and cooperation with the legislation of each country.

According to one of the most important principles of international law, namely – the independence of the country, each country is free to determine its national politic separately. It means that each country has its own experience in elaboration of the terminology concerning the implementation of the modern technology provided in automated vehicles.

To illustrate the integration of the international terminology at the national level, it is worth bringing the example of the Russian Federation. Recognizing the value of the harmonization of national legislation and international laws, there is a tendency according to which Russia acknowledges the international terminology. Thus, the Concept of on-road safety and security involving unmanned vehicles on public roads, adopted by the Decree of the Government of Russia No. 724-p (March 25, 2020) (hereafter – Concept on using unmanned vehicles), contains the link on the Resolution on the introduction of highly and fully automated vehicles in road traffic conditions, adopted on the 78th session of Global Forum on Road Traffic Safety (Geneva, March 25–29, 2019) (hereafter – Resolution on the introduction of highly and fully automated vehicles). Being a full member of the Global Forum on Road Traffic Safety, the Russian Federation is guided by the provisions of the Resolution on the introduction of highly and fully automated vehicles in order to promote road safety, mobility, and socio-economic progress. Therefore, the Concept on using unmanned vehicles operates with notions and their definitions recommended by the Resolution on the introduction of highly and fully automated vehicles. In particular, the notion "highly automated vehicle" prevails under the term "unmanned vehicle". The Concept of using unmanned vehicles states that the frequently encountered term "unmanned" is not so accurate, since it emphasizes that the driver (pilot) is not in the vehicle. However, this feature cannot always be realized at the current level of technology development. Moreover, the term "unmanned" also does not take into account the presence of intermediate levels of automation. In addition, an unmanned vehicle can be controlled remotely, through the commands of an external operator, which may mean that there is no automation of the vehicle as such. The most correct understanding of the notion "unmanned vehicle" is as a highly or fully automated vehicle operating in an unmanned mode, which means that during the use of this mode, the vehicle is under the control of an automated driving system.

According to this abstract, one can conclude that the Russian legislation follows the international recommendations. This statement is true to the extent that the mixture of terms is reserved in the Concept on using unmanned vehicles. Even the title of the Concept contains the collaboration "unmanned vehicle", though the context of the Concept confirms other terms as more appropriate. Moreover, there is also the linguistic aspect that might be taken into consideration. The Concept contains Russian terms and their equivalents in English. According to the translation into the Russian language, the notions "driverless car" and "unmanned vehicle" have the same translation of the words "driverless" and "unmanned". However, for the interpretation of international documents, the word "driverless" is appropriate only for the vehicles of Level 5, while the word "unmanned" can satisfy lower levels of automation. The difference in languages and the difficulties of translation can be also the obstacle to the harmonization of transportation legislation concerning the implementation of modern technologies including automated vehicles.

As the conclusion of the part concerning the terminology of modern technologies, it is substantial to consider the harmonization of the national and international legislations and the linguistic difficulties. These two aspects influence the quality of the transportation legislation devoted to modern technologies.

The value of automated vehicle development is great. Firstly, it is supposed to improve road safety. According to item 2.7 of Policy on Automated Vehicles 2017, over 90% of road traffic collisions are caused by human error that can be in different forms. Secondly, commercial transportation by modern automated vehicles is assumed to be more effective as vehicles are able to drive themselves restlessly. Thirdly, modern vehicles are recognized to be more friendly to the environment. This satisfies the modern strategies of building smart eco-cities expanded all over the world.

It will also be important to ensure that the legislation contains new rules for governing the implementation of automated vehicles. Nowadays, there are no rules that satisfy the needs of the transportation sphere that includes modern transport vehicles. On the international level, all international conventions contain rules regulating the liability of the carrier. For instance, item 1 of Article 17 of the Convention on the contract for the international carriage of goods by road (CMR), dated May 19, 1956, states that the person who carries goods is responsible for the total or partial loss of the goods and for their damage since he takes control over the goods until the time of delivery. The carrier is also liable for any delay in delivery. This rule is not accurate for the situation of the implementation of automated vehicles in carriage of goods. The reason for that is the following: the carrier may not be the person who is liable for the delivery of goods with the use of automated vehicles. The current provisions support the practical fulfillment of the carriage of goods, as the carrier undertakes goods and therefore is liable for their security. In the case of using automated vehicles, there is no clue (in theory and in legal documents) who will be the person who undertakes the goods and makes assignments in the electronic consignment notes. Moreover, if we presume that the carrier is the person who is liable for all losses, there might be cases when an automated vehicle comes out of his control because of a cyberattack or program fault. It is obvious that in such a case, the carrier is not the proper defendant due to the following circumstances: (1) goods are not under his control; (2) the vehicle is not driven by the carrier's driver; (3) the carrier is not responsible for the maintenance of the driving system. Therefore, there is a question: who will be liable for losses in case of goods and damages. We could assume that the new transportation legislation needs a new figure – an operator who will be responsible for losses. Moreover, the institute of insurance can be applied to make the transportation system more sustainable. Making obligatory the insurance of the operator's liability is the chance to support the changing system of transportation that depends on the modern smart vehicles. The developed insurance relationships are common practice in the USA and European countries, where flexible insurance conditions exist. However, there are still some countries that have a need to concentrate on the insurance market. A bright example of such a country is Russia. Far from everyone, transportation companies insure their responsibility. As a result sometimes, carriers fail to cover caused damages to consignors or consignees, whereas the latest may delay to perform their obligations against contractors, their partners in other contracts. Such situations promote unstable market conditions that have a negative influence on the economy itself. In conclusion, the compulsory insurance of liability will be a way out for developing the transportation sphere.

Thus, the development of the transportation sector in the age of automatization is a complex process. It combines technical, economic, juridical issues. Given how fast new technology is evolving, the regulatory framework must leave room to cater for further developments. Any changes should be limited to clearly identified problems for which feasible solutions exist.

The researched field is full of questions, answers to which have not been ready but will be given in a while. The process of creating technical, economical, and legal platforms requires time, mutual work of the globe. Therefore, several proverbs could be applied to the future complicated process as its slogans. The first one is that "In Unity there is Strength" which means that the problems of automatization in the transportation sphere must be solved not on the ground of competition, but in the atmosphere of harmonization. All nations can accomplish more as a group than they can by themselves. The second proverb "Haste makes waste" must be also taken into consideration while the process of automatization of transportation. All details of legal regulation are the subject of meticulous work.

As a result, the automatization of transportation in accordance with the stated slogans would contribute to the further economic growth of each country and the global economy itself, supporting the competitiveness. It will make our economy smarter and more efficient.

References

1. Dmitrieva O.A., Nikolaeva I.G., Rudakova E.N., Morkovkin D.E., Vlasov A.V. "Digitalization of the EAEU transport and logistics sector and its role in improving the Euro-Asian cargo transportation". *Advances in Social Science, Education and Humanities Research*. Volume 416. *4th International Conference on Culture, Education and Economic Development of Modern Society (ICCESE 2020)*, pp. 1296–1302.
2. Bryukhov R.B., Kovalenko K.E. "International legal regulation of road transportation (features of the legal consciousness of legislators)". *MATEC Web of Conferences 239*. TransSiberia 2018.
3. FIATA. https://fiata.com. Accessed September 21, 2021.
4. Transportist. "On the differences between autonomous, automated, self-driving, and driverless cars". 2017. https://transportist.org/2017/06/29/on-the-differences-between-autonomous-automated-self-driving-and-driverless-cars/. Accessed September 21, 2021.
5. ARGO. https://www.argo.ai/wp-content/uploads/2021/04/ArgoSafetyReport.pdf. Accessed September 21, 2021.
6. NHTSA. https://www.nhtsa.gov/technology-innovation/automated-vehicles-safety. Accessed September 21, 2021.
7. Takács Á., Rudas I., Bösl D., Haidegger T. "Highly automated vehicles and self-driving cars". *IEEE Robotics & Automation Magazine*, December 2018.

11

Big Data Analytics: The New Surveillance Discretion for the Criminal Justice System

Swati Kaushal and Chandrika Singh

CONTENTS

Introduction

The new technologies and information revolution has been a major breakthrough which has not just transformed but has even set new societal contexts for a modern progressive society. We are approaching toward and setting the trend for the new normal where technologies permeate our homes, offices, and even private spaces. In a complex society like ours, the implementation of such dynamics has also created new circumstances. The new nuances expanded the role of government, and it became all the more pertinent to maintain law and order to address such new diversions. The uses of information technology in law enforcement are the need of the hour to maintain national security and intelligence. The acknowledgment of the utility of technology in the legal domain is very recent. There is a vast variety of legal areas in which big data and its analytics can be put to use, such as predicting the decision of a case by artificial intelligence (AI), saving litigation cost, crime prediction, crime mapping, etc.[1]

The police have recently started using big data for criminal justice delivery in their vicinity for its various advantages.[2] It not only assists but also enhances their efficiency. It acts as a watchdog on crowd control and surveillance. It is frequently employed for facial identification, image enhancement technology, and scanning video footage for anomalies. Artificial intelligence-based technologies can play an instrumental role in deterring and solving crimes by data interpretation by producing outcomes based on human activities. However, there is a lack of research in investigating the potential applications to which

DOI: 10.1201/9781003215998-11

effective policing can be done through big data. Various private players have come up with innovative software and other technologies to contribute to the improvement of the criminal justice system.

Predictive policing with the use of big data combined with the machine learning process strengthens the rhetoric of using technology to put a stop to increasing crimes with increased efficiency and fairness of the law.[3] Big data collection has certainly helped in reducing the overutilization of police personnel's efficiency in doing menial tasks of data collection. The prowess of legal enforcement professionals can be utilized better by diverting the logical and mechanical tasks to automated systems such as data collection and calculation for preventing crimes by the usage of pre-crime technologies. Predictive policing has been used effectively in many jurisdictions across the globe for optimizing police deployment in predicting prospective offenders. These technologies have also been successful in identifying the offenders as well as predicting the type of crime which may be committed and its place of occurrence with precision.[4] The expertise of police personnel can now be diverted to provide solution-based outcomes related to crimes instead of engaging in menial tasks. The mapping of crimes is done on a very large scale as well as at the local level too. It further exposes the machine learning analytics to interpret the gathered data to forecast probable crime(s). As the technology is new, its pros and cons are still unexplored or incomplete. We must set some limits to its usage, especially in terms of its ethical connotations.

The collected data has been able to predict the probability of an offender committing the same crime again, but its authenticity is doubted. This theory might be predictable but its unbiases cannot be ensured. Prediction is not a dependable measure for determining the accuracy of the crime prediction. The biases reflected in its interpretation have to be excluded. It is suggested to develop another algorithm to remove or ignore the biases and prejudices while making pre-crime predictions through the analysis of big data on human behavior.[5] Human nature is not mechanical but rather reasonable, so its actions cannot be determined accurately, although its propensity may occur, but exactness is not certain. Pre-crime algorithms do not ensure the complete elimination of crimes. Only the likelihood of crime may be foreseen, but the probability of its non-occurrence cannot be ignored. The technology can be used in a more methodical way in identifying high-risk areas, which are frequent crime zones. The pre-crime technologies also take into account social media, which have become one of the most frequently used platforms for instigating criminal activities.

These systems are minefields of untreated data about the behavior, actions, and attributes of humankind from varied fields. The cause for their reactions to things may be influenced by their life experiences, which the data collection algorithms might fail to comprehend. Then, big data is at the disposal of police officials who may have their own perceptions and biases towards certain communities, which may reflect in their prediction of the data by identifying who are most likely to commit crimes. We are witnessing a loftier rate of search and seizure among certain racial or ethnic groups, constructed on them fitting a police officer's notion of "suspicious". The rigorous supervision of the law enforcement officials may be influenced by their ideologies, which will lead to manipulation of data to predict crimes based on their apprehensions.

The ethical implications of AI should not be ignored considering the ever-increasing demand and quantity of data. The usage of AI-based technologies in law enforcement has been quite common in developed countries. It has in fact led to the evolution of a new area of artificial intelligence and machine learning technologies to explore and generate hefty revenue. It has been used for controlling crimes both in private and public sectors.

The governments of various nations have been hugely investing in these technologies to upgrade the legal policing systems to reduce national security threats.

This chapter will feature the prospective applications of big data and its analytics for curbing crime. An attempt will also be made to identify the specific areas where it can be more effective. A comparative study with other nations where the technology has been propitious will also be discussed. The role of police in exercising the discretion to implement it to a particular group(s) of the population will also form an important part of the study. The new technologies will also be evaluated in relation to their technical, social, and normative nuances.

Big Data Policing

The society has grown infinitely to become more complex, multifaceted, and progressive with cutting-edge technologies that yield innumerous data reflecting the enormous change in societal structure and the way it works. In the age of the internet penetrating the lives of every individual, all their actions, engagements, day-to-day movements, and even grocery purchases get recorded and etched in the microelectronic territory. This data is stockpiled in very efficient centralized servers, then explored and investigated in countless ways. In the world of highly advanced technology, the most prominent power of big data is to capture huge reservoirs of data that are created knowingly and parenthetically. The applicability of big data has also intruded the legal system for various purposes.[6]

Police and Public Values

The task of police authorities is to reduce crime, big data analysis through various software can help in improving crime detection, building trust in such institutions, and reducing public fear. In this technology-driven world, reducing public vulnerability and taking timely action to ensure civility in public spaces has become a challenge. Not only this, the utilization of police authority and force impartiality and its measures to improve public trust and assurance in the police are counted as an integral part of the criminal justice delivery system. For providing an eminent service experience to citizens, the effectual and fair use of public funds in crime detection, apprehension of culprits, surveillance[7] tools are playing a major role.

Nowadays, big data is motivating an inclination in the direction of behavioral optimization and self-curated personalized law where all legitimate assessments, rules are augmented for paramount consequences and the law is personalized to distinct users depending upon the examination of past statistics. Possibilities for big data are many as it vows to deliver a systematical and evidence-oriented methodology in the direction of law. The machine analysis of big data can cut out the human bias,[8] maladroitness, and errors and also has the propensities to intervene independent legal judgment to produce inadvertent consequences in the whole legal system, and hence, moving ahead with caution is a requisite against using big data.

While law is abstract and semantic, big data is highly empirical and syntactic in nature or say fundamentally unrelated since it cannot deduce itself, nor can it detect the unclassified restrictions of legitimate values. Big data needs to be structured by someone through algorithms; since it's a difficult task, even for machine learning as procedures fail to tell what underlying dynamic forces might become pertinent in response to novel actions. By way of banking on big data, the policy-manufacturers take the risk of basing their inherently biased judgments on sheer connections recognized in the seemingly objective data in a highly indefinite and indiscriminate manner.

The challenge it carries with it is lacking a framework to systematically update its structure in harmony with progressing circumstances since big data possibly will weaken the democratic responsibility along with rule of law and may also obstruct judicious legal revolution.

If controlled with caution, big data can prove to be a prodigious tool to update judgment based on scientific evidence and generate "personalized" or customized law, but uncertainties are with reference to the assessments on questions of equity and natural justice. Limitations of predictive power in big data generate a vindictive reflexivity in the law, undermining the legal system's evolutionary capabilities and democratic accountability.

In order to intensify the competence and correctness of big data analytics, diverse categories of methodological approaches and tools have been established, for instance, social media analytics, social network analysis (SNA), text mining, sentiment analysis, visual discovery, and advanced data visualization (ADV). The rules and regulations with reference to data privacy may well impede the further expansion of big data analytics, making it the antithesis of this new-fangled technology. One of the purposes of this one-of-a-kind reverse salient is its impetus for incessant innovation in data interpretation semiotics, for the rules and regulations call technology designers to guarantee responsible usage of the collected data and processes applied.

The employment of big data analytics in law enforcement has indeed eased the transference of the policing attitude from responsive to defensive and precautionary, an attitude that is over and over again interconnected to information-seeking intelligence agencies.[9] The analytics executed by law enforcement organizations bank on the characteristics and individual information assigned on entities and believe in constructing a well-thought-out system of information on potential crime offenders.

Various issues are related to the use of big data that questions the need for maintaining public security at the cost of specific issues of privacy[10] by means of peripheral data that were collected unintentionally with the objective of delivering criminal justice. Examples of these can be electronic toll pass data, permanent address, and usage of the utility bill, grocery bill, pizza delivery order, and number plate recognition systems. These random data can bring innocent individuals with no criminal justice contact into the record of the law administration database putting them under the radar.

In such situations, the only atypical situations should be the ones where data on unsuspected entities should only be investigated and that it should only be examined for the purpose it was collected for. Although Palantir Technologies Inc., 2012 has given a response to the concerns of privacy and has embedded preemptive measures as regards privacy, but for the usage, it is still unclear and does not promise the accountability of the platform.

The balance between two essential rights is the essence of democracy. The police require the data on its populations to screen and identify possible criminals and wrongdoers to make sure the public is safe and secured. Not only peace and protection of public and property is a necessity but the residents have the fundamental right of privacy and personal freedom that must be guaranteed and well looked after, as both are the callings of the state. It is ostensible that the state must work to find the middle ground between two stakeholders that can be facilitated by the presence of a governing body to simplify the interface and create laws and put an end to hitches regarding several adverse consequences such as the infringement of privacy and civil liberties.

It is undeniable that in today's high-tech world, privacy has become a luxury since the presence of spyware software like Pegasus, sensitive data are actually being assembled that are usually nonconsensual and the authorities are breaking down and investigating the data to generally brand most of the innocent civilians as a danger to the peace and security of the general social order. The plight doesn't end here as the minorities, people of different race, color, creed, political parties, vulnerable sections like LGBTQ suffer prejudices the most. Notwithstanding the fact that they form a huge chunk of the population, hence a part of the society, whose safety ought to be guaranteed by the same police. Human analysts are the ones who do exploration of such gathered data and might be personally biased toward people from a certain ethnicity, race, caste, creed resulting in unfair analysis and causing huge discomfort to the targeted population.

The lucidity of the prevalent policing system requires to be improved to earn the confidence and conviction of the general masses, and it should be a pre-requisite to evaluate data for predictive policing to safeguard against prejudiced investigation. In the backdrop of all these apprehensions, one conceivable resolution is to augment a real-time assessing procedure executed by artificial intelligence to evade any kind of human intervention, and the procedure must be developed by the government as a replacement for a private company in order to avoid the dispute over differences of interests. The advanced role of big data can be to further increase the benefit for society, as it can be of use in programs that are basically to ward off the crime that greatly emphasizes serving people by discovering decent professions by which they would not cling to commit crimes like theft, robbery, dacoity, forgery, etc.

International Implications of Surveillance and Predictive Policing

The recent evolution in the long line of computer-based tools is the usage of big data in the field of law enforcement mechanisms. The introduction of big data has changed the existing practices and also requires a technological framework. The acceptance of the new technological tools is also something to ponder upon by law enforcement officials and its impact on its application. It is also required to foresee whether the new tools are to change the already existing practices or the new tools are to adapt to the already existing practices. Although the new technologies such as artificial intelligence and its related machine learning analytics may prove to be a breakthrough in increasing the efficiency of legal tasks, their sudden introduction may turn out to be a hindrance rather than helpful. The introduction of legal analytics has to be gradual, and the professionals also have to be mentally prepared to adapt to such technologies.

Various developed nations have successfully introduced legal analytical tools for strengthening their legal enforcement systems. In the UK, in the early 1970s and 1980s, the usage of computers for legal decision making started with the development of "expert systems". Expert systems are basically those computer programs capable of performing multifaceted tasks at a level which if not at, then definitely very close to the level that is expected of a human expert. The process of decision making by a human being is very complex, and it cannot be always predicted. The advantage of a good system is that it can give a reason for the conclusions it provides, sometimes it's in the form of a variety of logical statements and at times it's through innumerable citations to relevant reliable sources. It can be inferred that a good system has not only the tendency to reveal the outcomes but also explains the details and reasons for arriving at such outcomes, which highlights the transparency of the system. The drawback remains in the fact that it cannot solve intricate and legal problems.

The applicability of artificial intelligence in the legal arena has been quite wide. It has been frequently used in the police systems. The police have to conduct vigilance of the society at large and also at local levels so as not to miss any chance to prevent any crime and also provide assistance in case of crimes which already happened. The moment a crime takes place, it is the foremost task of the police to reach the place of crime and secure the victim and provide any immediate assistance if required. Another major task of the police professionals is to track the offender in which the role of computer-generated technology becomes very important. Big data have been mostly utilized for DNA collection and storage, mass surveillance, and predictive policing. Millions of DNA records can be easily analyzed within a few minutes with accuracy. Predictive policing employs very large volumes of unstructured data in the form of images and videos and even messages, which poses lots of problems in identifying the right information as required in a particular case.[11]

The UK has been using predictive technology tools for a long time. It is demarcated as "taking data from disparate sources, analyzing them and then using results to anticipate, prevent and respond more effectively to future crime". Artificial intelligence and its analytical tools can be used to interpret the behavior of people in a particular vicinity to identify prospective perpetrators of crime. Collection of data is not difficult, but its interpretation is a Herculean task, which requires proficiency to make someone criminal in the eyes of law. The interpretation must be based on evidence and solid reasoning, which can be done by humans only. Writing a reason-based code to analyze the data of a large number of people will also require physiological expertise to formulate such effective analytical instructions. The predictive technology has been frequently utilized in the UK, but its interpretation of whom to target is the endeavor of police professionals who might also have their own perceptions about any particular community, which poses a question on its transparency and authenticity.

The USA has also adopted predictive technologies to reduce its crime rate. The adoption of CompStat brought about a drastic reduction in crime. Automated mapping and statistical analysis of crime data is now being regarded as a vital portion of policing in the States. It has been now implemented by almost all the states across the US. Hotspot policing has been instrumental in identifying crime-prone areas by the usage of "Spatial Analysis". It can be effectively used in strategizing crime prevention in varied environments and also in varied crimes.

A database is created in the UK named Police National Computer (PNC), which comprises more than 12 million records that are categorized as personal and other such records that are generally of vehicles and drivers. This is different from the Police National Database (PND), which is a national intelligence handling system holding police records that generally come under the local police. The London Metropolitan Police Service (MPS) runs an Automatic Number Plate Recognition (ANPR) network, which is noted to obtain roughly 38 million reads each day. The law enforcement agencies of Cambridgeshire and Durham constabularies have been using software which allows them to admit the custody images in just a jiffy with the use of an application on their smartphones. However, this smart technology is not shared with officials across the country when they too have similar requirements for policing purposes. Facial matching software is also used in a few places. An absence of unity and partnership is seen at a local, regional, and national level.

An effective and efficient system can only be ensured when a national infrastructure that is united is established for supervising all the data through big data technology.

Predictive tools have proven to be effective in curbing crime, but then too, it is not adopted uniformly across the UK. Traditional beat policing consumes a great deal of time of the stakeholders then too hot spot policing is not employed frequently as it should have been. It also requires a smaller number of resources for deployment to predict when and where a crime is expected to take place. Currently, policing especially mapping of a crime spot is mostly based on already available recorded data and phone calls from the public informing the probability of crime. Hotspot maps are still produced manually by analysts, which many a time becomes redundant by the time it is put to use. Automated predictive crime mapping tools are an answer to such lags, which is also cost effective.

Kent police started using software in 2014 which used "a well-planned and evidence-based methods to tackle the problems that were generally resident and [provided] anticipatory and targeted watches that were basically on foot". It was successful and held a hit rate that was 11% that led to the prediction of the crime location ten times more than random patrolling and more than doubled the probability to predict crime by using procedures that were intelligence-led.

The open-source data gathered from social media platforms is humongous. A significant amount is spent on purchase of such data but its analytical capability is not done effectively. It is due to technological incapability and organizational barriers. The manual analysis is a tedious task, which will consume much time, and in most cases by the time it is analyzed, the interpretation may not be required as it may become outdated. Moreover, the analysis of such data is mostly required for large events such as riots or protests over a wide geographical area. Its utility is very limited for local jurisdictions. The collection of social media data is mostly spread across a large geographical area which is of limited usage to local policing.

The data outcome from PredPol software is grid-based predictions, which are not very helpful for large geographical landscapes. There is a requirement to refine the model and integrate it within the existing predictive policing software rather than creating and introducing new software for the same purposes.

Canadian police agencies and global real-time agencies are using or preparing to device real-time analytics resolutions for their investigative and operations teams to comprehend as to how these tools affect decision making in real-time, compound environments. Mass amalgamation of data has been utilized to produce analytical solutions for effective policing. IBM and SAS software have been frequently used because of their profound analysis and visualization aptitudes and also for their coverage of wide geographical areas. Other competitors in the line include Motorola, NICE, IBI, and RapidDeploy that have a very detailed understanding of the civic safety space and have skillful analytics and visualization solutions that are precisely for the safety of the public. They have been engaged in monitoring, managing, and controlling information assets.

Predictive Policing – Proliferation of AI in Surveillance: Drones and Public Safety

George Orwell's vision of mass surveillance in *Nineteen Eighty-Four* seemed like science fiction when it was published in 1949, and likewise, when 1984 rolled around. Today, the stark similarity of many of its far-fetched concepts seems eerily close to reality. Technology has a valuable role to play in policing, but it still raises serious legal, ethical, and moral questions.[12]

In 1968, 911 was established as the universal emergency number, and after this, the next three decades have witnessed the tremendous intensification of community policing, mechanization, and a surge in DNA technology and computerization.

To have a better society, a robust police system is an indispensable thing. Predictive methods permit law enforcement agencies to work with more vigor but with limited resources. Predictive policing is low-cost and can work quite well as it is being used to forecast crimes, predict offenders, and people who could be victims in a given moment, predict perpetrators' identity, their whereabouts, sites of crime, times of increased danger, and similarly, these methods are also used to bifurcate groups or, in some cases, individuals who are likely to turn out to be victims of crime. The objective of the predictive methods is to improve real stratagems to detect and prevent the offense, hence lower the crime rate, or sometimes to ramp up the investigations and to make their struggles all the suaver. Nonetheless, it should be made clear that at all stages just applying speculative policing procedures is different from discovering with a crystal ball. It is important that policing strategies and plans are producing substantial results as then they can be reckoned as effective.

It is imperative to understand that at all levels applying these methods is not equivalent to finding a crystal ball. For a policing strategy to be considered effective, it must produce tangible results. For example, crime rates should minimize, arrest rates for severe offenses should surge, there should be a discernible positive impact on social and justice outcomes and peace prevails. Tech giants like Amazon, Walmart have also been using new techniques to examine extreme weather patterns to decide what should be kept in stores accordingly, like say overstocking duct tape, ready-made meals, bottled water, antiseptic ointments, etc. as most of these are discovered through numerical and algorithmic analyses of erstwhile consumption data pertaining to customers' behavior during analogous serious weather conditions. Numerous akin interactions that can predict consumer behavior can be reconnoitered with predictive policing.

Investigation agencies are dependent on using automated analysis of data about crimes that happened previously, the conditions of the local environment, and other relevant acumen to "foresee" and avert any sort of offense from happening. Objectives behind this are to create awareness at the premeditated and planned echelons and to make progress in laying down schemes that foster additional competence and real-time monitoring, since with much awareness and well-advanced anticipation of human activities, police authorities can easily recognize and improve upon the approaches to thwart unlawful activities by habitual offenders against recurring victims of the crime. These techniques also allow the police departments to act more proactively with limited resources.

However, there is a sturdy bulk of proof to back the philosophy that law-breaking is foreseeable statistically largely because offenders have a tendency to function in their precincts of comfort, thereby meaning they generally commit the same crime that has been committed quite effectively in the past, normally close to the identical setting and period. Even though this is not unanimously accurate, it befalls with appropriate regularity to make these ways and means work realistically. These tendencies can be captured in data and utilized by police and investigation authorities to win through in dealing with crime.

Per Jeff Brantingham, an anthropologist, who helps in supervising the predictive policing project for the Los Angeles Police Department (LAPD) at the University of California, Los Angeles,

> The naysayers want you to believe that humans are too complex and too random— that this sort of math can't be done … but humans are not nearly as random as we think …. In a sense, crime is just a physical process, and if you can explain how offenders move and how they mix with their victims, you can understand an incredible amount.

These explanations are reinforced by different leading theses of criminal behavior, like on routine activity, rational choice, and crime patterns. Theoretically, the methods of predictive policing are applied to trace the patterns and aspects through analytics and then can navigate criminals' choices to check crimes with premeditated intermediations. Of all these theories, the blended theory suits "stranger offenses", like dacoity, robberies, chain snatching, burglaries, and thefts the best. Statistical regression, hot spot analysis, data mining, and near-repeat methods are some methods that are essentially used to decipher where a crime will come about over an identified time limit and as a result who is expected to be the target.

Temporal and spatiotemporal techniques and risk terrain analysis can be used to predict when a crime is most likely to happen and who is most likely to be the victim because they account for the surrounding as well as local residents and also discerns the geospatial influences that make up for crime threats. Some of these will look into physical settings that might be well suited for a particular category of crime.

In predictive techniques, the frequently used software are bit technical and as the software capabilities increase, the training requirements for law enforcement agencies also become inevitable. For more comprehensive analysis or data manipulation, statistical software packages and languages, such as SAS, IBM's Statistical Package for Social Sciences (SPSS), and R (open-source and freely distributed), dispense the much-needed prognostic ways and means that can be used in combination with police functioning in general. Another facet of the surveillance system is through the widespread use of cameras everywhere. From parks to shopping centers, to public places to closed-circuit television (CCTV) and dash cams, private organizations, smartphones, and body-worn cameras have rolled out almost ubiquitously. Cameras are even being adjusted with unmanned aerial vehicles (UAVs) or drones. People can also be spied on through mobiles and even scanned for armaments from a remote area.

Application and usage of drones have also increased in predictive policing since several law enforcement authorities and police organizations are utilizing drones that include Little Rock, Arkansas, Miami-Dade, Florida, and Arlington, Texas. These surveillance techniques are also adding to big data. Where some are restricted to car chases and siege situations, others are being used for general surveillance and they can fill up the gaps in CCTV coverage and offer police greater capability to track people. Now there's a speculation that big data analytics will introduce new crime alleviation strategies through predictive policing.

Numerous companies, like SeeQuestor, are working to consider this conundrum by offering software that enables law enforcement to deftly review people and faces in videos, but it still needs a review by human beings. The debatable question is whether these innovative technologies protect citizens more than ever, or just provide the governments a new-fangled set of eyes. The newly acquired AI capabilities were used by the police force to solve some critical and high-profile cases and were later awarded the SMART Policing Award this year by FICCI, an association of business organizations in India.

Smart Glass is a piece of software by Staqu which operates on a pair of smart glasses equipped with a built-in camera. The startup's facial recognition capabilities are used by the glasses to identify individuals in a crowd. And once the face is recognized from within a provided database, Smart Glass projects the results on the wearer's screen. The complete process occurs in real-time as officers simply look at the faces of people within the immediate vicinity. Smart Glass is also capable of both speech and image recognition. Moreover, the technology can stream its camera feed to a mini set top box like device placed nearby so that the video feed can be analyzed against criminal databases or known threats. Smart Glass could be a game changer in the field of personal security as it has the capability of helping security personnel to safeguard huge crowds at events attended by political leaders, high-risk celebrities, or visiting dignitaries.

When two Chicago criminologists, Clifford R Shaw and Henry D McKay, conducted research, on one of the first traces of predicting crime technology, which can be traced by exploring the persistence of juvenile crime in some neighborhoods of the city. They introduced the social disorganization theory, which states that location matters when it comes to predicting illegal activity. Since then, technocrats have started exploring data analytics within the purview of crime too by involving statistical and geospatial analysis to curb crime. It led to the creation of sophisticated mathematical models and regression analyses to predict the likelihood of future crimes.

Big Data Predictive Analytics in India

As per National Crime Records Bureau (NCRB)* data, crimes in India have risen in 2018 as compared to last year. To deal with these modern-day criminals, the impetus will have to give to technology in addition to advanced weapons. This is the age of big data, and analytics and can be a great help to law enforcement agencies.[13] When data analysis and predictive modeling can be used in fields like weather, stock markets, healthcare, etc., then why not to apprehend crime. In India, where there is a constant shortage of police force, data analytics can be a blessing. As per a report, National Crime Records Bureau (NCRB) is trying to apply crime data analytics software to deliver predictive policing so that crime can be stopped before it starts to happen. It is speculated that post-2018, predictive policing technologies will be available to five states – Kerala, Odisha, Maharashtra, Haryana, and Tripura – and then subsequently to all states and union territories. Data Science is influencing our lives in every possible way. Recently, crime hot spots have been identified to prevent the repetition of crimes in geographical areas by using a piece of software known as "CMAPS" (Crime Mapping Analytics and Predictive System). It was done with the collaboration of ISRO, which landed its space technology-based tools for ensuring internal security. The Crime Mapping Analytics and Predictive System (CMAPS) and ISRO's technology converged to make it possible. The software maps the clusters collected through ISRO's satellite imageries and capital's Dial 100 helpline to access real-time data to track crime hot spots. As per police officials, in a mega city like Delhi, myriad crimes can be predicted and can be curbed by applying big data analytics based on scientific and objective analysis of data. The new software has replaced the manual crime mapping, which used to happen after every 15 days. It also aids police in identifying the exact locations and spatial distribution of the crime over a particular geographical area.

On January 1, 2020, police created a "CCTV Surveillance Matrix" within the state of Himachal Pradesh by installing around 19,000 CCTV cameras for predictive policing strategy (Outlook India, 2020). Various states which are also working on the same line include Delhi, Telangana, and Jharkhand, which have fully functional predictive policing systems. The Open Group on E-governance (OGE), which is an initiative in collaboration with the Jharkhand Police and National Informatics Centre has been a forerunner in developing such projects related to the use of technology in making the legal enforcement system robust. The OGE began exploring the possibilities of using data mining technology to scan the records available online, i.e., the digitalized versions of the crime to develop predictive policing solutions. These surging crime trends have the potential to be a building block in the predictive policing project that the state police want to adopt.

The state has also laid emphasis on developing a large-scale Domain Awareness System for the collection of live data and creation of innovative ways to support the police officers on the ground instead of using predictive policing software. It is appreciative of mentioning the development of a Naxal Information System, Crime Criminal Information System, which is integrated with the CCTNS and a GIS that supplies customized maps which can project the exact area where Maoist groups can operate. The Crime Analytics Dashboard is also a very systematized software, which can present the incidence of crime based on various categories as and when required. It can provide information related to the type of crime and its location.

In 2015, Staqu Technologies pitched an artificial intelligence-centered tool to disentangle real-world difficulties of the law enforcement organizations. It developed a multilingual mobile app that empowers police officials to capture relevant facts and other details about lawbreakers, compute biometric information, recognize face, speech, and fingerprints also, while out on the patrol. On the same line, a pilot project was successfully run in the city of Alwar in group effort with its law enforcement agency known as "ABHED" (AI Based Human Efface Detection) and was a grand success as it helped police officials in apprehending approximately 1,000 criminal suspects. Such technologies can be extended to the areas where the crime rates are surely high.

Reasonable Suspicion of Predictive Policing and Human Right Violations

It is frightening to see that predictive policing is being used for purposes for which it was under no circumstances envisioned. The primary intention for the conception of the Palantir-like technologies was to break down the growing terrorism and thereafter the technology started to be used to arrest and apprehend dangerous criminals. The technology is also used to deter criminals.[14] Unfortunately, this new-fangled technology is now being used for multifarious operations that include capturing political dissenters or those that show political disloyalty and then sent to extralegal political education centers.

Considering the situation and issues allied to big data and predictive policing in surveillance, it has become a necessity to ensure increased transparency and public awareness of these programs, as well as measures must be taken to evade any kind of potential human rights abuses within the criminal justice system. It is undoubtedly true that nothing is inherently erroneous with using data to record crime and foretell criminal activity, but to some degree, it is erroneous if there is no transparency about how the data is being used, on what basis it is being shared to authorities. Considering the image of the Indian police that is tarnished for being casteist and communal, use of such technologies can lead to heightened abuse of power and harassment for a few. Affiliates of minority religions and lower castes are more likely to be in the crosshairs of law enforcement agencies, even if there is conclusive proof of incorruptibility.

Predictive policing has the propensity to dilute the citizens' rights. During the protests against the Citizenship Amendment Act, it was reported that the Delhi police were filming protestors and then running images through its Automated Facial Recognition Software (AFRS) to screen the crowd. This issue is more related to identification of the perpetrators and not directly related to predictive policing but the use of AI can put those faces in the databank that can be consequentially used to judge dissenters as imminent criminals.

In the very famous judgment of *Puttaswamy*, the Supreme Court in August 2017[15] unanimously maintained the fundamental right to privacy as enshrined under Articles 14, 19, and 21 of the Indian Constitution. It is a very landmark judgment and a central element of the legal battles that are yet to arrive regarding the state's potential to carry out strict surveillance. Unfortunately, there resides a gray area flanked by privacy and the state's requirements for maintaining peace and security.

A few months after the landmark decision, the central government set up a Data Protection Committee under retired Justice B.N. Srikrishna, and various public hearings were organized pan India that later produced a draft data protection law in the year 2018 and is still lying with the Parliament. Albeit experts have figured out some loopholes regarding the draft law as it does not deal passably with surveillance development, it can still be considered as a first baby step in the direction of protection of the right to privacy as enshrined in the Constitution of India.

World over it is being urged to discontinue working on predictive policing systems. Examples of Chicago,[16] Shreveport explains the stage of infancy in predictive policing that needs to be studied more.[17] Public and researchers have started realizing that it propagates structural racism.

The Way Forward: Conclusions and Recommendations

The praxis of big data to legal enforcement is certainly here to stay. These innovative technologies have transformed policing methods, especially predictive crime policing. Big data interpretation for legal enforcement has become an integral part of policing in the West. In India, it is still at the experimental stage. Currently, many of the officers lack the technological bend required to access or input data remotely, greatly restricting their operational capabilities. In India, technology innovation has started, but its acceptance by the officials will require more sensitization and training. The pragmatic approaches to the technology require professional enhancement of police officials. Most of them are reluctant to adopt and implement as their approach to blending their legal capabilities and technology is not guided. Although the future of predictive technology is promising, it requires time and effort in its acceptance in normal parlance of legal enforcement.

The execution of big data in the legal field also poses certain challenges as various states use different software or other technologies, which results in desynchronization of data gathered from various geographical areas. Many a time, it leads to overlapping of the same data and wastage of legal credentials. A uniform virtual platform should be established by the authorization of the central authorities to unify data collection across the country.

The police officials are often provided limited access to the data. The procedural hurdles to access data should also be resolved. An efficient infrastructure to manage huge data and its analysis will ensure effective application of the data.

One of the most famous cyberlibertarian, John Perry Barlow wrote in his "A Declaration of the Independence of Cyberspace", "Governments of the Industrial World, you weary giants of flesh and steel, I come from Cyberspace, the new home of Mind. On behalf of the future, I ask you of the past to leave us alone. You are not welcome among us. You have no sovereignty where we gather". He supported the non-interference of the government in the realm of the internet. He believed that the internet is beyond boundaries, and national laws should not create any barriers to it. He believed in the power of the internet but missed the human uptake on cyberspace, his belief was not pragmatic. The internet is ubiquitous, it has pervaded almost all the spheres of human life. But he failed to foresee the challenges it could pose in the future and what it can cause without any governmental interference on policy guidelines. Big data can be misused for ransomware attacks and has a tendency to create both illicit markets like AlphaBay, Silk Road, and licit market, which can embolden the data breaches.[18]

The perplexities of legal principles might not be interpreted by mechanical analysis of the data gathered. Scientists around the world are apprehensive about the rightful utility of big data by legal enforcement agencies.[19] We do not believe that a superior technology is a viable solution to it. We think that the implementation of big data technologies might be useful to some extent, but its repercussions cannot be ignored either. A legal system based on human values can only ensure a robust and trustworthy society where law and order can sustain by avoiding excessive intrusion into the private lives of every individual.

It is suggested that police should upkeep pre-meditated democracy ingenuities that provide people the chance to distinguish and discover the complexities of data-driven policing in-depth prior to coming to a well-thought judgment on the adequacy of police practice in the era of big data.

The focus must be directed toward privacy and ethics commissions, which ought to be established in the governance organizations of every single police force in the nation-state to address growing disquiet for the right to privacy about the dark side of surveillance technologies that are progressively becoming a major source of much police data. Guidelines must be brought for police forces where they must review procedures and actions with regard to data stewardship.

For data-driven policing schemes, the most important part has to be played by the central government in providing surplus capital to police officers training in numerous areas of priority. Hence, the need of the hour is for an innovative and synchronized approach to data precision in supervising structures of policing. Also, time requires that reviewing courts must mandate systematic and comprehensive explanations from officers for carrying out their search and seizure decisions to make the system robust.[20]

Efforts must be made toward updating the education system and conducting well-suited training programs to facilitate police officers and administrators considering the significance of accuracy and detailing when data is being captured. Along with these measures, modification of prescribed staff training programs by private corporations providing predictive and data-driven policing systems is now a necessity. These efforts have a tendency to profound the public trust in surveilling techniques adopted by police authorities, for maintenance of that public confidence is indispensable to the capability of the police service to track the kind of public value.

Notes

1. Chouhan, K.S., 2019. Role of an AI in legal aid and access to criminal justice. *International Journal of Legal Research*, 6(2), p.1.
2. Schlehahn, E., Aichroth, P., Mann, S., Schreiner, R., Lang, U., Shepherd, I., and Wong, B.L., 2015. Benefits and pitfalls of predictive policing, in *2015 European Intelligence and Security Informatics Conference*, pp.145–148. https://doi.org/10.1109/EISIC.2015.29.
3. Chan, J., and Bennett Moses, L., 2016. Can 'Big Data' analytics predict policing practice? (June, 2016). *UNSW Law Research Paper No. 1*, pp.20–82.
4. Shapiro, A., 2019. Predictive policing for reform? Indeterminacy and intervention in big data policing. *Surveillance & Society*, 17(3/4), pp.456–472.
5. Feldman, Y., and Kaplan, Y., 2019. Big data and bounded ethicality. *Cornell Journal of Law and Public Policy*, 29, p.39.
6. Backer, L.C., 2018. And an algorithm to bind them all? In *Social Credit, Data Driven Governance, and the Emergence of an Operating System for Global Normative Orders* (May 21, 2018). Entangled Legalities Workshop (Vol. 24).
7. Joh, E.E., 2016. The new surveillance discretion: Automated suspicion, big data, and policing. *Harvard Law & Policy Review*, 10, p.15.
8. Brayne, S., 2017. Big data surveillance: The case of policing. *American Sociological Review*, 82(5), pp.977–1008.
9. Brayne, S., 2018. The criminal law and law enforcement implications of big data. *Annual Review of Law and Social Science*, 14, pp.293–308.
10. Tene, O. and Polonetsky, J., 2013. Big data for all: Privacy and user control in the age of analytics. *Northwestern Journal of Technology and Intellectual Property*, 11, p.239.
11. Hardyns, W. and Rummens, A., 2018. Predictive policing as a new tool for law enforcement? Recent developments and challenges. *European Journal on Criminal Policy and Research*, 24(3), pp.201–218.
12. Strikwerda, L., 2020. Predictive policing: The risks associated with risk assessment. *Police Journal: Theory, Practice and Principles*, 20, pp.1–15.
13. Cukier, K. and Mayer-Schoenberger, V., 2013. The rise of big data: How it's changing the way we think about the world. Foreign Aff., 92, p.28.
14. Bachner, J., 2013. *Predictive Policing: Preventing Crime with Data and Analytics* (Washington, DC: IBM Center for The Business of Government).
15. K.S. Puttaswamy V. Union of India (2017) 10 SCC 1.
16. Joh, E.E., 2014. Policing by numbers: Big data and the fourth amendment (February 1, 2014). *Washington Law Review*, 89, p.35.
17. Selbst, A.D., 2017. Disparate impact in big data policing (February 25, 2017). *Georgia Law Review*, 52, pp.109, 109–195, 145.
18. Wall, D.S., 2018. How big data feeds big crime. *Global History: A Journal of Contemporary World Affairs*, 2018, pp.29–34, p.29.
19. Mugari, I., and Obioha, E.E., 2021. Predictive policing and crime control in the United States of America and Europe: Trends in a decade of research and the future of predictive policing. *Social Sciences*, 10, p.234.
20. Taslitz, A.E., 2010. Police are people too: Cognitive obstacles to, and opportunities for, police getting the individualized suspicion judgment right. *Ohio State Journal of Criminal Law*, 8, pp.7–78, p.77.

12

The Use of Inclusive Digital Technologies: To Improve Access to Justice

Sergey V. Zuev, Oksana V. Ovchinnikova,
Vera A. Zadorozhnaia, and Tatyana P. Pestova

CONTENTS

Introduction

The world's attention is focused on digital accessibility (Kulkarni M., 2018). Issues related to technical products, resources, and services provided by hardware and software are being actively discussed. However, the growing dependence on digital technologies can lead to even greater inequality, leaving behind vulnerable groups of the population, including persons with disabilities. At the same time, digital accessibility can equalize the position of such persons in all spheres of life (Kerry Dobransky, Eszter Hargittai, 2006).

We believe that the creation of an accessible digital environment for persons with disabilities when addressing law enforcement agencies will make it possible to protect their rights effectively. A comfortable digital environment for persons with disabilities should be built on the user's needs and based on maximum resource availability.

The scientific literature actively discusses the means by which persons with disabilities interact with law enforcement agencies and the court. The authors consider the indicators of the readiness of law enforcement and judicial systems to work with persons with disabilities in different countries in the context of their access to justice (Kremte, 2019; Panggabean, 2019; Kuosmanen & Starke, 2015; (Byrne et al., 2021; White & Msipa, 2018). The importance of providing high-quality inclusive public services, including language support, online consultations, etc., is noted in Twizeyimanaab J.D. and Andersson A. (2019). Flynn points out that disability suggests a new focus on the effectiveness of access to justice and the inclusiveness of the justice system in general (Flynn E., 2015). Norman believes that human needs should be put at the forefront, and technologies corresponding to these needs, behaviors, and resources should be adapted to them.

DOI: 10.1201/9781003215998-12

Previous studies have focused on the general communication problems of people with disabilities. The literature does not consider the potential use of digital technologies in specific communication processes arising when people with disabilities interact with law enforcement agencies.

To obtain practical results on this topic, we should narrow the field of research. We are experienced in serving in law enforcement agencies.

The purpose of the research is:

- To analyze the legal guarantees of the accessibility of justice when people with disabilities address law enforcement agencies, to determine ways to improve them;
- To analyze the possibilities of digital communication between citizens and government authorities;
- To develop basic requirements for the functioning of a digital portal to improve the accessibility of justice.

Examining these issues will improve the accessibility of justice and help people with disabilities learn and defend their rights.

Methods

In our study, we used the analysis of normative legal instruments regulating legal relations at the stages of proceedings for considering appeals of persons with disabilities to law enforcement agencies with proposals, complaints, statements, and reports of crimes, administrative offenses, incidents; a direct method for assessing the compliance of digital technologies used by law enforcement agencies when considering the appeals of persons with disabilities; a structural–functional method which allowed us to determine the physical and temporal boundaries of the conceptual model of the digital portal operation to improve the accessibility of justice; conceptual modeling of innovative activities of law enforcement agencies on working with persons with disabilities using modern digital technologies.

Results

Based on the results of our study, we concluded that a specialized public digital portal is needed to improve accessibility to justice. The main condition for the formation of such a portal is the availability of a regulatory framework and technological standards.

Our analysis of the literature allowed us to determine the main areas of increased access:

1) Improving legislation, which includes optimizing the national legislation on applying international standards of digital content and legal enshrinement of alternative methods of communication for persons with disabilities;
2) Improving the presentation of digital content, combining universal design and artificial intelligence.

We formulated the main requirements for interactive services. Portals should provide a unified standard for appealing to law enforcement agencies, use sections for particular life situations, rapidly analyze and control incoming appeals from citizens and the response time for considering the appeals and execution of decisions on them, accept feedback from the applicant regarding their satisfaction with the received response, and solve the applicant's problem.

When vulnerable populations file appeals, the process of interaction with law enforcement agencies should be adjusted dynamically. The software and interface of justice administration systems should be based on the principles of user-friendliness, simplicity, and intuitive use; meeting user expectations; minimizing the consequences of errors (tolerance for error); minimizing physical efforts (low physical effort); flexibility providing for several control methods; reliability of feedback; and possibility of clarification and repetition.

The main criteria for access to an interface include:

1. A chatbot or virtual assistant;
2. Compatibility with adaptive technologies;
3. Making it possible to confirm information about disabilities;
4. Designing the interface according to the standards of the Web Content Accessibility Guidelines (WCAG) 2.1.

The software and interface of justice administration systems should include a personal account which can verify the user's identity and can allow users to indicate their special needs; optional traditional or electronic methods of interaction; high-quality feedback with optional means of communication (electronic sign language interpreter from audio to sign language and vice versa, screen magnification software; speech synthesis); virtual voice assistant, chatbot, avatar (preferably appearing as a police officer or in other uniforms, imitating racial, ethnic, and cultural similarities); other uses of artificial intelligence; a service for holding online court sessions; flash notifications, vibrations, compatibility with hearing aids; convertibility of documents, video, audio, text; ability to sign a printed or electronic document; use of artificial intelligence for interrogations and other procedural actions; ability of the system to detect and correct errors, handle big data; system flexibility, error detection, user feedback.

Legal Guarantees of Accessibility of Justice

The UN Convention on the Rights of Persons with Disabilities (hereinafter referred to as the Convention), other international treaties, and the national legislation of several countries formulate legal guarantees of access to justice when persons with disabilities address law enforcement agencies. The most important provisions of the Convention in this field are Article 12 "Equality before the law" and Article 13 "Access to justice".

Equality before the law for persons with disabilities implies:

- Equal legal protection;
- Equal legal capacity;
- Providing appropriate support to persons with disabilities in exercising their legal capacity;

- Guaranteeing prevention of abuse (providing for support measures based on respect of rights; absence of undue influence or conflicts of interest in their implementation; proportionality of support measures to the circumstances; minimizing the time required for provision of legal support; supervision (control) by an impartial authority or judicial body).

Access to justice for persons with disabilities includes procedural and other accommodations, which simplify their participation in the legal process at all stages, which is ensured, inter alia, by training staff employed in the police and in the penitentiary system.

To implement the Convention, in August 2020, the UN Special Rapporteur on the Rights of Persons with Disabilities, the special-purpose committee, and the UN Secretary General's Special Envoy cooperated to develop a special guideline on access to justice for persons with disabilities: International Principles and Guidelines on Access to Justice for Persons with Disabilities (hereinafter referred to as the Guidelines), wherein a special section is devoted to the exercise of the right to address law enforcement agencies by persons with disabilities. According to Principle #8 of the Guidelines, their complaints must be investigated with the provision of effective remedies. To implement this principle, paragraph 8.1 of the UN Guidelines suggests that states should establish transparent and effective mechanisms for persons with disabilities to report complaints, including through the provision of individually adjusted remedies. In particular, states should (clause 8.2 of the Guidelines):

- Create mechanisms to handle complaints with the involvement of human rights organizations, administrative bodies, and specialized teams able to handle complaints of persons with disabilities and assign them legal remedies;
- Provide persons with disabilities with equal opportunities for filing complaints through the use of various methods and mechanisms (hotlines, e-services), taking into account gender and other features;
- Provide voluntary alternative dispute resolution methods (conciliation, mediation, arbitration, and restorative justice);
- Ensure that complaint filing and justice systems can detect and respond to gross, systematic, group, and major violations;
- Provide appropriate training for law enforcement and court personnel in dealing with persons with disabilities;
- Ensure, where necessary, the presence of an intermediary.

Russia, as a signatory to the Convention, should submit reports on the implementation of the requirements and recommendations to the UN Committee on the Rights of Persons with Disabilities. The UN experts noted positive shifts in Russia in the area of protecting the rights of persons with disabilities: the legislation provides for a direct prohibition of disability discrimination; currently, the country has adopted and is implementing the accessible environment state program, which has been extended until 2025.

The World Wide Web Consortium (W3C) defines the technological basis for the interaction of persons with disabilities with law enforcement agencies. The content conforming to WCAG 2.1 also conforms to WCAG 2.0. WCAG 2.2 is scheduled to be published in 2021.

Web content offers the following to the users:

- Several modes of access (for example, visual, auditory);
- Settings for the convenient perception of information (for example, magnification, color change);
- Control methods (for example, define the language, use sign language).

Digital content should be compatible with different browsers and helper tools (for example, alternative keyboards, screen readers). WCAG noncompliant interfaces create barriers for persons with disabilities.

UN experts noted that there are positive trends in Russia on protecting the rights of the disabled: the law explicitly prohibits discrimination on grounds of disability; the Accessible Environment program was adopted and implemented and extended until 2025.

Additional measures are necessary in the following areas of digital equality:

1) Improve legislation:
- Form a national legal framework for the application of international standards of digital content;
- Provide a legal framework for alternative means of addressing law enforcement agencies by persons with disabilities (for example, the right of persons with visual impairments to use sound recording equipment, including at the expense of the state, as it is a communication tool for them; the use of text to speech software).

2) Improve technical equipment and digital content delivery:
- In addition to the WCAG standards, implement the concept of universal design with its principles (The Seven Principles): Equitable Use—convenience, confidentiality, and appeal should be equal for all categories of users; Flexibility in Use—the design accounts for a wide range of user preferences and capabilities, Simple and Intuitive Use—eliminate unnecessary complexities, meet user expectations; Perceptible Information—effective presentation of information regardless of environmental conditions and user capabilities; Tolerance for Error—minimizing the adverse consequences of accidental or erroneous actions; Low Physical Effort—minimize repetitive operations, provide accessibility in any position (standing, laying, sitting), Size and Space for Approach and Use—direct visibility, convenient access regardless of the position of the user's body, sufficient space for using helper devices);
- Integration of auxiliary adaptive devices into the digital interface;
- The use of artificial intelligence to identify and compensate for communication disorders.

Digital Means of Communication of Citizens with Government Authorities

State-of-the-art digital technologies make citizens' appeals with statements, suggestions, and complaints a more mobile means of feedback between the state and society. The use of digital technologies to address government and administrative authorities is one of the components of the e-Government project.

The current level of e-Government development indicates the demand for an electronic form of addressing government and administrative authorities, and, consequently, an increase in the national level of informational support, computer literacy of citizens, Internet security, and trust in government authorities and their officials.

We can outline four main means of submitting an electronic appeal to government and administrative authorities, including law enforcement agencies, which have found practical applications: e-mailing; sending an appeal via the electronic front desk of official websites; via specialized government Internet portals (for example, government service portals); sending appeals via special electronic terminals installed in the offices of the government and administrative authorities; providing a recorded appeal on a removable medium (flashcard, CD, DVD).

The state must ensure digital equality among the population, which means the person's freedom to choose between traditional paper or electronic forms of communication (Bogdanovskaya I. Yu et al., 2009). If citizens choose the electronic format, authorities generally respond to citizens' appeals by e-mail through an electronic document or a message to the personal account of a registered user of a particular Internet service, although responses can also be sent in writing to the mailing address indicated in the appeal. The digital equality of the population when addressing government and administrative authorities should also be understood as part of equal access to digital technologies.

Access to appeal of government agencies (and their officials) and access to the response to such appeals should be ensured by e-Government technologies and by training the population in information and telecommunication technologies.

Citizens, as potential applicants, must be made aware of all possible forms and means of government appeal and response to increase the overall level of accessibility. This multivariance should be ensured by e-Government technologies and by training the population in information and telecommunication technologies.

From the standpoint of information security, accessibility is characterized as the presence of certain conditions which enable the use of information or services, the ability to connect to a data transmission network (and the operability of this connection), and the presence and operability of infrastructure to store and process requested information (Revnivykh A.V., Fedotov A.M., 2014).

Broadly speaking, accessibility means Internet access for all, but in the strict sense, it refers to taking into account people with disabilities (Jitaru E. & Alexandru A., 2008). According to Tim Bernes-Lee (W3C Director and inventor of the World Wide Web), accessibility means "taking advantage of the Internet. Accessibility to all people, regardless of their hardware, software, network infrastructure, native language, culture, geographic location, or physical or mental aptitudes" (Berners-Lee T., 2018).

Nevertheless, many states are unable to effectively use digital technologies to provide inclusive services (E-Government Survey, 2020).

Ensuring the accessibility of electronic appeals to the government and administrative authorities and the responses to such appeals should become a component of social inclusion. Government agencies should review the established forms of handling citizens' electronic applications, from receipt of the appeal to responding by e-document.

The digital transformation of government should be aimed at the promotion of digital inclusivity. This is essential for people with disabilities who are used to relying on the Internet to provide them with more independence, work, and social interaction opportunities (Jitaru E. & Alexandru A., 2008).

E-Government services should be used as the foundation for a Citizen Feedback Platform (hereinafter referred to as a platform), which should provide a uniform applicable standard for filing appeals using special sections listing certain life situations, prompt analysis and control of incoming appeals, prompt turnaround for considering the appeals and taking decisions on them, feedback from the applicant regarding their satisfaction with services provided, and response to user requests in social networks. This concept is based on the realization of several provisions.

First, creating an interactive platform will provide prompt solutions to citizen's problems. This platform will allow the authorities to receive objective information on pressing problems pertaining to citizens and take appropriate measures to solve them.

Notably, such platforms have already been created and operate successfully in other countries. For example, the Republic of Kazakhstan has a similar platform on the official public website Public Services and Information Online, which allows a citizen to file any appeal to any authority, provided that the person has registered a personal account and uses an electronic digital signature. Since March 2020, within the framework of the Digital Public Administration federal project of the Digital Economy national program, the Russian Federation has been conducting an experiment to introduce a feedback platform through the Unified Portal of Public and Municipal Services, covering more than 50 constituent entities of the Russian Federation. The interactive system includes four main components: citizens' appeals; public opinion polls and voting for the initiatives of government and local self-government authorities; incidents in social networks, including a search for problematic messages and response from the authorities; and government groups involving centralized social network account management and messengers with the function to moderate and build a content plan (Feedback Platform (FP), 2021).

Second, ensuring the accessibility of the platform according to the principles of the Web Content Accessibility Guidelines (WCAG) 2.1, namely:

1) Perceptibility: The information and components of the user interface should be presented only in a form perceptible by the users.
2) Controllability: The components of the user interface and navigation should be controllable.
3) Understandability: Information and operations of the user interface should be understandable.
4) Reliability: The content should be reliable to the extent needed for its interpretation by a wide range of different user applications, including assistive technologies.

Third, the use of the Internet resources of the platform should be provided by assistive technologies, i.e., hardware and software used by a person with disabilities separately or jointly with the main hardware and software complex to provide functionality which cannot be achieved through the use of conventional hardware and software means. Assistive technologies include:

1) Screen magnification programs, which generally contain other tools helping to better perceive visual information by visually impaired users (those with residual vision), users with impaired perception, and other sensory and physiological features, which impede reading a printed text.
2) Screen access programs allowing blind users to perceive text and other on-screen information using Braille, synthesized speech, vibration, sound, and other signals.

3) Alternative keyboards for users with restricted motion, which replace the conventional keyboard (keyboards using head pointers, simple switches, breath-based control systems, and other special input devices).

4) Alternative pointers for users with restricted motion, replacing the common mouse and allowing one to point and activate buttons (GOST R 52872-2019, National standard of the Russian Federation).

Fourth, to enable the user to confirm the information on their disability in their personal account on the specified platform, for example, in the section "My Documents and Data".

Fifth, the platform should provide a chatbot, which is designed to clarify information and ensure efficiency in solving the applicant's questions, for example, related to the structure of the appeal or the content of the response to the appeal, etc. Some countries are already experienced in this field. In March 2021, the format of interaction with users was changed in the new version of the Russian Unified Portal of State and Municipal Services. Visitors of the portal are greeted by robot assistant Max, who can be asked any question of interest. The chatbot explains how to find the necessary agency or service.

> The experimental version of Robot Max is already showing impressive results: it uses a "smart" search through the portal, receiving online answers to users' questions, and identifying their needs. This is an original solution that will raise the service level for portal users and relieve the burden on agencies. In the future, the robot will also be able to call the real portal operator for help if the user still has questions after a dialogue with artificial intelligence. This year the robot Max will gain a voice – it will be able to communicate with our users through voice devices, which are becoming familiar to many people. Work in this direction is already in progress, and the first tests are more than encouraging.
>
> **(Chernousov I., 2021)**

Sixth, to develop a mobile application of the Platform installable on tablets, mobile phones, and other devices for reading, entering, viewing, and reproducing digital information.

We believe that such an approach of government agencies to handling electronic appeals of citizens in general (and electronic appeals of persons with disabilities in particular) will correspond to key components of the transition of governments to digital technologies designated by the UN: ICT infrastructure, availability, and accessibility of technologies; public potential (to develop the potential at the level of society, so as not to leave anyone behind and to overcome the digital divide (E-Government Survey, 2020), and, therefore, to increase the level of trust in government agencies and their officials).

Conclusion

Access to justice can only be ensured by integrating e-participation activities with regular tasks and processes rather than carrying them out when necessary as an exception. Moreover, all processes should be personalized with consideration of the needs of a particular participant.

The high cost of innovative technological solutions is a factor in the appearance of digital inequality. The only way to ensure accessibility is to implement digital solutions to expand access to justice within the framework of government programs and further integrate them into digital e-Government platforms.

However, existing services do not provide the level of communication level necessary for the equal and effective participation of persons with disabilities in the administration of justice at all stages. Interaction with law enforcement agencies requires online services providing real-time audio and visual contact. This requires the creation of a specialized portal meeting the criteria of inclusiveness.

References

Berners-Lee, T. (2018). Web Accessibility Initiative Home Page. Access date Дата обращения 10.06.2021. http://www.w3.org/WAI/.

Bogdanovskaya, Y. I., et al. (2009). The right to access information. Access to open information. Moscow: Justicinform (in Russ.).

Chernousov, I. (2021). Max will find it. An updated portal of public services is presented. Access date 12.06.2021. https://rg.ru/2021/03/31/predstavlena-beta-versiia-obnovlennogo-portala-gosuslug.html (in Russ.).

Dobransky, K., and Hargittai, E. (2006). The disability divide in Internet access and use. *Information, Communication & Society* 9 (3), 313–334.

E-Government Survey 2020. (2021). Digital government in the decade of action for sustainable development. Access date 10.06.2021. https://publicadministration.un.org/egovkb/Portals/egovkb/Documents/un/2020-Survey/2020%20UN%20E-Government%20Survey%20(Full%20Report).pdf.

Feedback Platform (FP). (2021). Access date 12.06.2021. https://digital.gov.ru/ru/activity/directions/1058/ (in Russ.).

Flynn, E. (2015). *Disabled Justice?: Access to Justice and the UN Convention on the Rights of Persons with Disabilities*. Surrey: Ashgate.

GOST R 52872–2019. (2019). National standard of the Russian Federation. Internet resources and other information presented in the electronic-digital form. Applications for stationary and mobile devices, other user interfaces. Accessibility requirements for people with disabilities and other persons with physical dysfunctions. Approved and put into effect by the Order of the Federal Agency for Technical Regulation and Metrology of August 29, 2019 No. 589-st. Moscow: Standardinform (in Russ.).

Jitaru, E., and Alexandru, A. (2008). Content accessibility of web documents. Principles and recommendations. *Informatica Economica* 46 (2), 117–124.

Kulkarni, M. (2018). Digital accessibility: Challenges and opportunities. *II MB Management Review* 31 (1), 91–98.

Revnivykh, A. V., and Fedotov, A. M. (2014). Availability of information system resources. Vestnik Novosibirskogo gosudarstvennogo universiteta. *Informacionnye tekhnologii* (Bulletin of Novosibirsk State University. Information Technologies) 1, 55–63 (in Russ.).

13

The Use of Digital Technology for Enhancing Proof in Forensic Document Examination

Kipouras Pavlos

CONTENTS

Introduction

The 2009 National Academy of Sciences publication *Strengthening Forensic Science in the United States*, known as "The 2009 NAS Report", asserted that opinions in handwriting examination were based too much on subjective analysis rather than objective, science-based analysis (NRC 2009). However, court rulings require the use of a scientific evidence-based approach for deduction and presentation of opinions of Forensic Document Examiners (FDEs; Wakshull 2019). Subjective analysis (subjectivity) is a common problem in forensic science, but in Forensic Document Examination the situation is much more complicated, being one of the hardest kinds of examination to objectify (Stoel et al. 2010: 4). Subjectivity can have a severe negative impact on the expert's result (Morris 2021: 152–153).

The use of digital technology in Forensic Document Examination is a prerequisite to performing *objective*, accurate measurements of handwritten and other items in the questioned and indisputably genuine documents, as well as processing these measurements with objective, user-independent software. *Quantifiability* (measurability), *highly controlled conditions*, *clearly defined terminology*, *reproducibility*, and finally *testability* and *predictability* are the five essential components – requirements that characterize scientifically rigorous studies, according to standard analytic epistemology (Chalmers 2013); these requirements in Forensic Document Examination can only be satisfied by the use of digital technology. We should not underestimate the value of qualitative, empirical analysis undertaken by FDEs, leaving the expert's role be reduced to that of a simple "machine operator" (Allen 2016: 5). However, courts worldwide demand reliable, objective evidence in order to issue a decision that the forgery of a signature and/or handwriting was committed by a specific person (Kipouras 2021).

Using digital technology, FDEs can "cure" the imperfections of their human visual system (HVS), which consists of the limits of a human visual perception along with multiple optical illusions intrinsic to HVS. Digital technology allows FDEs to "see" beyond the limitations of their HVS. Furthermore, questioned documents scanned with digital technology can then be studied without time limitations and without fear of damaging the original documents. Careful digitization of relevant documents (scanning with standardized settings), along with detailed knowledge of the software tools that are appropriate for the tasks to be undertaken, and improved awareness of the limits and imperfections of HVS are essential for a science-based approach for FDEs, enhancing the quality of evidential proof in courts.

Elements of Human Visual Perception and Visual Illusions Relevant to Forensic Document Examination

In this section, only basic topics are treated: the reader who seeks more information should consult the textbooks and articles cited.

The limits of human visual perception capacity have been studied in detail since the 19th century, and the relevant studies are cited in textbooks of physiology, as well as in digital image processing books (e.g., Hall et al. 2021; Gonzalez et al. 2017) and related journals (Sinha et al. 2020; Purves et al. 2011; Kurki et al. 2009; Hendley 1948). There are also textbooks on visual illusions (e.g., Shapiro et al. 2017).

A crucial question for FDEs is the minimum required distance (MRD) between two dots on a document – linear spatial resolution – so they can be seen as two distinct objects with the naked eye and not be erroneously perceived as a single object dot. The answer to this question depends on the distance from the image to the eye. The resolution limit of the eye (and other camera-like devices) is given in textbooks in angular terms. The linear spatial resolution can be quantified by the angle subtended between the two closest dots that can be seen apart. "Normal" human vision, rated as 20/20 by the Snellen eye chart used by ophthalmologists, has a spatial angular resolution of about 30 cycles of light/dark per degree (angle), thus requiring at least 60 pixels per degree. So, for the two dots to be seen as two separate objects by a "normal" person, the angle between these two dots has to be at least 1/60 of a degree.

To calculate linear resolution from angular resolution, we need to know the distance from the image to the eye and use basic trigonometry. Assuming that the smallest pixel seen by the "normal" person is 1/60 of a degree and considering that the tangent of 1/60 degree is 0.000291 (about 1/3440), we conclude that if the viewing distance is 3,440 times the pixel pitch or more, the pixels are no longer visible and the two dots cannot be seen by a "normal" person as separate objects. The average reading distance (ARD) is about 15 in (38.1 cm). For this reading distance of 15 in, the minimum required distance (MRD) between the two dots, to be seen by a "normal" person as distinct objects, must be such that the product of the equation "MRD times 3,440" will not be less than 15 in (38.1 cm). Therefore, the MRD for a "normal" person must be no less than 15/3440 in = 0.0044 in (38.1/3440 = 0.011 cm or 0.11 mm), for an ARD of 15 in (38.1 cm), as depicted in Figure 13.1.

A) Two black dots (magnified) in an image seen from a reading distance of 15 in (38.1 cm) are perceived by a "normal" person as two separate black objects, if the distance between them is equal to or more than the minimum required distance (MRD), which is about 0.11 mm for a "normal" person and for a reading distance of 15 in.

B) The distance between the two black dots (magnified) is less than the MRD for the reading distance of 15 in, and so they cannot be seen as separate objects by a "normal" human observer.

C) When the distance between two dots (magnified) is less than the MRD, they appear as a single object to the "normal" human observer.

There are persons who have better visual acuity than the "normal": the best human visual acuity measured is about 2.25 times better than the "normal". So documents must be scanned with at least 600 dpi, with pixel diameter 1/600 in = 0.00423 in or 0.01075 cm. However, the scanned documents can be magnified and thus FDEs can observe them better. Therefore, FDEs must scan the disputed documents with a scanner using at least 600 dpi and then use magnifying lenses or digital tools to magnify the scanned document.

Another important limitation of the HVS is the restriction on the perception of shades of gray. This fact, known only by scientists in the past, became common knowledge for many ordinary people due to the bestselling book *50 Shades of Gray* and its related movie. In fact, the average human perceives fewer than 50 shades of gray, between "pure" white and "pure" black. In contrast, digital technology tools can measure up to 4,096 gray scales or more.

Furthermore, HVS perceives the shade of gray, the light intensity, or absolute brightness of an area, not by its absolute value, but through the perception of the relative intensity of light compared to the background. Figure 13.2 depicts this phenomenon, which is an optical illusion.

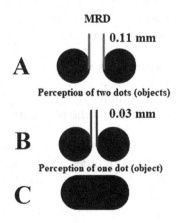

FIGURE 13.1
Visual demonstration of the MRD. Source: Author's creation.

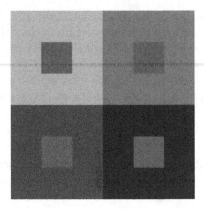

FIGURE 13.2
Different perceptions of the same shade of gray owing to different background contrast. Source: Author's creation.

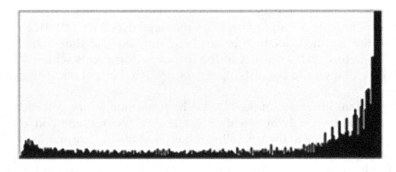

FIGURE 13.3
Histogram (from Adobe Photoshop). Source: Author's creation.

The small squares here are all exactly the same shade of gray, but they appear to have different gray-scale values (light intensity–absolute brightness).

All the inner small squares have the same absolute brightness–light intensity, but they are seen as progressively lighter as the bigger surrounding squares (background) become darker. This is a very impressive, well-known visual illusion due to an imperfection of the HVS. Digital technology tools do not make such mistakes: they measure the real, absolute brightness of an area in the image.

The absolute brightness of a shade of gray can be measured by commercially available computer programs for image processing (e.g., ACDSee, Adobe Photoshop, Capture One Pro, Corel AfterShot Pro, Fotor Photo Editor, GIMP, PhotoDirector, PicMonkey, Pixlr, etc.). Also, these programs can display and assess the histograms of shades of gray of an area in an image (Figure 13.3).

An image histogram is a type of histogram that is a graphical representation of the distribution of shades of gray, in absolute brightness, and in a digital image. The number of pixels for each tonal value is plotted. Examining the histogram for a specific image, we can judge the entire distribution of shades of gray at a glance.

Objective measurement of the properties of ink can be made by examining the histogram of an area of a scanned document containing this particular ink. Significantly different histograms for the ink in different areas of a disputed bank check mean the presence of different inks: this reveals the use of a different pen to produce a fraudulent check (Gorai et al. 2016), so the above is an objective method for proof of fraud in bank checks.

A second common visual illusion – a mistake made by the HVS – is the exaggeration of the length of vertical lines when compared with horizontal lines. This occurs in the *vertical–horizontal illusion*, where the two lines have exactly the same length, although the vertical line appears to be longer than the horizontal line (see Figure 13.4).

Observers overestimate the length of the vertical line, although the horizontal and vertical lines have exactly the same length.

This error does not occur with digital technology tools. In this example, measurement of the lines in scanned documents is accurate, while HVS perception is misleading.

A third visual illusion is the *Müller–Lyer illusion*, which consists of a set of arrow-like figures with parallel straight-line segments of equal length: observers perceive the straight-line segment in the middle as being longer than the other two, above and below it. The parallel straight-line segments comprise the "shafts" of the arrows, while shorter line segments (called the fins) protrude from the ends of each shaft. The fins can point inwards to form an arrow "head" or outwards to form an arrow "tail". The line segment forming the shaft of the arrow with two tails is perceived to be longer than that forming the shaft of the arrow with two heads (Figure 13.5).

The "shaft" of the arrow in the middle appears to be longer than the parallel "shafts" above and below of it, although they have exactly the same length.

A fourth visual illusion is the *Jastrow illusion*, which consists of two arches placed on top of each other. The arches are identical, but the lower arch appears to be longer, as depicted in Figure 13.6.

The lower arch appears to be longer, but the two arches have identical dimensions.

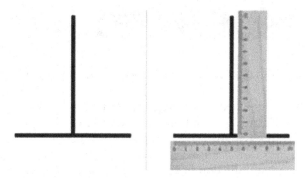

FIGURE 13.4
The vertical–horizontal illusion. Source: Author's creation.

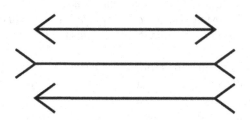

FIGURE 13.5
The Müller–Lyer illusion. Source: https://commons.wikimedia.org/wiki/File:M%C3%BCller-Lyer_illusion.svg.

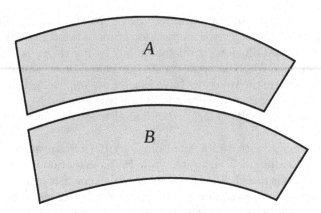

FIGURE 13.6
The Jastrow illusion (1). Source: https://commons.wikimedia.org/wiki/File:Jastrow_illusion.svg.

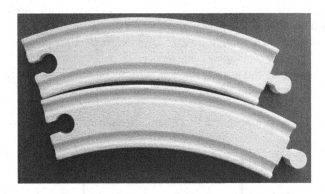

FIGURE 13.7
The Jastrow illusion (2). Source: https://commons.wikimedia.org/wiki/File:Jastrow-illusion-track.jpg.

This illusion also occurs when perceiving objects in the real world, not only figures in a paper document. In Figure 13.7, the human observer perceives the lower toy railway as being longer, although the two toy railway tracks are identical.

The two toy railway tracks are identical, but the lower one appears to be longer.

A fifth, impressive visual illusion is *Kanizsa's triangle* or the *subjective contours illusion*, which consists of falsely perceiving an edge, without a luminance or color change across that edge. Illusory differences in brightness of parts of the figure often accompany the illusory contours, as depicted in Figure 13.8.

An illusory (non-existent) contour of a bright white triangle is perceived, created by the mind of the observer by aligning Pac-Man-shaped parts of the figure. The illusory bright triangle seems rather brighter than the background, even though the various white parts of the figure have identical luminosity.

A sixth, well-known visual illusion is the *Ebbinghaus illusion* or *Titchener circles*. This is an optical illusion of false perception of two disks as having different diameters, while they are identical, as shown in Figure 13.9.

The two orange disks are identical, but the right circle appears larger.

A seventh visual illusion is the *Delboeuf illusion*. This is an optical illusion of false perception of the relative diameters of two black disks which are identical, but the disk

FIGURE 13.8
Kanizsa's triangle. Source: https://commons.wikimedia.org/wiki/File:Kanizsa_triangle.svg.

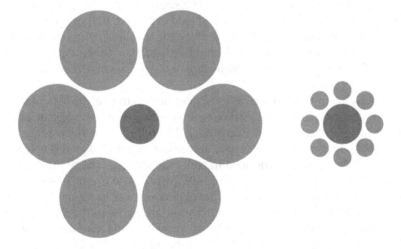

FIGURE 13.9
The Ebbinghaus illusion. Source: https://commons.wikimedia.org/wiki/File:Mond-vergleich.svg.

surrounded by a ring appears larger to the observer than the non-surrounded disc, as depicted in Figure 13.10.

The two black disks are identical, but the disk on the left surrounded by a ring appears larger than the non-surrounded disk on the right.

Knowing the above basic visual illusions, relevant for Forensic Document Examination, is essential for FDEs so that they can avoid false perceptions in the documents to be examined. Using digital tools for measurement of the dimensions and the (real) gray-scale values of figures, signatures, and/or letters in the documents is the "cure" for these visual illusions.

FIGURE 13.10
The Delboeuf illusion. Source: https://commons.wikimedia.org/wiki/File:Delboeuf_illusion.svg.

HVS imperfections also affect the *perception of color*s. There are many factors that affect color vision: the perception of the color of an area is not stable but depends on the properties of the neighboring areas (background), their luminance, etc. It is beyond the scope of this section to treat color vision in detail; the reader who seeks more information should consult the textbooks and articles cited.

FDE Evidence in Court

Before the presentation of digital technology tools, it is important to refer to the use of evidence in courts. In the case of expertise on FDE, the relevance of documents depends on their secure authorship (Huber 1999: ch. 2, p. 1). Expertise based on the ambiguous origins of comparative material cannot be considered safe evidence, first of all for the expert and much more for the judge. But, how do judges face the problem of evidence concerning the recognition of handwriting? What is the value they attribute to the expert's conclusion? Is it influenced by the general scientific community's consensus on the scientific rigor of this particular forensic branch?

Expert testimony in the courtroom is evaluated following several standards that have been established through different court decisions considered to be of crucial importance. The first point of reference derives from the 1923 Frye v. the United States (Frye) case, known as the *standard of general acceptance*, according to which any forensic technique or method used as evidence and introduced in court should meet the requirement of general acceptance by the relevant scientific community. In 1975, the *relevancy standard* was introduced, which required the qualification of the expert witness in terms of skill, knowledge, experience, training, and education. The third standard arose in the famous 1993 Daubert v. Merell Dow Pharmaceuticals case (Daubert), known as the *reliability standard*: this suggested falsifiability as a necessary condition for the admissibility of scientific expert testimony, with peer review, publication, and error rate as additional parameters (Zlotnick et al. 2001; NAS Report: 91).

FDE is the forensic branch that has been most often accused of lack of scientific method and subjectivity in the estimation of the evidence by the expert. The problem lies with the demonstrable reliability of the methods used within this expertise. For this reason, FDE needs to deepen the research on the validation of theory, method, and techniques of identification in order to ensure the objectivity and scientific nature of the sector, thus minimizing the problem of acceptance by judges (Page et al. 2011a, 2011b). Poor documentation of the analytical process, along with the lack of established and standardized procedures, remains the key to the acceptance of expertise. At the same time, the subjective nature of observation biases and the lack of control standards for exclusion of them diminish the guarantee of objectivity (Varshney, Bedi, 2019 : 19–23). Nevertheless, the bias factor (Sulner 2018 : 656–663) remains as far as the judgment of the court is concerned. Judges remain humans, and they can also be affected by several biases although they are experienced in the estimation of the exhibits and evidence of the case (Gigerenzer, Brighton, 2009 : 107–143, Mara Merlino, 2015 : 11). That does not mean that they can exclude completely the negative effects of biases. On the other hand, forensic experts are also experienced in their sector. Since we are comparing the subjectivity of judgment by experienced professionals (both judges and forensic experts), the danger of biases remains in both cases.

Is there any objective method or reliable technique according to which judges follow a particular procedure in order to decide, apart from the law's provisions? In an analogous way, there is a kind of abstract procedure that judges follow when articulating thoughts, estimating the reliability of exhibits and the application of law or jurisprudential rules, and evaluating evidence and testimonies. Is there a particular procedure or method of evaluation that a judge should follow? Is there any way of controlling the objectivity of a judge's decision? Is it impossible for two different judges to arrive at diametrically opposed decisions when examining the same case file? Are they superhumans with perfect and absolutely objective judgment? Has any research, regarding the objectivity of judges' judgments, indicated their "immunity" to the influence of biases? Are only forensic experts exposed to subjectivity and biases (Mussweiler 1999)? Are only forensic experts obliged to follow certain documented and commonly approved methods and techniques? Has there been any application of analogous standards deriving from the *Frey* or *Daubert* or *Kumho* cases and the subsequent theoretical discussions, achievements, and conclusions in this direction regarding the standardization of the logical and methodological procedure judges should follow? Are their reliability and objectivity technically demonstrable and commonly accepted in any way? Hiding beyond the mentality "judges cannot be judged" does allow the evolution of the humanistic sciences. Every kind of decision making, no matter where it derives from, is a challenge that should be exposed to criticism and judgment. The forensic expert is practically following the same procedure. He is a kind of technical or scientific judge, a forensic judge applying the same exact procedure in making evaluations and estimations, giving priority to one element over the other, interpreting the evidence according to his experience, knowledge, and, of course, by following the ultimate achievements and evolution of his scientific area, as he is obliged to do.

Returning to the aforementioned mentality regarding the application of standards of measurements of positive sciences in humanistic sciences, we have to repeat and underline that this is not always possible. This does not mean that forensic sciences should be excluded by default from criticism. I personally believe that every expert, independently of his sector, has his own range of scientific value, which is reflected in the quality of his investigative scrutiny and in the logical articulation of the evidence – based, of course, upon the methodological rules and mentality of his sector – in order to indicate and prove his conclusion. It is the same mental procedure that judges follow in order to decide. The legal mentality regarding the necessity of undeniable proof in order to reach the final verdict is completely understood and should be absolutely respected by all professionals dealing with the application of law and justice. The responsibility of judges is much heavier than that of any other professional involved in a trial. On the other hand, the judge has always the alternative of estimating much more evidence and exhibits than those analyzed by a forensic expert. In any case, he also has the possibility of cross-examination of evidence, which may reinforce his conviction or help him eliminate false impressions that lawyers, experts, or testimonies are voluntarily or not creating in order to deceive him. The judge's judgment is absolutely of utmost importance, but it is based on various and polysemous parameters. For this reason, there is the Latin apothegm *iudex peritus peritorum*, which means practically that the judge is the judge of the judges, considering the partial experts as minor technical–scientific judges within their particular areas and that the judge is the one above all who will decide – if necessary, even ignoring the experts' conclusions. For this reason, the legislations of many European countries (e.g., Italy and Greece) establish that the judge is not obliged to follow the expert's conclusion, even when this expert is nominated by the court, but he is free to decide – even to adopt in his decision the conclusions of just one expert out of many.

Apart from the jurisdictional and empirical requisites for decision making, the judge who has to decide upon handwriting analysis would do better to insist on an evidence line-up (Stoel et al. 2010: ch. 5, p. 1). Because the FDE's expertise is essential in practice, apart from following the law and doctrine's point of reference regarding the estimation of the expert's conclusion, judges should follow their common sense and the logical concatenation of evidence by the expert. Many of the exclusions of expert testimony result from deficits in the reasoning process and the logical (or not) correlation between methods, findings, and the facts of the case at hand (Joe 2005: 74; Shelton 2011: 17; Found et al. 2016: 59). This assessment can also indicate the reliability of the expertise (Federal Rule of Evidence 702, Linton Mohammed 2019 :131)).

The admissibility of forensic evidence by experts is of great importance because jurors are highly influenced by expert reports in their ultimate decision. Modern juries pay great attention to forensic evidence – possibly, owing to the increasing popularity of TV crime series. Such a tendency is commonly known as the "CSI effect". We have to point out, though, that jurors can also be influenced in their decisions by biases as well as by effects such as the "CSI effect". In addition, they do not have much experience in dealing with facts and testimonies, unlike judges. For instance, according to studies, they perceive DNA analysis as being almost absolutely accurate – considering it especially credible and reliable (NAS Report 2009, pp. 100–101). Practice suggests that judges share a similar opinion about DNA and tend to give priority to DNA evidence when there is conflicting forensic evidence. Modern jurisdictional procedures (Haack 2014: 4–5) of evidence are practically based on the degree of demonstration of proof. If this has reached the legally required point of reference, then the evidence is accepted as indicative of guilt or innocence and is sufficient to justify the verdict. But, what happens in the case of the incongruity of forensic conclusions from different specialties? Apart from a general understanding of the level of development of every forensic sector, judges should follow their common sense. Here, both judges and jurors are required to evaluate the expert's testimony (Haack 2014: 23). The problem arises from the use of specialized techniques and vocabulary, which demand special knowledge, comprehensible only to scientists with that expertise. Therefore, experts should try to explain quite complex matters in the simplest way: this is not always possible, even though the expert might be experienced in translating and condensing scientific proof in a simplified manner.

As stated in the NAS Report (NAS 2009: 87), the recognition of a scientific basis for a particular forensic method defines the reliability of evidence, mostly in criminal cases. The admission and reliance on a forensic report should be based on two requisites: (a) the scientific reliability of the methodology, which permits the accurate analysis and reporting of data; and (b) the parameters of control of this reliability by excluding individual interpretation, biases, weak protocols, and low standards of performance. Since law enforcement requires forensic contribution, the guarantee of scientific rigor is crucial. The basic principles of science that should characterize even the forensic sciences are the precise definition and repeatability of the methodological procedure; the identification of eventual sources of error, measurements, and their limits of inaccuracy; and validation studies regarding the percentages of false positives or negatives (National Research Council 2009: 121–122).

Additionally, we have to mention the importance of biases and the development of methods for avoiding them as much as possible, given the human nature of the scientists. These standards can improve the reliability of the forensic conclusion, which is not taken for granted by all forensic branches. In fact, the progress of forensic sciences has been boosted in the last century – and especially in recent decades – through the rapid evolution of technology. This has provided scientific equipment, increasingly more sophisticated and precise, enabling more profound investigations. Although some forensic sectors have been accepted as evidence in the trial – among them, even FDE (Huber 1999) – the "stain" of art rather than science remains in some cases. For this reason, the 2009 NAS report (p. 167) arrives at a conclusion that *"the committee agrees that there may be some value in handwriting analysis"* [my italics]. Of course, since then there has been a continuous evolution and a vast application of new methods and instruments, which have strengthened the scientific aspects of FDE. Digitalization, apart from becoming more accessible through declining cost, has accelerated in most of the countries of the world, even in less developed ones.

Digital Technology in FDE

Digitalization and Use of Software

Digitalization has empowered the search for truth in most of the diverse branches of forensic sciences. FDE is, fortunately, one of the most favored. Standardized scanning and processing of documents are essential, using easily available commercial software. It is crucial that scanned questioned documents can then be studied without time limitations and without fear of damaging the original documents. Unfortunately, indentations in the original paper document are lost as information in the scan. Therefore, the original document has to be inspected for indentations: There is a sophisticated technique to depict the indentations in the original document, using "laser profilometry performed in conoscopic holography" and other 3D imaging techniques (Dellavalle 2011, Dellavalle, Frontini 2020). Before referring to the technological aspect, we have to consider the problem of the subjective approach of an expert, according to not only his own personal sensorial conception but even educational preparation and professional experience, which could eventually mislead. We also have to take into consideration the lack of training in handling technological equipment which can distort results.

An important example of digitalization of a document, and subsequent analysis using digital tool technology and computer programs, is the following: Teulings et al. in 1993 used digitalization of handwritten documents for processing them in their digital form, which was indispensable for their study. The results of this study showed that the degree of similarity of strokes in handwriting is not the same for all strokes. Down strokes were more consistent than upstrokes in terms of vertical stroke size. However, unlike the vertical stroke size, the horizontal stroke size was not invariant. Both vertical and horizontal sizes showed substantial between-stroke correlations (Teulings et al. 1993). This was, and remains, an extremely important scientific discovery concerning the variation of the handwriting of the same individual (intra-writer variation).

Kipouras et al. (2021) have commented on the above findings:

We propose that the explanation for this inequality of similarity of strokes is based on the following:

a) The principal muscle generating horizontal strokes in handwriting is the extensor carpi ulnaris (ECU), a muscle of the forearm which shifts the right wrist so that the fingers bend toward the ulna bone. The ECU is controlled by a smaller area in the motor cortex than the intrinsic muscles of the hand, as shown in the "motor homunculus". Down strokes are generated mainly by muscles moving the thumb and the index finger holding the pen, and secondarily by muscles moving the wrist. The muscles moving the fingers are more controllable, as they are controlled by a bigger area from the motor cortex, as depicted in the motor homunculus.

b) Among the muscles contributing to down strokes are the flexors of the wrist and of the fingers, which are stronger, dominant and more controllable than the extensors. The above explanations are based on study of many relevant articles and books dedicated to anatomy and kinesiology of muscles (e.g. Kapandji, 2010).

Further analysis of online digitized handwritten items (handwriting research) is being performed by Teulings et al. and their coworkers from the International Graphonomics Society (https://graphonomics.net/) with the use of the specific software MoveAnalyzeR by Neurosript, which is used in 62 countries.

There are commercially available software programs with measuring tools for digitized documents – for example, measuring the distance between two dots, calculating an area in the document, etc. Optical character recognition (OCR) software applications are also used in FDE (e.g., Adobe Acrobat Pro OCR tools, Tesseract, Abbyy FineReader).

There are multiple software applications for signature verification. This can be done "offline", after the individual has signed. VerSign is an offline signature verification system which can be used to verify signatures on bank checks. SignatureXpert® by Parascript is used on many occasions requiring signature authentication, such as voting by mail. "Online" signature verification is performed when the individual signs on a special tablet where a software application is running (e.g., TechSign Biometric Signature verification with Samsung Tablet).

Artificial intelligence (AI) software for forensic document analysis helps FDEs and is a tool presented in courts; it enhances the objectivity of forensic results as it is independent of the user of the software (e.g., CedarFox, FISH, FlashID, Write-On, WANDA, Masquerade et al.). To date, the existing software solutions are helpful, but they cannot yet replace skillful and experienced FDEs. Software also exists for the reconstruction of shredded, torn, or fragmented documents, and software engineers can use the ideas and algorithms of software developed for the reconstruction of damaged archeological discoveries – torn to pieces, such as the reconstruction of the famous 1650 BC wall paintings [frescos] on Thera Island, Greece, developed by Prof. Papaodysseus and his coworkers (Papaodysseus et al. 2008). There are also "graphometry" tablets which run software applications analyzing the characteristics of signatures "written" on the writing surface of these tablets (e.g., Namiria Firma Certa, PRB, e-graphing, Euronovate).

The aim of this section is not the creation of an exhaustive catalog of technological equipment, methods, or software that can improve the efficacy of forensic demonstration. That would likely be unsuccessful, given the always-expanding technological innovations throughout the world. Our aim is, instead, focused on suggesting a way of taking advantage of digitalization to improve the objectivity of evidence: in the past, evidence could be based only on theoretical conceptions which remained as subjective interpretations without any possibility of certain proof.

Equipment and Methods in FDE

Digitalization technology has to be adapted to the standard equipment and methods of FDE. FDE was originally based on *"ictu oculi"* inspection of the handwriting. Later, magnifiers and photographic cameras were used, in order to capture the exact condition of the document and handwriting; this even included the use of specific sources of lighting through different angles or sides of the document. Methods soon emerged for the identification of paper quality and different kinds of pens, even monitoring the continuously changing chemicals used in various inks. The next step was the macroscopic and microscopic determination of documents (Nickel 1996: 127–145).

Scanning of disputed documents and the use of digital technology analysis of scanned documents is indispensable, *sine qua non*, in Forensic Document Examination in the 21st century. Scanners are also simple to use and easy to handle, creating a reproduction of the examined document quite precisely, although dependent on technical adjustments by the operator for the accuracy of image capturing (Kelly 2021: 248, 276; Dellavalle 2011: 133).

Forensic Document Examiners must follow the requirements of the SWGDOC Standard for Use of Image Capture and Storage Technology in Forensic Document Examination (available at https://www.swgdoc.org/index.php/standards/published-standards). Also, they must take into account the ASTM E2825-19, Standard Guide for Forensic Digital Image Processing (ASTM E2825-19, 2019).

Disputed documents (such as wills (testaments), bank checks, etc.) are usually kept and stored in facilities owned by Ministries of Justice, so Forensic Document Examiners have limited time – just minutes – to examine the original disputed documents. Hence, they need to scan the documents to have enough time to examine them in their offices using visual perception, and also to use tools of common, commercially available, or free Office programs (e.g., Microsoft Word, Libre Office) and use digital image processing tools, OCR software, and various software applications.

Macroscopic examination refers to reflected, oblique (Allen 2016: 206), or transmitted light to the document so as to make visible particularities, which cannot always be seen by the naked eye. While reflected light is the most usual method of examining a document, oblique light (Ellen et al. 2018: 186) – i.e., side light at a low angle striking the surface of the document – is a way of making more visible the indentation in the paper created by the pen. This technique can reveal possible erasures or other irregularities of the usual embossments of the paper caused by pressure exerted through the graphic instrument (i.e., the type of pen or pencil used). Transmitted light involves illuminating the back side of the document, which permits the identification of possible erasures, watermarks, or other alterations of the paper's surface or ink.

Microscopic examination (Nickel 1996: 145–152; Belensky 2017: 53) gives the expert an opportunity to investigate beyond the limits of human vision using high magnification. Of course, exaggerated degrees of magnification may be misleading, since we are obliged to conceive the particular point of a trace in the whole frame of the graphic environment. The expert should always have the critical ability to adjust his inquiry on either focusing on deep analysis or seeing the totality of the document. He must have a mentality analogous to the lens zoom function by analogy. He should understand when to focus (zoom

+) or move in the opposite direction (zoom –), so as to have an investigative mentality able to contextualize detailed revelations holistically. The most common tools of magnification are magnifiers and loops, but the stereoscopic microscope is much stronger and very useful for detecting the characteristics of the pen and the particularities of the traces. Although stereoscopic microscopes (Belensky 2017: 54–55) are very reliable, their mass, weight, and dimensions limit their use in the laboratory. The modern digital microscopes connected to portable computers or operating autonomously are more comfortable and allow captures of the particular parts of the traces in digital form, which can be easily elaborated by image processing software. Some modern digital microscopes even possess macroscopic functions.

The most common and direct means of showing evidence has always been photography (Nickel 1996: 152). Either in its older film images or in modern digital form, photography has continuously evolved, which has given great possibilities of demonstration and proof to forensic experts. Digital photographs (Kelly et al. 2021: 247) are also more easily elaborated and have become a reliable and strong means of proof, giving direct and easily understood information about the document. Sensitive sensors, lenses, and hardware or cutting-edge technology facilitate the work of experts, even in conditions of low lighting, since very often the disputed documents are deposited in different locations with difficulty in examining them.

The next conquest of FDE in its historical evolution has been the application of spectral techniques (Nickel 1996: 155–166; Belensky 2017: 57–58) or spectroscopy (Allen 2016: 171–173). It consists of an analysis of ultraviolet (almost 200–400 Nm) or infrared (almost 700–1000 Nm) frequencies that lie at the opposite edges of the human's visible spectrum (or VIS, almost 400–700 Nm) by directing light of a particular frequency to the paper in order to identify the degree of absorption or reflection. This can provide evidence regarding the paper's surface or the ink's components and also indicate alterations or additions (Ellen et al. 2018: 122). Video spectral comparator (VSC) is also very useful equipment for FDEs, since it applies the above techniques of the non-visible spectrum and different sources of lighting for examination of the document, giving evidence of utmost importance to the expert. The digital form of the files deriving from photographing all the findings can be easily included in the expert's report in particular word processor software in common use (Microsoft Word, Libre Office, etc.). Digital images can be elaborated in software for image editings, such as Adobe Photoshop and other programs.

The electrostatic detection apparatus (ESDA) is used for revealing latent impressions on the surface of documents (Belensky 2017: 56; Ellen et al. 2018: 186–191). Although it is not a digital procedure, it can give useful forensic information. There are other specific methods of laboratory examination that can be very helpful to FDE, partially digital or not digital at all. Some of them have retained their pre-digital conceptualization but use new, digital procedures. The importance of showing findings remains the same, but the elaboration of data is easier to demonstrate and include in the expert's report.

Additionally, there are many special kinds of inquiries regarding specific cases of investigation. These include analysis of inks, investigation of crossed lines regarding the sequence of tracing lines, chromatography and Raman spectroscopy (Allen 2016: 166–175), chemical analysis of paper (Allen 2016: 180–186), the dating of ink, chemical analysis of inks, and numerous other destructive or non-destructive methods that can provide more objective evidence. However, we are talking about more objective – rather than completely objective –evidence, because every method or technology has its own limits of accuracy and statistical error. These limitations return within the frame of legal requirements, as derived from the Daubert case. Francesco Dellavalle (2011: 18) wrote an extended and detailed reference to such forms of inquiry, which demand laboratory equipment and specialized methods. Analysis of this kind provides FDEs with further objective data of demonstration and proof.

The most revolutionary achievement of the last decades, in the author's opinion, is the use of biometrics in examining handwriting. The study of the dynamic aspects of writing has become possible by analyzing its characteristics in real time. In the past, the expert had to identify the characteristics of handwriting using the fixed image of a written document, trying to identify dynamic elements using various theoretical rules. Being present at the moment of a document's production in order to observe several phenomena was quite impossible. Graphobiometry or graphometry in its modern form is another advantage granted to today's experts. Although the expert may have to examine a static handwritten document, graphometric software used on persons relevant to the case can give further diagnostic tools. In particular, pen pressure is perhaps the most important characteristic that can be revealed by measurement (Pugnaloni et al. 2013), along with velocity and rhythmic variation. When examining signatures already traced in digital form (Harralson, 2013 : 56–59) – current practice in developed countries – direct comparison of the samples may provide more objective evidence, taking into consideration the technical limitations of the relevant software and hardware. Such software interprets graphics and measures the potential graphic characteristics of handwriting: this provides the FDE with the opportunity to establish a stable scientific basis, thus enhancing the reputation of Forensic Document Examination within the forensic sciences (Mazzolini et al. 2021).

Digital Technology in Forensic Document Examination: Correlating Handwriting to Muscle Contractions and Electromyogram (EMG) – Future Perspectives

A major future perspective is the use of digital technology to correlate the action of muscles of the upper limb to handwritten letters and signatures, understanding the action of these muscles, and as a way to discriminate genuine documents from forgeries, through kinesiology studies of the handwritten items. The importance of the correlation of the action of muscles (kinesiology) with handwriting cannot be overemphasized: This was first noted in 1894 by Hagan in his famous book *Disputed Documents* (Hagan 1894). However, this is a difficult task, since there are, in total, 43 muscles which participate to a greater or lesser degree in the act of handwriting – as shown by electromyogram (EMG) studies (Chihi et al. 2020; Derbel 2020: 71; Mahmoud 2020, etc.). There are very few published articles on this fundamental topic. For every stroke in handwriting, there are some main muscles which generate it, and the remaining muscles stabilize the upper limb; even today the kinesiology analysis of handwriting is quite difficult but essential. Moreover, until 1970, there was no book explaining in sufficient detail the actions of the muscles of the upper limb: this changed with the first edition of Kapandji's *The Physiology of the Joints* in 1970.

Despite these difficulties, researchers in artificial intelligence – with little knowledge of anatomy – understood that clockwise strokes are generated with different muscles compared to counter-clockwise strokes (Sayre 1965). An additional problem for researchers was that EMG was performed using needle electrodes: this was, at a minimum, an annoying and sometimes painful experience for the individuals being examined. Subsequently, surface electrodes applied non-invasively to the skin of the subject were invented and then used extensively in practice for the last two decades or so. Consequently, there are few (<20) articles in PubMed concerning the correlation of handwriting with the action of the upper limb muscles of *healthy individuals* using surface EMG electrodes.

The author referred to articles (Kipouras 2021):

> on the reconstruction of handwriting and other meaningful arm and hand move-
> ments from surface electromyography (sEMG) signals. Using computer algorithms, the
> researchers solved the inverse problem – that is, deducing which specific letter or figure
> was written, by examining the recorded sEMG signals of the relevant muscles of the
> hand or forearm.
>
> **(Chihi et al. 2020; Mahmoud et al. 2020; Abdelkrim 2019; Okorokova et al. 2015;**
> **Chen et al. 2017; Huang et al. 2010)**

Based on the above EMG studies, research in robotics (e.g., Balasubramanian 2014, etc.) and
books and articles dedicated to the actions of muscles (kinesiology), in a recent study by
Kipouras et al. (2021) propose that:

> there is a unique sequence of muscle movements required to produce a specific signa-
> ture or handwriting (…); this is supported by recent research on the reconstruction of
> handwriting and other meaningful arm and hand movements from surface electro-
> myography (sEMG) signals. Using computer algorithms, researchers solved the inverse
> problem – that is, deducing which specific letter or figure was written, by examining the
> recorded sEMG signals of the relevant muscles of the hand or forearm. Four facts lead
> to this conclusion: a) that the forearm rests on the table, b) that the pen is held in a "tri-
> pod" manner, c) that the tip of the pen (more generally, the tip of the medium used for
> writing) moves in a plane, and d) that the signature or handwriting must be produced
> by moving the pen along a specific trajectory, the same as followed by the other person.
> These four facts restrict the mobility of the upper extremity. Applying standard knowl-
> edge of the anatomy, physiology, and kinesiology of muscles and joints, the conclusion
> is that there is only one sequence of specific muscles contracting that can produce a par-
> ticular signature or handwriting. Only rarely is there an equivalent sequence of muscle
> movements that can accomplish this task.

This correlation, of muscle actions to handwriting, is made using digital technology to
analyze handwritten items and with specific software. The digitized handwritten items
are segmented according to the ballistic hand movements called "strokes", and the strokes
are correlated to specific muscle actions using sEMG signals. In the future, lists will be
created containing the strokes needed to write all the letters of the alphabet and all the
numbers from 0 to 10, and the corresponding muscles which act to create the required
strokes together with the corresponding sEMG signal. The discrimination of the genuine from forged handwritten documents will be based on the kinesiology analysis of
handwritten items, and a method to perform it is by using sEMG signals and digital
technology.

In the same article (Kipouras et al. 2021), the authors also propose:

> a new test for the verification of authenticity of questioned signatures and/or hand-
> writing – the "proprioception test". This test consists of holding a pencil in a "tripod"
> grip and following the trajectory of a photocopy of the questioned signature, and then
> doing the same for the genuine ones. In so doing, we can "feel", by proprioception, if the
> sequence of muscle contractions is different in the suspect case, compared to the genu-
> ine signatures. The ability of human proprioception of very fine movements is indisput-
> ably proved by the performance of very difficult tasks innumerable times worldwide
> somatic cell nuclear transfer (SCNT) and rice writing.

This test can also be performed by the judge in a court. The proprioception test can be used in the future, combined with the use of special graphometry tablets – with special touch, pressure, and velocity sensors, and analyzing online graphometric signatures with specific software (e.g., Euronovate, e-graphing, PRB, Namirial Firma Certa, inter alia). The graphometric signature is the electronic signature in which the signature path is captured, using a tablet, and added to the electronic document. A paper photocopy of a disputed signature could be attached to the writing area of the tablet, and the FDE, using the special pen for these tablets, could trace the trajectory of the disputed signature along the photocopy. The same procedure could be repeated for a photocopy of a genuine signature. The calculations performed by the software for each tracing (questioned and genuine signature) will appear on the tablet screen (this test can be called the "tracing signature photocopy tablet test"). These calculations can be correlated with the different "feelings" of muscle motions perceived from the "proprioception test" of the photocopies of the signatures and also to sEMG signals from surface electrodes applied to hand and forearm muscles.

An example of kinesiology analysis of disputed and genuine signatures is shown below from a recent article by the author Kipouras (2021):

As shown in Figure 13.11, the initial part (on the left) of the questioned signature X begins with a stroke in the clockwise direction, in contrast with the initial part of genuine signatures which begin with an anti-clockwise direction. The forger could not suppress his motor habit and began the questioned signature with a stroke with the opposite direction from the initial stroke in genuine signatures. Also, the final part of the questioned signature X is generated with an obviously different sequence of movements, as the final part of the signature X is a stroke with direction away from the center of the right wrist, while the final part in genuine signatures is a stroke towards the center of the right wrist. Holding a pencil in a "tripod" grip and following the trajectory of a photocopy of the questioned signature X, and doing then the same for the genuine ones, we can "feel", by proprioception that the sequence of muscle contractions is very different in the suspect case, compared to the genuine signatures. Examining further the middle part of the suspect signature, which is a simple squiggle with few particular elements, we observe that the forger did not succeed in his effort to simulate the same characteristics (such as the different proportions in the dimensions of the loops in the middle part of the squiggle, and the different distances between them). Moreover, the axes of the loops are almost parallel in the genuine signatures, whereas in the suspect signature the axes of the loops converge in the upper part: this needs a different sequence of muscle contractions than was used in the genuine signatures.

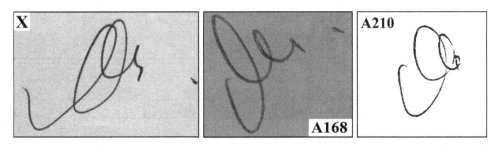

FIGURE 13.11
Suspect signature (X), and two genuine specimens. Source: Author's file.

The source of the above figure is the author's archive. In the above example, apart from the proprioception test, the aforementioned "tracing signature photocopy tablet test" can be performed, and the different calculations generated by the software for the disputed and the genuine signatures can be compared, proving that the disputed signature X is not a genuine one.

Future perspectives of applications of digital technology in Forensic Document Examination include improved digitalization, the development of better image processing and measuring tools, better OCR software, more robust AI software programs for document authentication, etc. These developments may require the collaboration of FDEs with computer engineers.

A Practical Case Study

The aforementioned parameters are shown in practice in the following case. We have to underline that digitalization allows and assures the support of the judge by turning personal theoretical judgments into demonstrable proof. The measurements have been obtained using the software *Firma Certa Forensic* of Namirial. The Greek language does not affect the principle of the case, since the important factor is the method of analysis and not the alphabetical script of the letters.

Historical Data of the Case

The disputed document is a receipt handwritten by the payer. Since the recipient did not take a copy of the document, the payer has written the whole text of the receipt, leaving enough space at particular points in order to add subsequent numbers that prove the payment of a much larger amount of money. How can digital technology prove this, since we have to hand only a photocopy of the document delivered in the case file? Moreover, other kinds of analysis which can be carried out only on the original are impossible, since the payer claims that the disputed document is lost. The part of the disputed document that is the main object of forensic interest is shown in Figure 13.12:

Note: Names have been erased leaving only the initial and final letters

It is obvious – even from the photocopy – that the last five words of the last row 12 present more intense traces. The photocopy has been made by one and only one reproduction of the original since the photocopier uses equally in all aspects of the document with the same parameters. The difference between the original and the photocopy is given by different scales of gray (Figure 13.13).

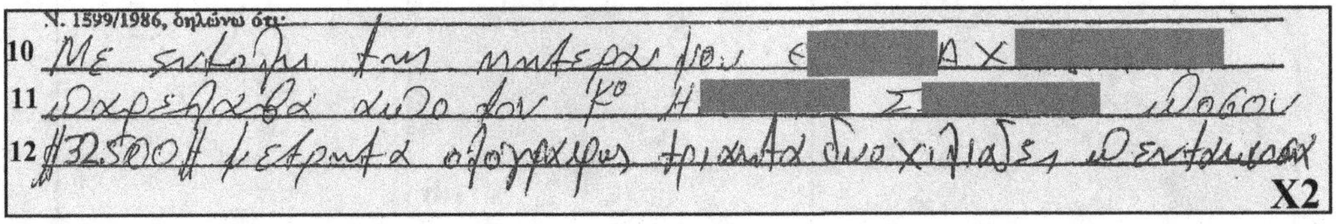

FIGURE 13.12
The disputed document. Source: Author's file.

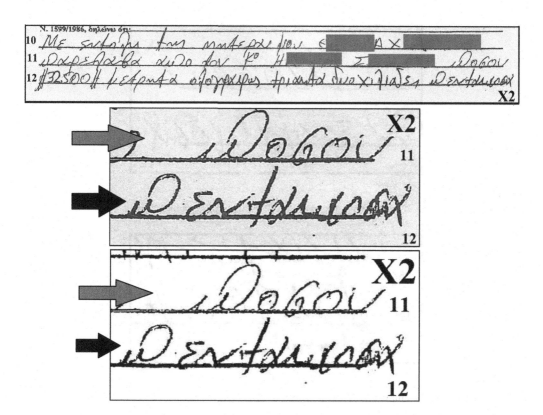

FIGURE 13.13
Different scales of gray of the traces in the last words of rows 11 and 12 (color scan in the first image; black and white scan in the second). Source: Author's file.

This phenomenon suggests the use of another pen in the added part or a different momentum of the writer using the same pen but in conditions of increased pen pressure. Different textures and the hardness of the writing surfaces could also be another reason. Figure 13.14 shows this with specific characters.

The same characteristic is found in the numerical reference to the amount, which means that we find common characteristics in the presumed added points of the text (Figure 13.15).

The text can be considered in two sectors, determined by their meaning. The first sector (called part A) goes up to the word "*μετρητα* " (meaning "in cash") of row 12: this sector is complete as a meaning. The concluding five words (called part B) contain a written reference to the amount as *ολογράφως* (written in full) *τριάντα* (thirty) *δ ύ ο* (two) *χιλιάδες* (thousand) *πεντακόσια* (five hundred) – an autonomous sector that confirms the amount written numerically in Part A. The scenario to be examined is that of an intentional blank space left before the number "2" in the reference "#2.500#" at first writing, in order to add subsequently the number "3" in front, changing it to "#32.500#". In order to complete this falsification, part B also required the presence of the written amount in the form *ολογράφως τριάντα δύο χιλιάδες πεντακόσια* – as opposed to what should have appeared on the first writing of the receipt, namely *ο λογράφως δύο χιλιάδες πεντακόσια*. How could this hypothesis of textual alterations be tested and confirmed?

Graphic variability of every individual is one of the most traditional theoretical achievements of handwriting analysis (Cristofanelli et al. 2004; Bravo 2005, pp. 33, 231; Welsh et al. 2014; D.Ellen et al. 2018, pp. 11–26). Nevertheless, this variability is highly individual

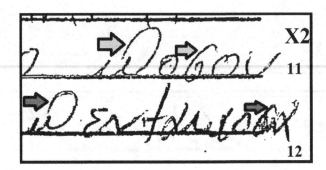

FIGURE 13.14
Lighter traces in row 11 and more intense traces in row 12. Source: Author's file.

FIGURE 13.15
More intense traces in the number "32". Source: Author's file.

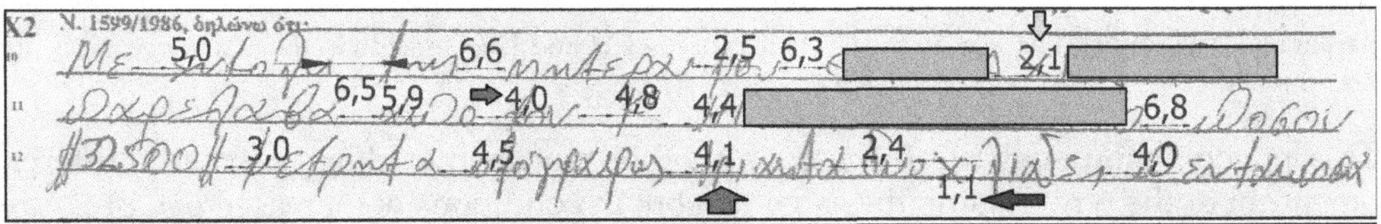

ROW	MAXIMUM (mm)	MINIMUM (mm)
10	6,6 («*της*» – «*μητερας*»)	**2,1** («*E......*» – «*X......*»)
11	6,8 («*.....O*» - «*ποσον*»)	4,0 («*από - τον*»)
12 (after the word "*ολογράφως*")	4,1 («*ολογραφως*» - *τριαντα*»)	1,1 («*δυο*» – «*χιλιαδες*»)

FIGURE 13.16
Maximum and minimum distances between words. Source: Author's file.

and refers, among other characteristics, even to the spaces left between consecutive words. Visual feedback regarding spacing can influence word spacing. Measuring the distance between words in the above-disputed documents presents a range of variation (in mm), as shown in Figure 13.16.

It is obvious that the maximum distance between words in part B is 4.1 mm, which is almost the minimum of the immediately superior 11th row (4.0 mm). The minimum value of the text is located in the investigated part B and amounts to 1.1 mm, which is about 50% less than the next lower value of the whole text (2.1 mm in the tenth row). The average value of 4.5 mm in part A is maintained at the same levels at the beginning of part B (4.1 mm between "ολογραφως" and "τριαντα" in row 12). The writer, being aware that the available space until the end of the row is not enough, gradually diminishes the distances, giving a value of 2.4 mm (between "τριαντα" and "δυο") and the minimum text's value of 1.1 mm (between "two" and "χιλιαδες"). Then, since there remains only one word to be added ("πεντακοσια"), the writer returns in the initial part of the word to his normal value (4.0 mm). Due to writing the letters in a reduced space, since he has to trace all the word in this row, he again reduces the inter-letter space in the final part of the word, even by tracing letters overlapping each other. The space under the last row 12 had been initially deleted by crossed lines, so that the space could not be used for writing (Figure 13.17).

The first three letters ("πεν") occupy 14.1 mm of space (average 4.7 mm), while the last seven ones ("τακοσια") 18.0 mm (average 2.57 mm). The same conclusion derives from the number #32.500#. We notice the following differences:

(1) Different distance between the initial and final symbol "#" with respect to the first number "3" and final number "0" (Figure 13.18).

(2) Different distances between "#" and initially the first number "2'" (2.2 mm) and final number "0" (0.9 mm). The dimension of the added "3" and "2" is almost identical (2.8 mm–2.9 mm) (Figure 13.19).

(3) Since the blank space was 2.2 mm, the numbers "3" and "2" are necessarily slightly overlapping, which does not happen in any other combination of numbers (see Figure 13.20).

The combination of all these findings confirms the alteration of the disputed document subsequent to its original writing. The handwritten addition of the five words in the last row justifies the transformation of "2,500" to "32,500".

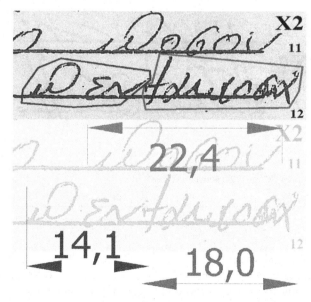

FIGURE 13.17
Normal inter-letter spacing vs diminished spacing. Source: Author's file.

FIGURE 13.18
Different distances between "#" and initial–final number. Source: Author's file.

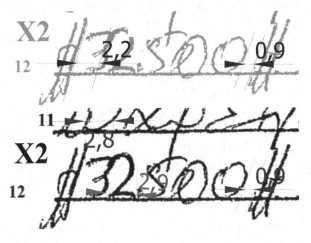

FIGURE 13.19
Distances between "#" and initials (A-part "2", B-part "3") and final number. Source: Author's file.

FIGURE 13.20
Slightly overlapping numbers "3" and "2". Source: Author's file.

Conclusions

Until now, artificial intelligence (AI) software cannot replace FDEs. However, AI software is very helpful, and generally digital technology tools are indispensable for Forensic Document Examination to perform *objective*, accurate measurements of the handwritten and other items in the questioned and the indisputably genuine documents, as well as processing these measurements with objective, user-independent software. Thus, the use of digital technology in FDE enhances proof in courts. Moreover, scanned documents can be studied without time limitations and without fear of damaging the original documents. The human visual system (HVS) has a limited ability to perceive small details, also limited ability to perceive gray-scale differences, and has multiple intrinsic optical illusions. Using digital technology, FDEs can see details beyond the limits of the naked eye, and measurements performed with digital technology "cure" the optical illusions that cause misperception of the examined documents. It is crucial that FDEs should follow the relevant SWGDOC and ASTM standards. Future perspectives of digital technology in FDE are very promising, and among them is the application of digital technology in FDE using kinesiology and electromyography so that the genuine handwritten items can be authenticated by confirming that they have been created by the unique sequence of muscle contractions, which generates similar genuine items.

Acknowledgment

The author expresses his gratitude to Dr. Dimitrios E. Skarpalezos MD, PhD, for his scientific contribution to this essay.

Bibliography

Abdelkrim A, Benrejeb M (2019), Conventional and non conventional body motions modelling and control. Application to the handwriting process, *Asian Journal of Control*, 21(11), available at https://onlinelibrary.wiley.com/doi/full/10.1002/asjc.2127.

Allen M (2016), *Foundations of forensic document analysis, theory and practice*, Willey Blackwell.

ASTM E2825-19 (2019), *Standard guide for forensic digital image processing*, ASTM International.

Balasubramanian R, Santos VJ (eds) (2014), *The human hand as an inspiration for robot hand development*, Springer International Publishing.

Belensky R (2017), *Forensic examination of signatures*. Amazon Digital Services LLC - KDP Print US.

Bravo A (2005), *Variazioni Naturali e Artificiose della Grafia*, Libreria Moretti Editrice, p. 33, referring: Girolamo Moretti (1942), *Perizie Grafiche su base Grafologica*, L'albero.

Cecil JS (2005), Ten years of judicial gatekeeping under Daubert, *American Journal of Public Health*, 95(S1), S74–80.

Chalmers AF (2013), *What is this thing called science*, 4th edition, McPherson's Printing Group.

Chen Y, Yang Z (2017), A novel hybrid model for drawing trace reconstruction from multichannel surface electromyographic activity, *Frontiers in Neuroscience*, 11, p. 61.

Chihi I, Kamavuako EN, Benrejeb M (2020), Modeling simple and complex handwriting based on EMG signal, Chapter 6, in *Control theory in biomedical engineering* (pp. 129–149), Academic Press, London, Elsevier Inc.

Cristofanelli P, Cristofanelli A (2004), *Grafologicamente, Manuale di Perizie Grafiche*, CE.DI.S. EDITORE, p. 231.

Dellavalle F (2011), *La strumentazione per l' analisi documentale in ambito forense*, Sulla Rotta del Sole Editore.

Dellavalle F, Frontini S (2020), A preliminary study of 3D depth measurement of the grooves generated by three different pens for handwriting, *Journal of the American Society of Questioned Document Examination*, 20(2), pp. 37–53.

Derbel F (ed.) (2020), *Communication, signal processing & information technology*, Walter de Gruyter GMBH, Berlin and Boston, p. 71.

Ellen D, Day S, Davies C (2018), *Scientific examination of documents, methods and techniques*, 4th edition, CRC Press.

Found B, Bird C (2016), Evidence evaluation and reporting procedures (module 10), *Journal of Forensic Document Examination*, 26, p. 59.

Gigerenzer G, Brighton H (2009), Homo heuristicus: why biased minds make better inferences, *Topics in Cognitive Science*, 1(1), pp. 107–143, available at https://onlinelibrary.wiley.com/doi/10.1111/j.1756-8765.2008.01006.x.

Gonzalez R, Woods R (2017), *Digital image processing*, 4th edition, Pearson.

Gorai A, et al. (2016), Document fraud detection by ink analysis using texture features and histogram matching, in *2016 International Joint Conference on Neural Networks (IJCNN)*.

Haack Susan (2014), *Evidence Matters, Science, Proof and Truth in the Law*, Cambridge University Press.

Hagan EW (1894), *A treatise on disputed handwriting and the determination of genuine from forged signatures. The character and composition of inks, and their determination by chemical tests. The effect of age as manifested in the appearance of written instruments and documents*, Banks & Brothers, available at https://archive.org/details/disputedhandwrit00haga.

Hall J (2021), *Guyton and hall textbook of medical physiology*, 14th edition, Saunders.

Harralson H (2013), *Developments in handwriting and signature identification in the digital age*, Anderson Publishing, Elsevier, pp. 56–59.

Haselton MG, Nettle D, Murray DR (2016), The evolution of cognitive bias, in DM Buss (Ed.), *The handbook of evolutionary psychology: integrations* (pp. 968–987), John Wiley & Sons, Inc.

Hendley CD (1948), The relation between visual acuity and brightness discrimination, *Journal of General Physiology*, 31(5), pp. 433–457.

Huang G, Zhang D, Zheng X, Zhu X (2010), An EMG-based handwriting recognition through dynamic time warping, in *Annual International Conference of the IEEE Engineering in Medicine and Biology Society*.

Huber RA, Headrick AM (1999), *Handwriting identification: facts and fundamentals*, CRC Press.

Kapandji A (2010), *The physiology of the joints*, Churchill Livingstone.

Kelly JS, Angel M (2021), *Forensic document examination in the 21st century*, CRC Press.

Kipouràs P (2021), Evidence for a 3-stage model for the process of free-hand forgery of signatures and/or handwriting, *IJISET - International Journal of Innovative Science, Engineering & Technology*, 8(1), available at http://ijiset.com/vol8/v8s1/IJISET_V8_I01_23.pdf.

Kipouràs P, Kosmidis C, Skarpalezos ED (2021), Why two genuine signatures are never identical in practice: evidence and scientific explanation, *IJISET - International Journal of Innovative Science, Engineering & Technology*, 8(4), available at http://ijiset.com/articlesv8/articlesv8s4.html.

Kurki I, Peromaa T, Hyvärinen A, Saarinen J (2009), Visual features underlying perceived brightness as revealed by classification images, *PLoS One*, 4(10), p. e7432, available at https://journals.plos.org/plosone/article?id=10.1371/journal.pone.0007432.

Mahmoud I, Chihi I, Abdelkrim A (2020), Adaptive control design for human handwriting process based on electromyography signals, *Complexity*, 2020, Article ID 5142870, available at https://www.hindawi.com/journals/complexity/2020/5142870/.

Mazzolini D, Mignone P, Pavan P, Vessio G (2021), An easy-to-explain decision support framework for forensic analysis of dynamic signatures, *Forensic Science International: Digital Investigation*, 38, p. 301216.

Merlino M (2015), *Validity, reliability, accuracy and bias in forensic signature examination*, p. 11, available at https://www.ojp.gov/library/publications/validity-reliability-accuracy-and-bias-forensic-signature-identification.

Mohammed LA (2019), *Forensic examination of signatures*, Elsevier, Academic Press.

Morris R (2021), *Forensic handwriting identification, fundamental concepts and principles*, 2nd edition, Elsevier Academic Press.

Mussweiler T (1999), Hypothesis-consistent testing and semantic priming in the anchoring paradigm: a selective accessibility model, *Journal of Experimental Social Psychology*, 35, pp. 136–164, available at http://www.idealibrary.com.

National Research Council (2009), *Strengthening forensic science in the United States: a Path Forward*, The National Academies Press (The NAS 2009 Report) available at https://www.ojp.gov/pdffiles1/nij/grants/228091.pdf.

Nickel J (1996), *Detecting forgery, forensic investigation of documents*, University Press of Kentucky.

Okorokova E, et al. (2015), A dynamical model improves reconstruction of handwriting from multichannel electromyographic recordings, *Frontiers in Neuroscience*, 9, p. 389, available at https://www.ncbi.nlm.nih.gov/pmc/articles/PMC4624865.

Page M, Taylor J, Blenkin M (2011a), Forensic identification science evidence since Daubert: part I--a quantitative analysis of the exclusion of forensic identification science evidence, *Journal of Forensic Sciences*, 56(5), pp. 1180–1184, available at https://pubmed.ncbi.nlm.nih.gov/21884119/.

Page M, Taylor J, Blenkin M (2011b), Forensic identification science evidence since Daubert: part II--judicial reasoning in decisions to exclude forensic identification evidence on grounds of reliability, *Journal of Forensic Sciences*, 56(4), pp. 913–917, available at https://pubmed.ncbi.nlm.nih.gov/21729081/.

Papaodysseus C, Exarhos M, Panagopoulos M, Roussopoulos P, Triantafillou C, Panagopoulos Th (2008), Image and pattern analysis for 1650 B.C. wall paintings study and reconstruction, *IEEE Transactions on Systems Man and Cybernetics, Part A: Systems and Humans*, 38(4), pp. 958–965.

Pugnaloni M, Federiconi R (2013), Forensic handwriting analysis: a research by means of digital biometrical signature, https://www.researchgate.net/publication/271825986.

Purves D, Wojtach WT, Lotto RB (2011), Understanding vision in wholly empirical terms, *Proceedings of the National Academy of Sciences of the United States of America*, 108 Suppl 3, pp. 15588–15595.

Sayre KM (1965), *Recognition: a study in the philosophy of Artificial intelligence*, p. 301, University of Notre Dame Press.

Shapiro A, Todorovic D, et al. (2017), *The Oxford compendium of visual illusions*, Oxford University Press.

Shelton ED (2011), *Forensic science in court, challenges in the twenty-first century*, Rowman & Littlefield Publishers Inc.

Sinha P, Crucilla S, Gandhi T, et al. (2020), Mechanisms underlying simultaneous brightness contrast: early and innate, *Vision Research*, 173, pp. 41–49.

Stoel R, Berger C, Fagel W, Heuvel E (2010), *The shaky criticism of forensic handwriting analysis*, Netherlands Forensic Institute (NFI).

Sulner A (2018), Critical issues affecting the reliability and admissibility of handwriting identification opinion evidence—How they have been addressed (or not) since the 2009 NAS report, and how they should be addressed going forward: a document examiner tells all, *Seton Hall Law Review*, 48(3), Article 5, available at https://scholarship.shu.edu/shlr/vol48/iss3/5.

Teulings HL, Schomaker LR (1993), Invariant properties between stroke features in handwriting, *Acta Psychologica*, 82(1–3), 69–88.

Varshney VP, Bedi M (2019), Lacunae in forensic handwriting examination: scope for exploitation, *Medicolegal Section*, 5(1), pp. 19–23.

Wakshull M (2019), *Forensic document examination for legal professionals: a science based approach*, Q9 Consulting, Inc.

Welsh MB, Delfabbro PH, Burns NR, Begg SH (2014), Individual differences in anchoring: traits and experience, *Learning and Individual Differences*, 29, pp. 131–140.

Zlotnick J, Lin JR (2001), Handwriting evidence in federal courts - from Frye to Kumho, *Forensic Science Review*, 13(2), pp. 87–99.

14

Legal Analysis in Forensic Investigation

Galina Rusman and Julia Morozova

CONTENTS

Introduction

The realities of today imply the widespread introduction of digital technologies in all types of legal activities. For example, in Russia, during the pandemic, participants in criminal proceedings can take part in court sessions and submit petitions, including participation in a court session remotely, via messengers (for example, WhatsApp), video communication (for example, Skype), and video conferencing. Courts of various instances are looking for ways to help participants exercise their procedural rights, including remotely.

These opportunities are not provided for by the criminal procedure legislation. On the one hand, these measures can be considered procedural encouragement, since the court allows the participants in the process to exercise their rights, including the right to health and the right to access to justice. On the other hand, it is the creation of procedural opportunities to ensure the implementation of the rights of participants during the period of restrictions imposed due to the lack of other opportunities. This is necessary for the implementation of the procedural deadlines established by law.

Throughout the existence of forensic examination as a procedural action that solves certain research tasks, scientific progress contributes to its development. Forensic examination is the most effective source of introducing the latest achievements of science and technology into the field of legal proceedings.

The use of modern technologies in forensic research significantly helps in solving crimes and establishing various circumstances. Modern technologies make it possible to examine objects received for examination from new sides, which previously could not be studied due to the insufficient level of scientific development for this and the lack of necessary methods and techniques. Thus, forensic examination is in the process of constant development due to the emergence of new technical means, methods, and research methods. It is worth noting that technologies borrowed from other fields of activity (physics, chemistry, information technology, and so on) are used in the initial forensic examination. This requires determining the specific properties of expert technologies. Technology is a set of rules (stages, techniques, methods) of applying scientific knowledge to obtain the desired result. As part of the implementation of judicial activities, the technology must comply with the form established by the legislator.

The introduction of technologies requires a significant reform of the system of forensic expertise in many countries. The leading position in the reform of forensic expertise is occupied by the United States. Back in 2009, the report "Strengthening Forensic Science in the United States: A Path Forward" (Strengthening Forensic Science in the United States: A Path Forward, 2009) was published in the United States. The European Network of Forensic Science Institutes (ENFSI) notes the need to improve the quality of forensic examination in Europe.

In order to achieve a qualitatively new level of expert research, States need to introduce modern technologies in a timely manner and ensure their legal admissibility in a timely manner.

The purpose of this study is to determine the features of the legal regulation of expert activity in individual states and to identify the problems with the legal admissibility of new expert technologies.

Literature Review

There are a large number of scientific articles and reviews that consider hotel expert technologies. Various biometrics methods include DNA analysis (Moiseyeva, 2014, Bessonov, 2019, Amelung & Machado, 2021, Zapico, 2021), examination of the iris (Ammour et al., 2017; Ammour et al., 2020), and finger marks. Some works reveal the features of the use of nanotechnologies in establishing individual circumstances of the case (Chauhan et al., 2017, Kanodarwala et al., 2019) and the use of artificial intelligence (AI) technologies (Stepanenko et al., 2020, Hall et al., 2021). Some authors consider the concepts of forensic examination (Zinin & Mailis, 2002), expert technologies (Khlopotnoy & Shaikova, 2019), and the problems of legal regulation of individual issues of expert research in individual countries (Trefilov, 2018, Horsman, 2018, Losavio et al., 2019, Neuteboom & Shakel, 2016). But there are no scientific studies devoted to the analysis of the legal admissibility of expert technologies and the results of their application. This shows the need for a comparative analysis of the legal norms regulating the possibility of using modern technologies in forensic examination.

Results

Legal Regulation of Forensic Examination and Expert Technologies

An analysis of the legislation of some countries shows the lack of a single, unified approach to regulating the introduction of new technologies in forensic examination. There is no single approach to the concept of forensic examination.

US scientists point out that the term forensic examination covers a wide range of studies, each with its own set of technologies and practices. Thus, there is a significant difference between expert disciplines, methods, methodology, reliability of results, the number of potential errors, and the admissibility of research results as evidence (Strengthening Forensic Science in the United States: A Path Forward, 2009, p. 6–7). The Russian legislator defines forensic examination as "a procedural action consisting of conducting research and giving an expert opinion on issues whose resolution requires special knowledge in the field of science, technology, art or craft".

The essence of forensic examination consists of the usage by an expert in accordance with the law, of special knowledge to conduct research and give an opinion on issues relevant to the case, put before him by an authorized, and in some cases by an interested person by an authority (plaintiff, defendant, court, person conducting an investigation).

In turn, systems of logical and instrumental operations, methods, and techniques, that is, methods of expert research, are used to obtain the information necessary to solve the questions posed to the expert (Zinin & Mailis, 2002, p. 31).

Various material means and devices belong to the technical part of the expert method. The actions performed with the research materials and the methods of implementing the method are included in the operational part. The justifying part of the method is the expected result of the method application and its scientific base. It follows that, together with the substantiating and operational parts of the method, material and technical support make it possible to implement the method in practice (Rossinskaya, 2019b, p. 115).

The development of methods, means, and techniques of forensic examination mainly depends on the development of methods and technical means of research of various sciences. The achievements of these sciences and their methods undergo certain changes due to the research objectives, a specific expert task, and, finally, the conditions to which experts are forced to adapt the application of the method (Averyanova, 2009, p. 278).

The process of forensic research often requires painstaking work from an expert, especially in the framework of criminal proceedings. Modern reality, with the widespread use of digital technologies opens up new opportunities, which helps to facilitate the work of an expert, makes the research process more operational, and also helps to reduce the number of errors (Kamalova, 2019, p. 180).

The study shows that only those methods and means that meet certain requirements can be used in forensic research.

1. Their application should not contradict the principles of legality and ethical norms of society. The methods used should not infringe on the rights of citizens, humiliate their dignity, lead to a violation of the norms of procedural law, and should exclude threats and violence. It is also necessary to comply with the procedural requirements for the expert's opinion.

2. Of great importance is the scientific validity of the methods used in the production of forensic examinations, as well as the reliability of the results obtained with their help.

3. The necessary criteria are the accuracy of the results, their reliability, the possibility of their verification, and the possibility of repeating the study.

4. When choosing a method, the decisive factor is its effectiveness in solving the problem facing the expert. The method should be most effective in achieving the set goal, which should be achieved in the shortest possible time. At the same time, the ratio of the forces and means spent on conducting the study should be commensurate with the value of the results obtained.

5. The methods and means used must be safe. The safety of the method should be achieved through the use of more advanced technical means and research methods.

Also, the methods of expert research must meet the requirement of preserving the object in the form and condition in which it was submitted for examination. Complete or partial destruction of objects, changing their appearance or basic properties, is allowed only in extreme cases in agreement with the person who appointed the examination.

The results of the use of technical means in the production of forensic examination should be obvious and visual both for experts and for all other participants in the proceedings.

Forensic expert activity is a complex system of such elements as regulatory regulation, the status and functions of subjects of expert activity, technical means, scientific foundations, methods, and techniques for conducting expert research (Khlopotnoy & Shaikova, 2019, p. 457).

The legal basis for the use of technical means and methods in forensic expert activity in Russia is the Criminal Procedure Code of the Russian Federation, the Civil Procedure Code of the Russian Federation, the Federal Law "On State Forensic Expert Activity in the Russian Federation" and the Federal Law "On Police", as well as departmental regulations of the Ministry of Internal Affairs of the Russian Federation, the Ministry of Justice of Russia.

According to Article 57 of the Criminal Procedure Code of the Russian Federation, an expert is "a person with special knowledge and appointed to conduct a forensic examination and give an opinion". Appointment and production of a forensic examination are carried out in accordance with Articles 195–207, 269, 282, and 283 of the Criminal Procedure Code of the Russian Federation. The rights and obligations of the expert are fixed in parts 3 and 4 of Article 57 of the Criminal Procedure Code of the Russian Federation. The expert is responsible for giving a deliberately false conclusion in accordance with Article 307 of the Criminal Code of the Russian Federation.

In accordance with Article 85 of the Civil Procedure Code of the Russian Federation, the expert is obliged to accept the examination entrusted to him and conduct a full study of materials and documents. At the same time, the law establishes the obligation of the expert to ensure the safety of the materials and documents submitted for research.

The criminal procedure legislation of some countries does not regulate the use of special knowledge and forensic technologies. This situation follows from the analysis of the relevant legislation of the USA and Canada Criminal Code (1985).

Currently, the application and use of information technology is an integral part of the practical activities of a forensic expert.

In accordance with part 2 of Article 2 of the Federal Law "On Information, Information Technologies and Information Protection", information technologies are "processes, methods of searching, collecting, storing, processing, providing, distributing information and ways of implementing such processes and methods".

According to Article 4 of the Federal Law of 31.05.2001 "On state forensic expert activity" state forensic expert activity

> is based on the principles of legality, respect for human and civil rights and freedoms, the rights of a legal entity, as well as the independence of an expert, objectivity, comprehensiveness and completeness of research conducted using modern achievements of science and technology.

Rossinskaya E.R. (2019a) notes that the digitalization of forensic expertise includes a variety of information technologies, which the legislator defines as processes, methods of searching, collecting, storing, processing, providing, distributing information, and ways of implementing such processes and methods necessary, including for solving forensic tasks (Rossinskaya, 2019, p. 570).

In most developed countries, the use of certain technologies in forensic examination is regulated by the relevant legislation.

The system of organizing the quality of forensic activities in most countries, as well as in Russia, does not differ in uniformity. Thus, the practice of the European Network of Forensic Science Institutes (ENFSI), which requires its members to accredit laboratories in accordance with ISO 1720 or ISO 17025 standards, indicates that the adopted system of validation of methods (standard operating procedures) does not fully meet the tasks of quality control of examinations by process participants (Neuteboom & Shakel, 2016, p. 21–25).

Undoubtedly, a single national regulator would more effectively solve the issues of the introduction and uniform application of technologies in the field of forensic expertise. Kazakhstan, Lithuania, the USA, Estonia, the countries of the European Union, and other countries are currently implementing the idea of creating unified judicial expert centers.

Common standards for forensic service providers relating to such sensitive personal data as DNA profiles and dactyloscopic data have been developed to unify forensic expertise in the European Union COUNCIL FRAMEWORK DECISION (2009).

In the Republic of Lithuania, the Council for the Coordination of Forensic Expertise has adopted a Professional Code of Ethics for an Expert, according to which an expert is obliged to conduct research in accordance with established principles, namely in accordance with the law, using scientifically proven research methods, necessary equipment, and materials.

Thus, "the implementation of the increase in conditions of scientific and technological revolution of judicial examination largely depends on its legal status, how regulating its function and production of rules of procedural law to help turn these opportunities into reality" (Belkin, 1988, p. 21).

For quality implementation of new technological advances in forensics, the Russian legislator is necessary to develop new methods, methods of obtaining information in the course of expert research with the help of digital technology.

Despite the fact that the term "technical means" is repeatedly found in the Criminal Procedure Code of the Russian Federation, there is no definition of them in the article that reveals the basic concepts. This can lead to confusion, since the term "technical means" depends on the meaning of a particular article. So technical means are perceived as:

- Funds intended for the investigation of criminal case materials (Article 58);
- As a means of control (Article 107);
- As a means of recording (Article 166);
- As a means of recording (Article 170).

The lack of a common understanding of the term "technical means" may lead to uncertainty of their legal status and the order of use in expert activities. Despite the fact that the essential features of technical means are not legally defined, evidence obtained with the help of such means is permissible if the provisions of the procedural law governing the collection, investigation, and evaluation of evidence are not violated. Consequently, there should be no obstacles to the use of technical means in forensic activities in Russia.

Let's consider the general principles of the admissibility of the use of technical means. These include legality, compliance with moral standards, scientific validity, the accuracy of the results of their application, effectiveness, and safety.

Since the expert's opinion relates to evidence, the requirement of admissibility applies to it. Thus, the methods and technologies used in the course of expert research must meet this requirement.

The list of permissible technical means in procedural legislation cannot be exhaustive. Legislatively, it is necessary to provide only the conditions for their application.

The main criterion for the use of technologies in forensic examination is legality and admissibility. Permissibility is the conformity of the application of a means, technique, method, and technology to the norms of the law. At the same time, the law cannot cover all continuously developing means, methods, and technologies. The law should define only general conditions for the admissibility of technical means and technologies and general rules for their application.

The legality of the use of modern technologies and technical means should be determined based on the general principles of the admissibility of technical means and methods in legal proceedings. Taking into account the multiplicity, ambiguity, and continuous improvement of technical means, it is impossible to place an exhaustive list of them in the criminal procedure legislation (Rossinskaya, 2019, p. 112).

A similar rule is contained in one of the paragraphs of the Basel Commentary to the Swiss Code of Criminal Procedure ("Kein numerus clausus der Beweismittel" – "Not an exhaustive list of evidence"). According to paragraph 1182 of the Swiss Code of Criminal Procedure, "all conceivable evidence is suitable, even if it is not fixed or not yet fixed in the procedural law" (Trefilov, 2018, p. 276). In this case, the legislator considers it impossible to anticipate the subsequent achievements of scientific and technological progress.

Thus, the legal status of technical means and technologies used in the production of forensic examination should be more qualitatively regulated in the legislation regulating expert activity. At the same time, it is worth limiting ourselves to general signs of technical means that are essential for evidentiary value, since it is impossible to list all existing technical means in the law in view of the continuous development of modernity, the emergence of completely new and obsolescence of existing technologies.

Modern Technologies in Expert Activity

Currently, among the most common areas of application of information technology in forensic activities, one can distinguish:

- According to the data under study – biometrics;
- According to the technique used – scanning;
- By type of information – graphic information.

These directions do not exist in their pure form since they can be used in combination with other technologies in the process of expert research.

Thanks to the expert high-tech investigation of traces using biometric technologies, it has become possible to investigate and solve criminal cases of crimes committed in conditions of non-obviousness (Nedorostkov, 2018, p. 58).

In accordance with the Federal Law of Russia «On Personal Data», biometric personal data is

> information that characterizes the physiological and biological characteristics of a person, on the basis of which it is possible to establish his identity... and which is used by the operator to establish the identity of the subject of personal data.

Today, such data include human appearance (including the anatomy of the retina and iris), the venous pattern of the hands, fingerprints and palms, voice, handwriting, smell, and DNA.

Biometric data have a number of characteristic features that allow them to be used for forensic identification of a person. These include:

1) Uniqueness of biometric characteristics of each person;
2) The prevalence of biometric characteristics, which makes it possible to collect, accumulate, and process them;
3) Relative stability of biometric characteristics over a large amount of time;
4) The ability to classify biometric characteristics.

Biometric identification methods are divided into two groups: statistical, investigating physiological characteristics, and dynamic, investigating behavioral characteristics of the individual.

Displays of physiological and behavioral characteristics of a person are objects of identification in forensic examination. An example of the use of biometric data is the conduct of genotyposcopic, fingerprinting, habitoscopic, phonoscopic, and handwriting examinations. By using biometrics, an expert can verify the reliability of the research results obtained by him.

Automatic processing of research results makes it possible to identify hidden patterns and interpret them correctly. But still, data processing by itself cannot serve as a guarantee of the quality of the expert study performed if the examination itself is carried out methodically incorrectly or on the basis of erroneous data (Dmitrieva, 2017, p. 497).

Genotyposcopic examination is based on DNA analysis methods. They allow you to identify a person at the level of an individual-specific identity. Forensic DNA phenotyping (FDP) is a bundle of DNA analysis technologies that have emerged for the prediction of human physical characteristics, including externally visible traits such as the color of eyes, hair, and skin, as well as biological age and biogeographic ancestry (Amelung & Machado, 2021).

DNA analysis methods are used in different countries, and the prospects of their application in investigation and consolidation in legislation are actively discussed by scientists (Wienroth, 2020; Samuel and Prainsack, 2018a, 2019; Amelung & Machado, 2021; Zapico, 2021).

In Russia, the use of DNA methods in the examination, in addition to the relevant procedural legislation, is regulated by the Federal Law "On State Genomic Registration in the Russian Federation".

The new EU legislation requires every EU Member State to establish a forensic DNA database and to make this database available for automated searches by the other EU Member States. As DNA profiles are regarded as personal data, national privacy legislation derived from the European Data Protection Directive 95/46 also applies to forensic DNA databases DNA DATABASE MANAGEMENT REVIEW AND RECOMMENDATIONS (2016).

Thus, in 2018, in the South German state of Bavaria, forensic DNA phenotyping technologies were included in the Police Law in order to implement preventive police actions

(Bayerische Staatskanzlei, 2018). The amendments to the said law of the Bavarian State allow the use of the following technologies of forensic DNA phenotyping – eye, hair, and skin color as well as age and biogeographic ancestry, "to investigate people they deem an 'imminent danger' ... persons who have not necessarily committed any crimes but might be planning to do so" (Vogel, 2018). The Bavarian legislator justified the need to use forensic DNA phenotyping technologies with counter-terrorism measures to prevent potential mass shootings or terrorist attacks, which emphasize the concept of security (Amelung & Machado, 2021).

Iris identification is also one of the methods based on biometrics.

The results of using a multimodal biometric identification system showed special reliability and speed of recognition. This system is based on the fact that the face and iris work in identification mode. The merge is applied at the level of the corresponding score with different normalization and the merge rule (Ammour et al., 2017) and consists of modules for reading, feature extraction, and matching (Ammour et al., 2020).

Scanning of the iris can be carried out in the visible or infrared zone of the spectrum, which allows you to obtain high-quality identification signs.

Scanners, regardless of the principle of operation, provide remote receipt of input information. The absence of direct human contact with the equipment, since the capture of the iris image, is made simply by looking into the scanner lens. The scanner analyzes the image quality of the eye in the frame and determines the center of the pupil, the center of the iris, and its boundaries.

The most common biometric method at the moment is the recognition of papillary patterns. It is based on the individuality and uniqueness of papillary finger patterns in each person.

Despite the fact that automated fingerprint information systems do not belong to the technologies directly used during the production of examinations, they help to significantly facilitate the process of expert research and reduce the expert's work time.

Automated fingerprint information systems (ADIS) ensure the processing of significant arrays of fingerprint records, identify persons who left handprints at the scene of the incident, and establish the facts of the formation of handprints seized at the sites of several crimes by one person.

Automated fingerprint information systems (ADIS) work with two types of fingerprint objects – fingerprints and palm prints and finger and palm prints. The trace is encoded and checked in the array of fingerprint cards available in it. As a result of the check, a recommendation list of fingerprint cards is formed, which contain the most similar fingerprints and palm prints. Such technologies allow, among other things, to avoid possible errors in the production of research Report on the Erroneous Fingerprint Individualization in the Madrid Train Bombing Case (2005).

Image analysis methods based on the use of special software have not only simplified the technology of comparative study of traces of papillary patterns but also significantly improved their images, making blurred dynamic traces sufficiently clear and suitable for fingerprinting. It became possible to complete the elements of the papillary pattern that were not displayed in the trace by statistical forecasting. The use of image analysis methods made it possible to identify finger marks by micro-signs, allowing for both a comparative morphological study, taking into account possible distortions during abandonment and mathematical processing of their images (Moiseyeva, 2014, p. 72).

In the biometric handwriting recognition method, the writing process itself acts as an object of comparison. The basis for identifying a person by handwriting is his stability and uniqueness of handwriting.

The parameters of the signature process itself are recorded in real time by the method of dynamic verification: the speed of movement of the hand in various areas, the force of pressure, the angle of inclination of the writing instrument, the sequence of drawing lines, and the duration of the different stages of the signature. The use of this method completely excludes imitation since it is absolutely impossible to copy the hand movements of the signature author (Dmitrieva, 2018, p. 58).

Currently, the simulation environment software has been developed. The technology is based on the user reproducing a word in his own handwriting several times.

It is the means of biometric identification of a person's identity by handwriting work with signatures that the user enters on a graphic tablet. Since handwriting expertise works with handwritten notes that were already written at the time of the study, it is necessary to scan the document under study with a verifiable signature and a number of documents with signature samples before the study.

After training a neural network based on handwriting samples, an examination can be carried out. An artificial neural network analyzes 480 parameters inaccessible to a human expert, but it is able to generalize only the current data in the "original" and "fake" databases (Ivanov, 2016, p. 253).

In addition, neural networks are being developed to conduct a preliminary analysis of handwriting in order to identify signs of forgery of signatures made without the use of technical means (Ushakov, 2020, p. 65).

The use of biometric parameters for personal identification is of paramount importance. Although the forgery of biometric characteristics is significantly difficult, the probability of this is not excluded.

The use of biometric technologies in forensic examination is possible if a unified approach to the algorithmization of signal processing technologies, their orientation, and scope of application is developed, as well as under conditions of confidentiality and improvement of current legislation (Dmitrieva, 2017, p. 500).

Thus, biometrics and technologies based on its principles are gradually becoming a means of criminalistic identification of persons and can be widely used in forensic examination as one of its methods.

The use of nanotechnology in forensic examination is developing significantly. These technologies make it possible to identify hidden traces (evidence), for example, finger marks, when other methods do not bring results (Kanodarwala et al., 2019). In addition to fingerprint detection, nanotechnology is used to detect narcotic drugs, measure alcohol levels in drunk drivers, detect explosives, nerve gas, saliva, and identify inorganic pigments in road accidents and other cases (Chauhan, et al., 2017).

Methods based on the use of various scanners are becoming increasingly popular due to their high information content.

The most common are scanners that transfer graphic and textual information from paper to digital format. For further work with the information obtained as a result of such scanning, programs for processing graphic information and optical text recognition are used.

Scanning technologies are also the basis of scanning electron microscopy, during which the object of study is scanned by an electron beam (probe) of an extremely small cross-section (several angstroms). Scanning electron microscopy allows not only to photograph the microinclusion but also to determine its composition.

The use of electronic fingerprint scanners allows you to accurately capture images of fingerprints and palms of hands, thanks to automatic recognition and encoding of images in automated fingerprint information systems.

Non-contact, non-invasive surface scanning methods make it possible to obtain a high-quality digital image of a fingerprint and a comparative sample in order to establish the prescription of the trace (Merkel, 2021).

Automated Ballistic Identification Systems (ABIS) allows you to obtain digital images of the bullet surface, the bottom surfaces, and the shell casing. Also, the ABIS identifies primary and secondary traces of rifling fields on the bullet, the position of the traces of the idle and combat facets on the secondary traces, and on the sleeve – traces of the striker and cartridge stop, reflector, sender, ejector hook, receiver window or shutter cover, magazine bend, and so on.

Among the scanning methods, it is worth highlighting 3D modeling based on the results of scanning using a 3D scanner. Its application opens up broad prospects for the development of many types of forensic examination.

3D scanning technology allows you to conduct a "virtual" expert experiment and solve situational and diagnostic expert tasks under difficult modeling conditions. 3D models are stored electronically and the expert will be able to refer to this or that object at any time, and find out all its characteristics, including dimensions and a set of particular features.

3D scanning technology is gradually being applied in some types of examinations. Thus, during fire-technical expertise, this technology helps to determine the location of the fire source and the direction of the spread of gorenje; during explosive expertise – to determine the epicenter and mechanism of the explosion; during automotive expertise – to determine the technical condition of the vehicle, the mechanism of a traffic accident, the technical condition of the road and the like; during engineering and technological expertise – to determine the condition of equipment, devices and mechanisms, their suitability for performing routine operations, identify defects and malfunctions, reconstruct the mechanism of accidents and accidents; during construction and technical expertise – to determine the validity of projects and the mechanism of accidents and so on (Bystryakov et al., 2017, p. 20).

Using the virtual reconstruction of the fire place, get a virtual model of a fire, which determines the area of damage, the area of the smoke, the zone of thermal damage, and the area of maximum heat damage, to obtain a spatial picture of the fire (Yarmak, 2011, p. 335).

Construction and technical expertise, the technology allows to carry out an expert study, even in the absence of project and other technical documentation of the object (Kharchenko, 2019, p. 110).

3D modeling in the course of automotive expertise is used to reproduce the mechanism of road accidents. This is due to a change in the structure of road traffic situations preceding road accidents (the speed of cars increases, the density of traffic flow increases, etc.) (Kudryavtsev, 2019, p. 310).

> A special place in the study of the paintwork of the car is occupied by the study and analysis of mechanical damage such as scratches, dents, chips that appeared as a result of accidents, collisions, the impact of solid objects from the road surface. Laser scanning can fix up to 100% of such damage.
>
> **(Snatkov, 2017, p. 125)**

3D modeling will be able to combine all available data about the event and visualize the material situation. The visibility of such a virtual copy of the scene of the incident will allow you to build, check, and evaluate various expert versions and then solve identification and situational and diagnostic tasks.

Among the main advantages of using 3D scanning, it is necessary to highlight a reduction in data collection time, a reduction in the complexity of the information collection procedure, an increase in the accuracy of the result, and long-term storage of results. In addition, the productivity and quality of expert's work increase, the reliability of the results of expert's research is ensured, and the possibility of accessing the research object (its model) an unlimited number of times without prejudice to the object (Kharchenko, 2019, p. 110).

Scanning technologies, 3D modeling based on them, allow solving expert modeling problems, thus recreating the spatial characteristics of the objects under study.

Modeling in the process of certain types of examinations allows an expert to solve situational problems, thus reconstructing the course of events and identifying possible causes.

Graphic information is the most visual and informative type of information found in the field of forensic examination. It includes various visually perceived data, for example, tables, diagrams, photographs, videos, visible displays of finger marks, shoe marks, vehicles, burglary tools, and so on. Methods of working with images are effective and in demand in forensic examination.

The graphical way of presenting information in forensic examination allows you to analyze information. Image analysis systems make it possible to carry out diagnostic and identification studies, for example, handwriting, fingerprinting, tracological, ballistic, and others.

With the help of measuring photography, photographs are obtained from which it is possible to determine the dimensions of the photographed objects and the distances between them. It can be carried out by the method of stereophotogrammetry (using stereo pairs) and monophotogrammetry (using a single camera).

The application of the rules of measuring photography with deep and linear scales allows you to measure objects and the distances between them recorded on video. For example, photogrammetric software "PhotoModeler Scanner" sets the dimensions of objects on such video recordings by applying landmarks (reference points) with known coordinates by direct measurement (Egorov, 2018, p. 155).

During the production of an expert study, an expert sometimes needs to determine the color of the object of study. Color is subjective information and can vary significantly depending on the biological characteristics of the color perception of the expert. To objectify the process of determining the color characteristics of research objects, experts need to use technical means.

In research photography, the colorimetric characteristics of an object are primarily determined, which include three specific values of the primary colors: red, green, and blue. And only after that, tools are used to automatically improve the image or auto-adjust its brightness and contrast when processing in a graphic editor. Prior to the study, such processing is unacceptable, since it can change the color and tone balance and give an incorrect picture of the color of its surface.

Ivanov A.V. and Protasov K.V. (2019) note that after obtaining an image of the object of study with the correct reflection of its color gamut, it is necessary to determine the color characteristics expressed in the generally accepted numerical format – the color code of various systems: RGB, CMYK, and others. The next stage of the work is the introduction of the received color code into a special program for selecting colors and generating color schemes, as a result of which the color name from the reference book will appear (Ivanov & Protasov, 2019, p. 72).

The most common means of working with images are various graphic editors. Experts quite often resort to photo processing in order to improve the image with the help of photo editors. In order to comply with legal requirements, the expert's conclusion should include a description of the devices used during the study, their characteristics, lighting parameters, characteristics of the source file with the image, and the graphic editor used, in which image processing procedures were carried out. Actions to correct the image must be performed with a copy of the file, the original must remain unchanged.

The need to use graphic editors in the process of expert research is not to improve the image as a whole but to improve the distinctiveness of the identification features of appearance, in order to ensure the subsequent possibility of identifying the studied persons.

Graphic editors are used by the expert in order to prepare illustrations for the expert's conclusion. Graphic editors have a user-friendly interface, a wide range of functions for simultaneous editing of several digital images, and superimposing one image or its fragment on another, using different settings (transparency, contrast, filters, masks, etc.). At the same time, it is possible to translate discrete information into its analog form for subsequent comparative research.

If analog images (photographs, negatives) have been received for the study, they are digitized. If it is necessary to examine fragments of video films, then they are storyboarded, after which freeze frames are selected in which the person under study is in the most successful position (pose of the person under study, camera position, tilt, head rotation, lighting, etc.).

For example, for fingerprinting and tracological examinations, software is used to improve the quality of images of objects, their analysis, and mathematical measurement of the surface, allowing diagnostic and identification studies.

Such software allows you to convert the traces into the required image format, identify faint signs, change the color correction and brightness of the image, mirror the trace, enhance the clarity (contrast) of the traces.

For example, the system "Papillon Rastr" uses non-destructive transformations of digital images in order to improve their visual perception and identify hard-to-distinguish details, examining images, with the mandatory preservation of the original image and the history of its modification (Papillon Systems, 2021).

In addition, graphic editors are used in handwriting expertise by converting different densities of the dye strokes into different colors to compare the pressure in the strokes, as well as in the technical and forensic examination of documents for the identification study of prints by an optical overlay.

During the production of forensic phototechnical and videotechnical examinations, photographic images are examined, as well as dynamic images (video recording) and static (a separate frame of video recording). Using various software tools, the expert applies the following methods to improve the quality of video images: sharpening, changing perspective, increasing the size of the image under study, cropping, noise removal, distortion correction, and converting omnidirectional cameras into panoramas (Askerova, 2017, p. 49).

During the preliminary stage of studying photos and videos, developments from the field of biometric technologies can be used, which allow the use of algorithms for recognizing people's faces on them.

Within the framework of expert research, the question sometimes arises of identifying the fact of using retouching when processing photos. So, the "About Face" program developed by Adobe, which is based on a convolutional neural network, analyzes each pixel of the image and identifies places with traces of editing in Photoshop. This program can restore the original image based on the results of the analysis Adobe Inc. (2019).

Methods of working with images in the production of forensic examinations are often used to improve the quality of the image under study and to present meaningful information in a more visual form. Thanks to the use of such methods, the expert receives the most important information that is difficult to detect with the naked eye without image processing (Askerova, 2017, p. 50).

However, common graphic editors are of little use for the work of an expert since they are intended for a wide range of users whose tasks do not include compliance with strict rules of image processing, thus allowing irreversible arbitrary change and distortion of the original information.

In order to identify the prospects for the use of new technologies in the course of expert research, it is necessary to consider some individual types of information technologies that are gradually finding application in forensic activities.

Thanks to the further development of genetic research technologies and the study of the human genome using computer analysis of the results obtained, the next step in the development of genotyposcopic examination capabilities (along with the traditional task of identifying a person by biological traces left) may be the drawing up of a portrait of an unknown criminal by DNA traces left by him.

The potential of using the achievements of genetics in this case is as follows: firstly, the further development of genetic research technologies represented by genome-wide sequencing, and secondly, the existing achievements of geneticists around the world in the study of the human genome using computer analysis of the results (bioinformatics). The development of sequencing technologies based on the scientific achievements of genetics and bioinformatics, based on the understanding of the connection of a larger number of different variants of the genetic text with a specific state of an externally observed and hidden trait, will allow an unknown criminal to make a portrait of him with specific search signs, containing comprehensive information about his ethnogeographic origin, eye and hair color, size and shape of body and face parts, various diseases, etc. (Bessonov, 2019, p. 23; Amelung & Machado, 2021; Zapico, 2021).

Currently, virtual reality (VR) technology is becoming widespread in forensic examination to increase the level of visibility of research results. For example, an experiment was conducted at the University of Zurich using VR technology to recreate a shooting scene (Badzyuk & Ermakov, 2019, p. 100).

Such an application of VR in forensic ballistics can find application for the method of visualizing the trajectory of bullets in three-dimensional space, and participants in the trial will be able to visually perceive the results of the expert study.

Information technologies form the basis of the process of digitalization of forensic expertise. At the same time, many researchers consider digitalization as a starting point for the introduction of artificial intelligence technologies in various fields of activity.

In Russian law, artificial intelligence is

> a set of technological solutions that allows simulating human cognitive functions (including self-learning and searching for solutions without a predetermined algorithm) and obtaining results comparable, at least, with the results of human intellectual activity when performing specific tasks. The complex of technological solutions includes information and communication infrastructure, software (including those that use machine learning methods), processes and services for data processing and solution search Decree of the President of the Russian Federation (2019).

In modern science, there are two approaches to understanding artificial intelligence:

1. The top-down type implies applied modeling of individual components (processes) of human thinking in order to solve highly specialized, particular tasks.

2. The ascending type assumes full-fledged thinking, i.e., a comprehensive assessment of incoming messages and making informed decisions based on them in conditions of incomplete, fragmented information and subsequent maximally rational behavior (Stepanenko et al., 2020, p. 209, Hall et al., 2021). Such artificial intelligence is characterized by situational versatility, that is, full-fledged adaptive qualities, and not narrow functionality.

To date, all existing artificial intelligence systems are of the top-down type. An artificial intelligence system of this type is aimed at solving a specific, rather narrow task.

The most common technology in artificial intelligence systems is artificial neural networks, which are the most important component of machine learning technologies.

Machine learning assumes that artificial neural networks train themselves, which makes it possible, due to high resistance to statistical noise, to process large amounts of information in order to identify patterns that are not obvious to human perception.

To date, artificial neural networks are able to recognize (determine the necessary features in the data under study), predict (determine the future state of a certain information system), and classify (distribute data into groups according to specified parameters) (Stepanenko et al., 2020, p. 212).

In many ways, the rapid development of artificial intelligence technology based on neural networks was caused by the emerging opportunities for collecting big data (Big Data).

In such cases, digital forensics tools are often used. Digital forensics tools are now used on a daily basis by examiners and analysts within local, state, and Federal law enforcement; within the military and other US government organizations; and within the private "e-Discovery" industry (Simson, 2010, p. S64).

To conduct a forensic examination, it is necessary to compare a lot of materials and data, as well as analyze them, in this case, Big Data analysis methods will help an expert work with an array of information, provide an opportunity to quickly obtain data, calculate a number of necessary indicators, and process the available data.

These tools can remove duplicate information, allow several specialists to work simultaneously, and provide the possibility of remote work with documents.

Artificial intelligence technologies are aimed at collecting and processing big data. They can be used to conduct a preliminary analysis during an expert study of handwriting, handprints, and others.

Thanks to artificial intelligence systems based on artificial neural networks, the use of computer vision technology has become possible in the processing of Big Data.

Computer vision can be used to solve problems in portrait, photo, or video forensic examination by identifying a person's identity from his image, as well as when examining documents.

For example, with fingerprint identification of ADIS "Papilon", the algorithm for selecting recommendation lists using neural networks allows you to reach a new level of identification of scarred, deformed, uninformative traces.

Improving the efficiency of the processing of significant arrays of fingerprint records will help to significantly facilitate and speed up the process of checking traces and subsequent forensic fingerprint examination.

Artificial intelligence can only solve narrowly focused tasks, but it cannot carry out a comprehensive assessment of incoming data, and therefore it is not able to make decisions in conditions of incomplete information.

There is a problem with legal regulation of the introduction of artificial intelligence technologies in forensic examination. Modern law recognizes the result of the work of artificial intelligence as a means of activity. Consequently, the responsibility for the process and the result of the research performed, which is recognized as evidence in the case, is borne by the forensic expert as a subject of activity and not technology.

Discussion and Conclusion

The analysis shows the readiness of states to develop expert technologies and introduce them into expert activities. At the same time, the activities of a forensic expert are not regulated by procedural legislation in all countries. In some countries, this is a specialized legislation that applies to licensed experts. At the same time, the requirements of the procedural or profile law impose additional restrictions on the software used in expert practice.

Currently, the primary task of the legislator is to improve the legal regulation of such technologies as artificial intelligence, to determine the status of artificial intelligence, and to determine the possible areas of its application, taking into account the need to comply with ethical and economic aspects.

Thus, modern technologies provide significant assistance in the production of forensic examination, facilitating the process of research and comparison, speeding up the processing of information, and increasing the accuracy of the results obtained.

The use of modern technologies by an expert during research allows for a more thorough study of the object under study, which makes the results more reliable and reliable. The expert's conclusion, based on the results of such a study, becomes more objective.

References

Adobe Inc. (2019), available at: https://theblog.adobe.com/adobe-research-and-uc-berkeley-detecting-facial-manipulations-in-adobe-photoshop/ (accessed 2021.07.14).

Amelung, N., Machado, H. (2021) "Governing expectations of forensic innovations in society: The case of FDP in Germany", *New Genetics and Society*. https://doi.org/10.1080/14636778.2020.1868987

Ammour, B., Boubchir, L., Bouden, T., Ramdani, M. (2020) "Face–Iris Multimodal Biometric Identification System", *Electronics*, 9(1), pp. 85. https://doi.org/10.3390/electronics9010085

Ammour, B., Bouden, T., Amira-Biad, S. (2017) "Multimodal biometric identification system based on the face and iris", 2017 5th International Conference on Electrical Engineering - Boumerdes (ICEE-B). https://doi.org/10.1109/ICEE-B.2017.8191981

Askerova, L.F. (2017) "The significance of the use of image analysis methods in the production of forensic examinations", *Science, Education and Culture*, 6(21), pp. 48–51.

Averyanova, T.V. (2009) *Forensic examination: The course of general theory* (in Russ.). 480 p.

Badzyuk, I.L., Ermakov, A.R. (2019) "Virtual and augmented reality technologies in the forensic activity of law enforcement agencies", *Actual Problems of Criminalistics and Forensic Examination* (in Russ.), pp. 100–103.

Bayerische Staatskanzlei. (2018) "Gesetz über die Aufgaben und Befugnisse der Bayerischen Staatlichen Polizei (Polizeiaufgabengesetz – PAG) [Law on the Tasks and Competences of the Bavarian State Police]", available at: https://www.gesetze-bayern.de/Content/Document/BayPAG/true (Accessed 2019.08.15).

Belkin, R.S. (1988) "Forensic examination: Issues requiring solutions", *Soviet Justice*, 1, pp. 21–22.

Bessonov, A. (2019) "Some promising directions for the further development of Russian criminalistics", *Academic thought*, 3(8), pp. 22–27.

Bystryakov, E., Usanov, I., Evgrafova, K. (2017) "Prospects of using the visualization method in the production of forensic examinations", *Expert-Criminalist*, 2, pp. 19–21.

Chauhan, V., Singh, V., Tiwari, A. (2017) "Applications of nanotechnology in forensic investigation", *International Journal of Life Sciences Scientific Research*, 3(3): 1047–1051. https://doi.org/10.21276/ijlssr.2017.3.3.13

Civil Procedure Code of the Russian Federation was accepted on November 14, 2002 (in Russ.), available at: http://www.consultant.ru/document/cons_doc_LAW_39570/ (accessed 2021.07.09).

COUNCIL FRAMEWORK DECISION 2009/905/JHA of 30 November 2009 on Accreditation of forensic service providers carrying out laboratory activities. https://eur-lex.europa.eu/eli/dec_framw/2009/905/oj#document1

Criminal Code (R.S.C., 1985, c. C-46), available at: https://laws-lois.justice.gc.ca/eng/acts/C-46/page-1.html (accessed 2021.07.11).

Criminal Procedure Code of the Russian Federation was accepted on December 18, 2001 (in Russ.), available at: http://www.consultant.ru/document/cons_doc_LAW_34481/ (accessed 2021.07.09).

Decree of the President of the Russian Federation was accepted on October 10, 2019 «On the development of artificial intelligence in the Russian Federation» (together with the «National Strategy for the development of Artificial Intelligence for the period up to 2030») (in Russ.), available at: https://www.zakonrf.info/ukaz-prezident-rf-490-10102019/ (accessed 2021.07.12).

Dmitrieva, L. (2017) "The possibilities of biometric research in the production of forensic examinations", in *Proceedings of the conference: Theory and practice of forensic examination: International experience, problems, prospects*, pp. 137–141. "https://www.elibrary.ru/item.asp?ysclid=l6p1cnw15k174298932&id=30532184"https://www.elibrary.ru/item.asp?ysclid=l6p1cnw15k174298932&id=30532184.

Dmitrieva, L. (2018) "The use of biometric identification of an individual in the production of forensic examinations", *Bulletin of Economic Security*, 1, pp. 56–58.

DNA DATABASE MANAGEMENT REVIEW AND RECOMMENDATIONS. April 2016. https://enfsi.eu/wp-content/uploads/2016/09/final_version_enfsi_2016_document_on_dna-database_management_0.pdf (accessed 2021.07.14).

Egorov, N. et al. (2018) "On the integration of methods and techniques of forensic photography and video recording", in *Collection of Materials of the International Scientific and Practical Conference: Modern criminal procedure law-history lessons and problems of further reform*, pp. 154–157. https://elibrary.ru/download/elibrary_36437149_30149972.pdf.

The European Network of Forensic Science Institutes, available at: https://enfsi.eu (accessed 2021.07.20).

Federal Law «*On Information, Information Technologies and Information Protection*» (in Russ.) was accepted on July 27, 2006, available at: http://www.consultant.ru/document/cons_doc_LAW_61798/ (accessed 2021.07.12).

Federal Law «*On Personal Data*» (in Russ.) was accepted on July 27, 2006, available at: http://www.consultant.ru/document/cons_doc_LAW_61801/ (accessed 2021.07.12).

Federal Law «*On State Forensic Expert Activity in the Russian Federation*» (in Russ.) was accepted on May 31, 2001, available at: https://base.garant.ru/12123142/ (accessed 2021.07.12).

Federal Law «*On State Genomic Registration in the Russian Federation*» (in Russ.) was accepted on December 3, 2008, available at: https://base.garant.ru/12163758/ (accessed 2021.07.11).

Federal Law «*On the Police*» was accepted on February 7, 2011 (in Russ.), available at: http://www.consultant.ru/document/cons_doc_LAW_110165/ (accessed 2021.07.09).

Hall, S.W., Sakzad, A., Choo, K.-K.R. (2021) "Explainable artificial intelligence fordigital forensics", *Wiley Interdisciplinary Reviews: Forensic Science*, e1434. https://doi.org/10.1002/wfs2.1434

Horsman, G. (2018) "A forensic examination of the technical and legal challenges surrounding the investigation of child abuse on live streaming platforms: A case study on Periscope", *Journal of Information Security and Applications*, 42, pp. 107–117. https://doi.org/10.1016/j.jisa.2018.07.009

Ivanov, A.I., Gazin, A.I., Kachaykin, E.I., Andreev, D.Yu. (2016) "Automation of handwriting expertise based on the training of large artificial neural networks", *Models, Systems, Networks in Economics, Technology, Nature and Society*, 1(17), pp. 249–257.

Ivanov, A.V., Protasov, K.V. (2019) "Determination of color characteristics of the object of research by computer means during forensic examination", *Society and Law*, 4(70), pp. 69–74.

Kamalova, G.G. (2019) "Digital technologies in forensic examination: problems of legal regulation and organization of application", *Bulletin of the Udmurt University. The Series "Economics and Law"*, 29(2), pp. 180–186.

Kanodarwala, F.K., Moret, S., Spindler, X., Lennard, C., Roux, C. (2019) "Nanoparticles used for fingermark detection – A comprehensive review", *WIREx Forensic Science*, 1, e1341. https://doi .org/10.1002/WFS2.1341

Kharchenko, V.B. (2019) "Features of the use of 3D laser scanning in forensic engineering and technical expertise", *Legal science*, 10, pp. 109–110.

Khlopotnoy, A.V., Shaikova, M.V. (2019) "The use of modern technologies in the production of forensic examination through the prism of the works of Professor E. R. Rossinskaya: theory and practice", in *Development of criminalistics and forensic examination in the works of Professor E. R. Rossinskaya. To the anniversary of the scientist, expert, teacher: Materials of the international scientific and practical conference*. Moscow: Prospekt (in Russ.), pp. 457–460.

Kudryavtsev, Yu.A. (2019) "Computer modeling in forensic expert activity", in *Materials of the II All-Russian Youth Scientific and Practical Conference: Investigative activity: Problems, their solution, prospects for development* (in Russ.), pp. 308–311. https://elibrary.ru/item.asp?id=39539694&ys clid=l6p1rq9mt9606327791.

Losavio, M.M., Pastukov, P., Polyakova, S., et al. (2019) "The juridical spheres for digital forensics andelectronic evidence in the insecure electronic world", *WIREs Forensic Science*, 1, e1337. https://doi.org/10.1002/wfs2.1337

Merkel, R. (2021) "Non-contact technologies and digital approaches to (latent) fingermark aging studies", in J. De Alcaraz-Fossoul (ed), *Technologies for fingermark age estimations: A step forward*. Cham: Springer, pp. 85–111. https://doi.org/10.1007/978-3-030-69337-4_4

Moiseeva, T.F. (2014) "New methods and means in the formation of new types of forensic expert research", *Bulletin of the O. E. Kutafin University (MSLA), Moscow: O. E. Kutafin Research Institute (MSLA)*, 3, pp. 69–75.

Nedorostkov, V. (2018) "On possible measures to prevent extremism and terrorism in Russia", *Truth and Law*, 4, pp. 56–59.

Neuteboom, W., Shakel, N.V. (2016) "History of cooperation between forensic institutions in the framework of ENFSI", *Forensic Examination of Belarus*, 2(3), pp. 21–25.

Papillon Systems. (2021), available at: https://www.papillon.ru/ (accessed 2021.06.22).

Report on the Erroneous Fingerprint Individualization in the Madrid Train Bombing Case. January 2005, 7(1). https://archives.fbi.gov/archives/about-us/lab/forensic-science-communications/fsc/ jan2005/special_report/2005_special_report.htm?__cf_chl_jschl_tk__=pmd_6c08ea51d426341 aa13a2a77b473542f53321290-1627990031-0-gqNtZGzNAo2jcnBszQg6

Rossinskaya, E.R. (2019a) "Digitalization of forensic and forensic activities interrelations and delineations", *Questions of Expert Practice*, S1, pp. 569–574.

Rossinskaya, E.R. (2019b) *Forensic expert activity: Legal, theoretical and organizational support: A textbook for postgraduate studies*, ed. by E.R. Rossinskaya, E.I. Galyashina, Moscow: Norma: INFRA-M, (in Russ.), 400 p.

Samuel, G., Prainsack, B. (2018b) "Forensic DNA Phenotyping in Europe: Views «on the Ground» from Those Who Have a Professional Stake in the Technology»", *New Genetics and Society*, 38(2): 119–141. https://doi.org/10.1080/14636778.2018.1549984

Samuel, G., Prainsack, B. (2019) "Civil Society Stakeholder Views on Forensic DNA Phenotyping: Balancing Risks and Benefits", *Forensic Science International: Genetics*, 43, pp. 102157. https:// doi.org/10.1016/j.fsigen.2019.102157

Simson L. Garfinkel (2010) "Digital forensics research: The next 10 years", *Digital Investigation*, 7, pp. S64–S73. Published by Elsevier Ltd. All rights reserved. https://doi.org/10.1016/j.diin.2010 .05.009

Snyatkov, E.V., Dorokhin, S.V., Charkin, I.I., Esaulova, A.N., Savinkov M.A. (2017) "Improving the objectivity of the examination of paint coatings using the 3-D scanning method", *Voronezh Scientific and Technical Bulletin*, 1(19), pp. 123–128.

Stepanenko, D.A., Bakhteev, D.V., Evstratova, Yu.A. (2020) "The use of artificial intelligence systems in law enforcement activities", *All-Russian Journal of Criminology*, 14(2), pp. 206–214.

Strengthening Forensic Science in the United States: A Path Forward. (2009) Committee on Identifying the Needs of the Forensic Sciences Community, National Research Council, Washington, 352 p., available at: http://www.nap.edu/catalog/12589.html

TEISMO EKSPERTŲ VEIKLOS KOORDINAVIMO TARYBA. SPRENDIMAS DĖL TEISMO EKSPERTŲ PROFESINĖS ETIKOS KODEKSO PATVIRTINIMO, 2014 m. vasario 24 d. Nr. B-16. Vilnius, available at: https://www.e-tar.lt/portal/legalAct.html?documentId=944af830a82e11e38e1 082d04585b3dd

Trefilov, A.A. (2018) "Peculiarities of proof in criminal proceedings Switzerland", *Judicial Power and Criminal Procedure*, 2, pp. 274–280.

Ushakov, R.M. (2020) "Technology of big data as a vector of development of forensic technology and application prospects in the context of their legality", *Ural Journal of Legal Studies*, 2, pp. 54–70.

Vogel, G. (2018) "In Germany, controversial law gives Bavarian police new power to use DNA", *Science*, available at: https://www.sciencemag.org/news/2018/05/germany-controversial -law-gives-bavarian-police-new-power-use-dna (accessed 2021.07.16).

Wienroth, M. (2020) "Socio-technical disagreements as ethical fora: Parabon NanoLab's forensic DNA Snapshot™ service at the intersection of discourses around robust science, technology validation, and commerce", *BioSocieties*, 15, pp. 28–45. https://doi.org/10.1057/s41292-018-0138-8

Yarmak, K.V. (2011) "About the possibilities of using 3D technologies in forensic examination", in *Informatization and Information Security of Law Enforcement Agencies*, pp. 335–336.

Zapico, S.C. (2021) "Latent fingermarks and DNA recovery", in J. De Alcaraz-Fossoul (ed), *Technologies for fingermark age estimations: A step forward.* Cham: Springer, pp. 285–308. https://doi.org/10 .1007/978-3-030-69337-4_10

Zinin, A.M., Mailis, N.P. (2002) *"Forensic examination"*: Textbook, Moscow, Law and the law, Yurayt-Izdat (in Russ.), 320 p.

15

Data Biases and Predictive Policing System in New Delhi

Baidya Nath Mukherjee and Bhupinder Singh

CONTENTS

Introduction

"When in doubt, predict the present, trend will continue"

- Merkin's Maxim

PredPol, Inc., one of the leading US companies, working on artificial intelligence (AI) and machine learning (ML) in 2012 announced to devise software that can predict future crime based on identifying patterns and analyzing past criminal records (Jain & Chopra, 2021). The idea of "predictive policing" gradually changed the face of criminal administration in the United States. This idea has now been adopted by many countries including India. Law enforcement agencies throughout India now have initiated developing access to big data storage platforms like Hadoop, NoSQL, etc., to store structured and unstructured data and thereby analyze the prediction of the crime pattern within their jurisdiction (FICCI & Ernst & Young, 2020). As of 2021, many states, including Kerala, Maharashtra, Uttar Pradesh, Orissa, and Tripura, have adopted machine learning (ML) as a tool to curb the crime rate in their state. However, in India, Delhi Police pioneered the adoption of machine learning (ML) in

DOI: 10.1201/9781003215998-15

enforcing law and order. In the year 2015, Delhi Police first announced its intention to adopt predictive policing by implementing the Crime Mapping, Analytics, and Predictive System (CMAPS) (Delhi Police Commissioner's Desk, 2015). The software was operationalized with the partnership of the Indian Space Research Organization (ISRO) under the patronage of "Effective use of Space Technology-based tools for Internal Security Scheme", which will be used for live spatial hotspot mapping of crime, patterns in criminal behavior, and suspect analysis. However, the influence of AI is not only limited to policing. The former Chief Justice of India, Justice S. A. Bobde, while addressing on the occasion of Constitution Day, accentuated the importance of artificial intelligence (AI) in enhancing the efficiency of the judiciary (*The Print Team*, 2019). However, he made it clear that AI can never take the place of a judge. However, the use of AI in the judiciary may take some time, but the technology is gradually holding on in the Indian Police System. In November 2019, a Gurugram-based start-up Staqu launched an AI software named "JARVIS", which will help to analyze CCTV footage and flag activities involving violence, pickpocketing, and intrusion besides analyzing crowds (Sen, 2020). The technology mines CCTV footage by producing short real-time alerts and thereby considerably reduces the time to come up with actionable data. Presently, Staqu is working with eight states and UTs of India, including Uttar Pradesh, Punjab, Haryana, Rajasthan, Bihar, and Telangana. Similarly, Innefu Lab, in collaboration with Delhi Police, came up with an AI technology, which offers gait and body analysis (Sen, 2020).

On the one hand, artificial intelligence has the potential to ameliorate many existing structural inefficiencies; on the other hand, without sufficient oversight it may potentially cause harm. The danger and limitations of predictive policing are not unknown. The report by a leading research group, The Partnership on AI, discovered potential flaws in the way predictive policing tools are sketched. Most of the defects are in the data on which it runs. Erroneous and unfair data leads to inaccurate outcomes. Racism being a long-standing concern in policing, there is a probable chance of repeating those existing inequalities. A frail link between the variable and outcome may lead to an inaccurate algorithm, and an inaccurate algorithm means biased results (Human Rights Watch; NAICS: 813311, 2019). India needs to explore such deployments at this stage, where there are continuous efforts to adopt artificial intelligence (AI) in every sphere. Science fiction teaches us how giving power to machines without reasonable oversight does not bring good results for humans. Extreme caution should be adopted while implementing AI in Law enforcement to avoid life-changing mistakes (Human Rights Watch; NAICS: 813311, 2019).

Through this chapter, an attempt has been made to study the various kinds of biases present in the Delhi Police's data-collecting practices and how they can affect the criminal justice administration in the country. Further attempts have been made to suggest a solution against the prevailing opacity of the policing data.

Crimes and Predictive Policing

Predictive policing is only effective for crime types where (1) secured indicators for the risk of the crime are known and (2) data of these indicators can be collected beforehand. Hence, predictive policing is suitable for crimes such as theft, robbery, snatching, eve-teasing, and rioting (Perry et al., 2013, p. 128). The conditions that are ideal prerequisites are (1) inclination towards reporting of crimes by the victims and registration of those crimes by the police, (2) precise determination of time and place of crime, and (3) presence of sufficient data needed to link the relevant indicators of the new crime event with that of the predictive model. The presence of these prerequisites will have a positive impact on the predictive model and will thereby enhance the accuracy and meticulousness of the model. However, there are no such real standards making the crime data unsuitable for predictive models. The considerable amount of data fetched by the police department is analyzed to create threshold data parallel to the preciseness and accuracy of the risk prediction. Quality assessment and evaluation of big data are a must when it comes to predictive policing, as misnomer data will lead to ineffective and substandard predictive models. One of such usual defects is not mentioning the exact crime location while reporting a crime event.

Predictive Policing Model in New Delhi

The Delhi Police Commissioner has notified in the public domain about the adoption of artificial intelligence as a means to predict and prevent crimes (Delhi Police Commissioner's Desk, 2015). The Crime Mapping, Analytics, and Predictive System (CMAPS) has been implemented by the Delhi Police for resource allocation. The technology has been instrumental in saving lives and arresting the perpetrators, as reported through several national news portals (Singh, 2020). Though the overall impact of this technology and its influence on the Delhiites are yet to be known, the technology accesses the Delhi Police Dial 100 data and utilizes ISRO's satellite imageries to spatially locate the call and envisage them as cluster maps to identify crime hotspots (Singh, 2017). The predictive models and algorithms help police to apprehend where the next crime is likely to occur. This concept of policing may be considered a "Paradigm Shift" in the policing pattern in the national capital that stretches across 1,483 km split into 13 districts, and each district is headed by a Deputy Commissioner of Police. CMAPS automatically updates the data every three minutes and replaces the traditional crime mapping pattern of the Delhi Police, where data were manually gathered at an interval of 15 days. The software plots the information gathered on the geospatial map of Delhi, enabling the police to spot the exact location and spatial distribution of crime by identifying the call patterns. Suppose there are too many instances of robbery reported over call to Dial 100 from a particular location that has a bar nearby; those calls are then matched with the past criminal records of that area and analyzed whether the bar is the particular reason behind the instances of crime and whether crime is again going to occur. The data that are inputted in this system are from the Dial 100 database and First Information Report (FIR) data saved in the Crime and Criminal Tracking Network System (CCTNS).

Though CMAPS is an automated predictive policing technology, previously manual hotspot-mapping initiative was also prevalent in the Delhi Policing System. Unlike artificial intelligence, the data were manually inputted by the Digital Mapping Division (DMD) of the Delhi Police. CMAPS, which updates the data every 3 minutes, has replaced the mechanical crime mapping of Delhi Police, where data were updated at an interval of 15 days (Singh, 2017). Now all the district police commissioners are provided with the login ID and password of the CMAPS, who in turn use them to brief the station house officers to decide on resource management and enhance police allocation in problem areas. The technology comes with enhanced features, which help the police to analyze crime by optimizing for a variety of considerations. The technology is equipped with filters for areas like railway/bus terminals, schools, and marketplace, and also has filters, which can point out areas like bars, migrant colonies, minority settlement areas, and so on.

Data Collection and Data Creation

Data collection and analysis carried on by the Digital Mapping Division (DMD) are very crucial components of the predictive policing system because (1) these data are useful for the CMAPS and helpful in building the geospatial maps and infrastructural layers, and (2) the lacunas in the manual mapping indicates the institutional constraints within which CMAPS functions. The first instance of predictive policing by the Delhi Police occurred in 2007 to detect patterns and curb instances of car-jacking in the city, which set the grassroots work for the complete mapping system of Delhi. The mapping structure was built by DMD after it surveyed the city's popular landmarks, metro pillars data, police station boundaries, and upgraded the capital's landscape given important changes. The police station boundary layers are helpful in determining the crime jurisdictions and also significant in determining which areas need more policing. The mapping process was not consistent. Initially, the personnel of the mapping division used to put dots on the digital map of the city built on ArcGIS, indicating crime spots. It did not last long, as the license of ArcGIS expired in 2017. And, secondly, the Delhi Police database consists of around 500,000 addresses, which are inaccurate and full of errors. The data neither had latitudinal–longitudinal coordinates, nor it consisted of detailed information about the geographic structure of the city, which led the officers to plot the locations of crime onto the pertinent jurisdictional police stations. Later on, the mapping officers adopted the technique to point the location of the crime at the nearest possible point to the actual location of the crime. Such inconsistency in the methodology of mapping crime and lack of accuracy raises serious questions about its reliability. These lacunas and inconsistencies are intended to be corrected by the CMAPS, but having an inaccurate precedent upon which the automated mapping will work casts doubt on the credibility of CMAPS. Another drawback of the AI-involved mapping technique is that as the quality of the outcome is directly proportional to the inputted data, sometimes grievous crimes committed in non-selected areas fall out of the radar (Innes et al., 2005, p. 39).

Source of Data

The Dial 100 call center is the main source of big data for predictive policing in Delhi. The emergency calls are routed through 40 channels, with every channel attended by a call executive bearing its own unique ID. While the details of the crime are briefed over call, the call taker notes the information in the "PA 100 form", which records the information and categorizes it into 130 categories including one as miscellaneous when it is difficult to spot the incident accurately. In the case of more than one offense committed in a transaction, the crime of the highest magnitude is considered. This subverts the accuracy of the data in indicating the frequency of data across the spectrum, while the rest of the remaining details are mentioned in the notes section of the form. The location of crime fetched is not always the exact location of crime as the complaints generally fail to mention the exact address due to the semi-planned structure of the city. Due to the irregular structure of the city, the complaints fail to mention the exact location from where they are calling and mention the local landmarks while mentioning their locations as if they are reporting the crime to the local police stations. For instance, a complaint stated that "I am calling from the crime spot which is just opposite to the pink house", making it difficult for the call taker to track the location and enter the exact address details. Due to skewed information by the complaints, the call takers mark the police station as a crime spot instead of the actual location. Such errors lead to an awful data structure. As the call takers are incentivized to conclude calls without reasonable delay, they resort to asking standard questions about the location of the crime and abstain from any kind of detailed inquiry.

Dispatch and PCR Van

After the details in the "PA 100" form are filled, it is forwarded to the dispatch section through the PA 100 software. The dispatch section is divided into 11 zones representing the 11 districts of Delhi (now revised to 13). The call is forwarded to the concerned zone, wherefrom it is further transferred to the concerned Police Control Room (PCR). The officer of the concerned dispatch section provides the details manually to the officer of the PCR van. The form is designed in such a way that a message consisting of all the call details reaches the handheld device of the PCR van, as soon as the call takers close the PA 100 form. The police officers after a proper investigation send the investigation report from the handheld devices which also contain the exact location of the crime. Sometimes, the PCR van officers argue that they are facing problems in handling the devices as they are inadequately trained to operate them. Investigation has revealed that the PCR van officers try to avoid investigation by avoiding visiting the crime scene. It has also been found that they often negotiate with the dispatch personals to assign them with less investigation during their shift.

Heinous Crimes to Go to Green Diary

Thereafter the records of the odious crimes such as rape, robbery, eve-teasing, and theft are forwarded to the dispatch command room after receiving from the call center. If the halaat report confirms the crimes, the same is recorded in the green diary. The categorization of the crime requires a minimum understanding of legal provisions and interpretative skills, which is missing among the officers in the dispatch department. For instance, one day a shopkeeper called to complain about two men who approached the shop in demand of two bags of ghee. Once the shopkeeper put the ghee on the counter, they demanded a bottle of Chawanprash to which the shopkeeper turned around to get it. Taking the opportunity, the offenders ran away with the bags of ghee. After a big debate and applying their on-ground policing experience, the officers concluded that the crime was snatching and the same was recorded in the green diary (Marda & Narayan, 2020, p. 321). The designation of crimes is arbitrary and based on lay man's understanding and absurd logic. Moreover, the officers have their own perspectives and cultural inclination while categorizing crimes. Crimes against women are pre-judged on the basis of their clothes and time of the crime, i.e., if any offense against women happens in daylight and without the women's fault, the crime is acceptable. Else various spurious questions like "Why do women wear such clothes?" or "Why are women out during such times?" arises. These also have an adverse effect on the predictive policing system of Delhi.

The police records depend absolutely on the whims and caprice of the police and whether the officer believes the incident was a crime or not. The major hypothesis behind that is the distrust of the police officers towards the specific complaints. The police officers argue that if the calls to the number 100 would have been chargeable, the frequency of calls from the slum areas would have been much lesser. Furthermore, the police officers quote that most of the offenses against women reported are false on the ground. Most often, the girls implicate their boyfriend in rape accusation over a small quarrel, as told by many police officers in the dispatch command room. As per the "Dial 100" call history, the callers from the slum areas are more in comparison to other areas. The number of calls may not be the indicator of high crime rate, but there is lack of access to other segments of the criminal justice administration for these poor sections. Hence, the crime call needs to be objectively scrutinized and classified as most of them reach a compromise without being registered as FIR, though recorded in green diary and spotted on hotspot maps. Therefore, using the "Dial 100" database to map crime may be problematic as it may lead to a measurement error as all calls to the Dial 100 do not indicate real crime every time (Klinger & Bridges, 2006).

In case of crimes reported during odd hours at night, due to slow-going process of investigation and delay in haalat report, it is entered in the green diary draft with a "pending" remark. It is sent for mapping by 6 a.m. by the next morning and is updated if the haalat report is received before time and if the halaat report does not come, the same call is mapped without corroborating the same with the haalat report. It leads to a mapping error as any spurious call may mark the area as criminal, though the same could be false or misleading. Hence, it cannot be concluded whether the crime calls were in fact what they claimed. Lack of standard and coordination between the officers in charge of the dispatch room and those preparing the green diary ensures that the hoax calls are often recorded as true calls and mapped as crime spots.

Analysis of Data

A segment argues in favor of implementing AI in criminal justice administration and advocates that those algorithms can be fairer and more bias-free in comparison to human decisions (Heaven, 2020). But there exist some genuine problems. The data used to run the predictive policing tools does not at all provide an accurate image of criminal activity as the data mostly comprises the call data, which is a weaker reflection of the actual crime patterns. Every call does not necessarily lead to a crime and is further converted to conviction. Feeding these colored data allows the past to shape the future. The foundational data of CMAPS is a result of a series of irregular police practices, asymmetric reporting, and subjective decision making. This chapter highlights some common threads identified through research, which emerges from the foundational aspect of CMAPS and extends to data practices at the division level, which reflects the institutional culture and has significant effect within the predictive policing system to be introduced by the Delhi Police in future.

Types and Origin of Biases

Data biases are a challenge to every machine learning (ML) system and the existing biases in the Delhi predictive policing system can be outlined with the help of the framework suggested by Suresh and Guttag (2021).

Historical bias: The practice of collecting information in policing is an age-old phenomenon, starting from a compilation of "badmash registers" to maintaining the track records of the criminal tribes; the act has always been a discriminatory one, with a greater onus befalling mostly on the minority groups (Singha, 2015). It is not the case that crime is more probable to occur in slums or poor parts of Delhi, but when it comes to choosing and mapping the place of crime, the apathy of the officers in the mapping rooms towards the people from the minority and poor areas and more forgiving tendency towards the posh parts of Delhi are apparent and pertinent. Secondly, the fact that the policing system in India carries with itself the controversial record of being discriminatory and vicious with vulnerable individuals, which clearly indicates that the bias is actively validated and installed into the data.

Representation bias: The data on which CMAPS operates is basically the data generated through Dial 100 calls and Crime and Criminal Tracking Network System (CCTNS) data. This data sample underrepresents the population from the privileged class in the society due to its poor sampling method. During our visit to the DMD division of the Delhi Police, few officers in the division bear the preconceived notion that the majority of the calls to Dial 100 are from the slum or ghettos, whereas the people from the posh area barely call. This simply indicates that the prospect of crime is markedly higher in the slums and ghettos, leading to tight security in those areas and eventually leading to more arrests and reports.

Measurement bias: It occurs when collecting, computing, or choosing features and labels are used in a prediction problem (Suresh & Guttag, 2021). It is present in the predictive model of the Delhi Police due to the inaccurate spatial distribution of the city and the inability of the individuals to engage with the system and their peer group. The call takers in the Delhi Police headquarters say that the complaints over call often could not mention the accurate address of the crime scene or from where they are calling, and most of the population being the women. Officers further mentioned that the women are mostly unaware of their surroundings or the actual location of the locality. Hence, in most cases, if any such woman calls over Dial 100 to report a crime, the call takers find it difficult to locate the same on the map.

Data Discrimination

India is at the basic stage of building technological proficiency in fully implementing machine learning (ML) for law enforcement. At this point, India needs to develop a mechanism for refining the data used in predictive policing systems. Predictive policing legitimizes discrimination on the pretext of mathematical analysis. The historical data on which predictive algorithms work does not actually indicate who is more likely to commit a crime, rather it indicates who is more policed. The Indian Police is both communal and casteist (Darapuri, 2020), though it is probably the attitude of a considerable section of the police in India. It is very much evident from this research that the way in which data is collected and analyzed in the Delhi police predictive system has an inordinate impact on the vulnerable section of the society. Despite protections guaranteed to the minority and vulnerable groups under the Indian Constitution, these groups have been constantly subjected to violence and discrimination. Despite being innocent, the members of the lower caste groups and religious minority groups are mostly within the reticle of the police with no reasons. As per 2018 data, the fact that the vulnerable and minority class are overrepresented in the Indian jail exhibit discrimination within the criminal justice administration (Thakur & Nagarajan, 2020). Similar discrepancies can be witnessed in the predictive policing system in New Delhi. Crimes reported from organized sectors are more likely to be recorded with an actual address, in comparison to crimes reported from minority areas, thereby causing an imbalance in the crime hotspot mapping. Furthermore, the extensive selective enforcement and discrepancies of the police officers in the DMD leads to overpolicing of the localities inhabited by minorities and vulnerable groups. This leads to racial bias in the crime data and thereby results in a discriminatory feedback loop – the more policing over a certain group, the more likely the algorithm identifies that group as potential criminals. A study conducted by the UK Government's Centre for Data Ethics and Innovation has suggested that pinpointing certain areas as hotspots induces police officers to stop and detain people from those areas due to prejudice and without any fair reason. Furthermore, the device of "layers" in crime mapping analytics and predictive systems can be used to filter ghettos and minority settlement areas, which are a probable extension of the irrational belief that crimes are committed by the people residing in these areas. Such variable used while processing and analyzing data is a protected attribute under Articles 14 and 15 of the Indian Constitution.

The stages of algorithmic discrimination are:

1. *Inaccurate or incomplete training data*: The use of non-reflective and incomplete data has implications on the manner in which the algorithms will analyze the data. A specific problem that arises in such cases is "overfitting" (Danks & London, 2017). In the case of supervised learning systems that require labeled data sets, the problem is more serious. The functioning of the algorithm models depends on the availability of suitable databases. As rightly argued in "Weapons of Math Destruction – how big data increases inequality and threatens democracy" by Cathy O'Neil, in case of type 2 crimes, i.e., vandalism or carrying a small quantity of drugs can be found only if in that area there is an enhanced police surveillance. As there is usually oversurveillance of police in minority populated areas or ghettos, the training data is likely to suggest that minority populations or slum populations commit more crime than they do (O'Neil, 2016).

2. *Algorithmic processing*: The solutions driven by artificial intelligence are an unstructured problem – like the risk profile of individuals. Expressing the amorphous question in source code enables the AI to assess the vast tract of data. The initial process is the assigning of value to the questions that the AI can understand. By means of applying the hidden layers, the AI generates an output value used to assign the desired risk profile to an individual. This assignment of values is termed as a scored society by Citron and Pasquale (Citron & Pasquale, 2014). As per Article 14 of the Indian Constitution, there exists unintentional indirect discrimination as certain classes of people are treated disparately by an apparent neutral algorithm.

3. *Interpretation of output*: Another bias interrelated with the abovementioned algorithm bias is the probability of misinterpretation of algorithm outputs. As the system functions through hidden layers, there is a high probability of disparity between the information the algorithm produces and the output required by the users. Due to such disparity, the user elucidates the quantitative result and applies the same in a qualitative manner.

The crime data that are used as the input for the predictive policing algorithms are not necessarily an accurate representation of the actual criminal activity as it is naturally limited to what police officers record and what the victims choose to report. Furthermore, most of the crime remains uncounted. The National Crime Records Bureau data only records the principal offense, i.e., when an FIR is lodged against an offense of murder and theft, theft may remain uncounted. Such inaccurate data aggravated by unknown bias may lead to inaccurate targeting.

Arbitrariness and the Predictive Model

Equality is antithetic to arbitrariness and also the cornerstone of every civilized society. On the other hand, arbitrariness is the prime concern while recording crimes by police officers. The calls over Dial 100 are interpreted by the officers based on their general understanding and knowledge. It does not follow any standardized format or specific parameters while categorizing the crime in PA 100 form and thereby making way from green diary to CMAPS. Categorization of data and making it algorithm ready is a powerful semantic and political intervention, which once instituted is treated with reverence by the algorithms (Gillespie et al., 2014). Crime categorization is mostly arbitrary and most often works against the vulnerable groups of the society and in turn implants the predilections within the predictive technology. Further, the way PA 100 form is designed implies the social understanding of the institutional notion of crimes. The forms represent the very essence of bureaucratic institutions, and the investment in such forms has been a cultural project to exclude residual categories; for example, genders are expressed in terms of male/female and fail to identify the third gender. The PA 100 form, which is already defined with 130 categories, often limits the capacity of the call takers to record the nuance of every call and forces categorization into accepted norms which are inadequate at best.

Transparency in Predictive Policing: Demand of Contemporary Society

Predictive policing appeals to provide black box solutions to crimes, and the related problem with such a solution is that it lacks sufficient clarity and transparency. The opaqueness in data collection and data analysis demands an effective design of predictive models that can ensure that predictive policing can live up to its promise (Ferguson, 2017, p. 1165). But such an effective model cannot be ensured with reasonable oversight. The questions that arise in contemporary society, with respect to the transparency in the predictive services, are to *whom* the predictive practices are transparent and *how* should the transparency be achieved. Opacity exists at every level of the predictive structure presently. If the keys to the vault of data are held by the law enforcement agency and private developers, then how to achieve better transparency, whether by handing those keys to any designated third-party agency or by keeping the vault open for any curious passerby. Assorted institutions such as Civilian Review Boards, Law enforcement, court, or legislature can be entrusted with the duty to oversee predictive policing. Furthermore, defect with any of these establishments leaves public surveillance as the best option (BAKKE, 2018).

Lack of transparency shadows accountability. Through our research, we found that a thick layer of opacity exists around predictive policing and related big data. It is kept aloof from the purview of Right to Information under the pretext of safety, security, and strategic information of the state. Transparency and accountability are the cornerstones of every democracy, and the unique feature of predictive police demands the need for transparency. Though absolute transparency may not be practicable in the case of predictive policing, a qualified transparency can obviously be ensured to retain public trust and confidence. An autonomous statutory institution free from any political and bureaucratic pressure can be established, which can keep a tap on the predictive policing being exercised in India. India needs to ensure an effective policing structure and not overpolicing the minority segment of the society.

Conclusion and Suggestions

The main aim of this article was to give insight into the institutional approach toward data and the biases present in the predictive policing system of New Delhi. While attempting to dispel the prevalent myths about predictive policing, an attempt has been made to identify the biases within the Delhi police predictive structure. The existing predictive model is not structured, textured, and pervasive. Despite the presence of ample evidence reflecting the inefficiency of predictive policing, the lure of predictive policing is too good to resist. The algorithms of predictive policing are as good as the data they utilize. Biased data institutionalize crimes against minorities. It is proposed that any predictive technology must be assessed through the lens of the institutional limitation and culture within which it functions. Keeping into consideration the nation-wide acclaim for predictive policing systems and artificial intelligence, the government should pass an algorithmic accountability law to ensure better implementation. Based on these important recommendations, it is proposed to make efforts to hold the significant systems in the public domain by means of focusing on a few aspects surrounding the system: (1) extensive research focusing on the various aspects of predictive policing should be promoted. (2) The procurement procedure of the predictive policing model must be transparent by means of proper notification and available for public scrutiny. (3) Standard operating procedures, clear guidelines for discriminatory actions, grievance reporting formats, and redressal procedures should be formulated at the state level and notified to the public. (4) An autonomous public authority should be appointed at the central level, who will keep a tap on the quality of data used in the predictive policing system, and lastly (5) a proper mechanism for appealing, fixing, and correcting should be parallelly developed. It gives a framework for the police authorities to think through the effectiveness of the predictive policing system and help the researchers and the society at large to examine the impact of the predictive policing system without any concrete evidence of the outcome. Law enforcement agencies should make predictive policing more inclusive by making it accessible and open to the masses and addressing the issues of bias. Bias is what is deeply embedded in the policing system in India. Hence, the solution does not lie in finding an impartial policing system; instead, the solution lies in ensuring a proper mechanism of checks and balances by means of maintaining transparency and accountability.

References

Bakke, E. (2018). Predictive Policing: The Argument for Public Transparency. *Annual Survey of American Law*. Retrieved July 11, 2021, from https://annualsurveyofamericanlaw.org/wp-content/uploads/2019/08/74-1-Predictive-Policing-The-Argument-for-Public-Transparency.pdf

Citron, D. K., & Pasquale, F. A. (2014). The Scored Society: Due Process for Automated Predictions. *Washington Law Review*, *89*(1), 1–32. https://digitalcommons.law.umaryland.edu/cgi/viewcontent.cgi?article=2435&context=fac_pubs

Danks, D., & London, A. J. (2017). *Algorithmic Bias in Autonomous Systems*. 26th International Joint Conference on Artificial Intelligence. https://www.cmu.edu/dietrich/philosophy/docs/london/IJCAI17-AlgorithmicBias-Distrib.pdf

Darapuri, S. R. (2020, September 09). The Police in India Is Both Casteist and Communal. *The Wire*. https://thewire.in/caste/police-casteist-communal

Delhi Police Commissioner's Desk. (2015). *From the Commissioner's Desk*. Delhi Police. Retrieved May 25, 2021, from https://www.delhipolice.nic.in/CP%20Forword2015.pdf

Ferguson, A. G. (2017). Policing Predictive Policing. *Washington University Law Review*, *94*(5), 1112–1188. https://openscholarship.wustl.edu/cgi/viewcontent.cgi?article=6306&context=law_lawreview

FICCI & Ernst & Young. (2020). *Predictive Policing and Way Forward*. [Virtual]. India. Retrieved May 24, 2021, from https://ficci.in/spdocument/23009/FICCI_EY_Predictive%20Policing_.pdf

Gillespie, T., Boczkowski, P. J., & Foot, K. A. (2014). *Media Technologies Essays on Communication, Materiality, and Society*. MIT Press. https://mitpress.mit.edu/books/media-technologies

Heaven, W. D. (2020, July 17). *Predictive Policing Algorithms are Racist: They Need to be dismantled*. MIT Technology Review. Retrieved June 30, 2021, from https://www.technologyreview.com/2020/07/17/1005396/predictive-policing-algorithms-racist-dismantled-machine-learning-bias-criminal-justice/

Human Rights Watch; NAICS: 813311. (2019, May 08). AI in Law Enforcement Needs Clear Oversight. *Financial Times*; London (UK), 10–11.

Innes, M., Fielding, N., & Cope, N. (2005, January 01). 'The Appliance of Science?': The Theory and Practice of Crime Intelligence Analysis. *The British Journal of Criminology*, *45*(I), 39–57. https://doi.org/10.1093/bjc/azh053

Jain, G., & Chopra, R. (2021, April 1). *AI tech is increasingly being used by police worldwide: Here's why India needs to regulate it*. Scroll.in. Retrieved May 24, 2021, from https://scroll.in/article/989094/ai-tech-is-increasingly-being-used-by-police-worldwide-heres-why-india-needs-to-regulate-it

Klinger, D. A., & Bridges, G. S. (2006). Measurement Error in Calls-For-Service as an Indicator of Crime. *Criminology*, *35*(4), 705–726. https://doi.org/10.1111/j.1745-9125.1997.tb01236.x

Marda, V., & Narayan, S. (2020). *Data in New Delhi's Predictive Policing System*. New York: Association for Computing Machinery. https://doi.org/10.1145/3351095.3372865

O'Neil, C. (2016). *Weapons of Math Destruction*. New York: Crown. 9780553418811

Perry, W. L., McInnis, B., Price, C. C., Smith, S. C., & Hollywood, J. S. (2013). *Predictive Policing: The Role of Crime Forecasting in Law Enforcement Operations*. Rand Corporation. https://www.rand.org/content/dam/rand/pubs/research_reports/RR200/RR233/RAND_RR233.pdf

The Print Team. (2019, November 27). AI Can Improve Judicial System's Efficiency' – Full Text of CJI Bobde's Constitution Day Speech. *The Print*. https://theprint.in/judiciary/ai-can-improve -judicial-systems-efficiency-full-text-of-cji-bobdes-constitution-day-speech/326893/

Sen, S. (2020, March 21). How AI Can Be Used in Policing to Reform Criminal Justice System. *The Print*. https://theprint.in/tech/how-ai-can-be-used-in-policing-to-reform-criminal-justice -system/384786/

Singh, K. P. (2017, February 27). Preventing Crime Before it Happens: How Data is Helping Delhi Police. *Hindustan Times*. https://www.hindustantimes.com/delhi/delhi-police-is-using-pre- crime-data-analysis-to-send-its-men-to-likely-trouble-spots/story-hZcCRyWMVoNSsRhnBN- gOHI.html

Singh, V. (2020, March 12). 1,100 Rioters Identified Using Facial Recognition Technology: Amit Shah. *The Hindu*. https://www.thehindu.com/news/cities/Delhi/1100-rioters-identified-using -facial-recognition-technology-amit-shah/article31044548.ece

Singha, R. (2015). Punished by Surveillance: Policing 'Dangerousness' in Colonial India, 1872–1918. *Modern Asian Studies*, 49(2), 241–269. https://doi.org/10.1017/S0026749X13000462

Suresh, H., & Guttag, J. (2021, June 15). *A Framework for Understanding Sources of Harm throughout the Machine Learning Life Cycle*. Cornell University. https://arxiv.org/abs/1901.10002

Thakur, A., & Nagarajan, R. (2020, January 17). Why Minorities Have a Major Presence in Prisons. *Times of India*. https://timesofindia.indiatimes.com/india/in-a-minority-but-a-major-presence -in-our-prisons/articleshow/73266299.cms

16

How Far Data Mining Is Legal!

Jayanta Ghosh and Vijoy Kumar Sinha

CONTENTS

Introduction

The words "big data" and "big data analytics" are initially based on artificial intelligence, intelligence, and corporate analysis. These concepts were employed during the 1950s, 1990s, and 2000s [1]. While many contend that big data is an imprecise word for various ideas [2], most of these definitions have a theme often described as huge pools of information that may be collected, transmitted, aggregated, stored, and analyzed [3].

Smart algorithms can recognize and anticipate target group behavior and provide insight into occurrences in real time. This allows decision makers to use big data evidence rather than intuition [4]. However, in addition to the advantages of big data, the usage of (personal) data is also expanding. The research report of the European Union stated that while using a web browser, the users left their electronic identification by digital cookies without providing their personal data, which helps to gather information about an individual [5]. Internet surveillance is not restricted. Mobile phones, cameras, and payments are all the tools to collect information and monitor, which also include biometrics, social media networks, loyalty cards, and interactive services. Inspite of having these issue, the big data tool has an advantage in the field of production, health care, communications and research but presently is being misuse in other fields [6].

Data mining is the process in which current data search for new information. The primary challenge faced by data mining involves converting low-level data into higher, more compact, more abstract, or more valuable businesses, which are generally too vast to grasp [7]. The use of analytical data and finding algorithms to list and extract patterns is essential to the data mining procedure [7].

Data mining is "the non-trivial extraction of implicit, previously unknown, and potentially useful information from data" [7]. In every element of this definition, we can realize that data mining can differentiate from the earlier known "data processing" and "database enquiry technologies".

Implicit data extraction indicates that data mining outputs are not already present in database data items [7]. Traditional database retrieval information provides data arrays consisting of individual (or complete) record field data from the existing knowledge in answering the defined or predefined datafile inquiry. The query is an expression of a traditional database, i.e., the answer to a query itself is a data item in the database [7]. It is an array of many objects. However, the techniques of data mining pull information implicitly from the database – knowledge that "there is usually no previous data" is disclosed [7]. In general, data mining finds patterns or correlations between data items or records not discovered earlier but revealed in the data itself (and are not data objects by themselves). Data mining thereby pulls previously undiscovered information. In other words, data mining uses complicated procedures to answer those questions which were not asked before. To initiate data mining for the unearthing procedure, the data mining phase must thus be constrained such that outcomes can achieve an appreciation of the original objective.

This paper focuses on issues, such as privacy and ethics, which individuals are vulnerable to because of the reality of data mining. This paper is divided into five parts, with the first describing the concept and origin of data mining and how data mining works. The second part highlights the dynamics of the legal system with the data mining concept. The third part studies the issues of data mining that conflict with the law. The fourth part discusses the various laws of the Indian legal system to tackle data mining. The fifth section of the paper explores the judicial position of India on data mining and privacy. The paper concludes with a set of suggestions on how to deal with this situation.

Stages of Data Mining

The concept of "data mining" mentions a specific phase in the process of finding knowledge. The steps which make up the process of finding knowledge include [7]

- Pre-processing;
- Data mining;
- Post-processing.

Pre-processing

In "pre-processing" – identification of objectives – the first and maybe most essential stage consists of comprehending the field within which discovery methods are used and determining desired results. In the case of national security, the first task is to identify the possible terrorists from the large population with the help of electronic traces in their personal and transactional databases. The main objective should be to achieve that while safeguarding principles of privacy and civil liberty in a larger context of the use of these technologies in a free democracy.

The next step is to gather, filter, and store data during pre-processing and to assemble data to be extracted into a single data file for future processing. Applications for database mining usually involve collecting data in one database, commonly known as a database storage facility. However, existing R&D initiatives attempt to build "virtual" data accumulation approaches in which the access of a target query to many remote data files may be negotiated on local terms. A "prospecting agent" will be able to access dispersed data files across a network and adjust to the local knowledge circumstances and needs, both for database processing and for data access. The data mining process does not need a single, large database (which is crucial for retaining privacy safeguards in domestic security applications). "Provided that certain (very low) size thresholds are exceeded in providing statistical validity, data mining techniques can be applied to databases of a wide variety of sizes". A significant technological benefit of virtual accumulation (in contrast to a single, large database) is that it allows privacy rules to be enforced on additional data files or specific types of data before accessing it.

Once data has been acquired, conventional data mining techniques need to be cleaned or converted – removing or correcting unnecessary, inaccurate, or useless data and standardizing data for processing. In general, cleaning and transformation are regarded to be separate stages. However, there are certain exceptions. Transformation entails normalization of data-type conversion and attributes selection, whereas cleaning entails removing noise and dealing with missing data. In other words, cleansing entails cleaning up the database, which will help in order to match better the intended processing or algorithm utilized.

In order to achieve usable results from existing data mining tools, data cleaning and transformation are essential. Data mining does not require "clean" data, although it does benefit from it. Statisticians can correct known data issues, like missing or noisy data, using procedures already in existence.

Data Mining

In the data mining procedure, one of the essential elements is the use of certain algorithms in order to extract, uncover, and identify some completely undiscovered features from the cleaned data. Data mining also includes developing predictive and descriptive models, and judgment is a part of post-processing and involves applying the framework to fresh data.

An important part of the data mining process is to use certain data cleaning algorithms to extract, identify, or discover some completely unknown features of the data. Strictly speaking, data mining involves the development of descriptive or predictive models, while evaluation is part of post-processing that involves applying models to new data.

For expository purposes, there are two primary algorithm types: *clustering and association rules*. Classification and unsupervised clustering are included in the term "clustering." Data classification involves classifying data into which was before categories, while unsupervised clustering involves mapping data to the new classification generated during the data analysis procedure and deduction. In addition, association rules may be used to uncover relationships between data characteristics and incorporate ways to define dependencies between data, locate linkages between data, and represent sequential patterns in data, among other things.

In both clustering and association rules, within the limits of the algorithm and according to its rules, higher-level data gets exposed by lower-level data. Comprehensive domain-based knowledge and data familiarity are needed to build and pick algorithms to prevent unnecessary, deceptive, or trivial attribute correlations. We must analyze even the most relevant correlations to determine whether or not they are relevant and beneficial to our initial objectives. It is possible that the application of data mining technologies blindly might lead to the identification of meaningless and erroneous patterns.

Generally speaking, there are two primary methods or methodologies for data mining – "top-down" and "bottom-up" approaches. Approaches from the top-down start with a hypothesis and aim to validate it. As a result of the bottom-up data mining, the hypothesis may be formed, or it can be developed from real-world experience. As a result of this bottom-up method, the data are analyzed, and patterns are found to build a hypothesis or model. If you know what you are searching for, you may use a bottom-up method that is either supervised or directed.

Post-processing

The first step in post-processing is to analyze and evaluate the found patterns and determine their relevance within the domain context. For example, subcategories of clusters can be generated, weak linkages can be shown, or specific patterns that are deemed useless in the domain context might be removed. The crucial stage in the knowledge discovery process is applying the newly acquired information to fresh knowledge to unearth more correlations, identify discrete content, or forecast upcoming behaviors.

Notably, it is not possible to evaluate the results of data mining in isolation from the knowledge discovery process as a whole. The extra processes related to pre-processing, including relevant pre- and post-domain data in the development process, as well as post-processing in itself, are essential to ensure that meaningful conclusions can be drawn from the data. The mechanized part of the knowledge discovery process converts data into information. Post-processing experts analyze and apply the data to the field of judgment and transform the data into knowledge as the basis of action.

Analytical decision making by humans cannot be replaced by data mining techniques. Statisticians may use data mining as a vital computer tool for supporting the analysis of humans in combining new information, formulating and making predictions, and constructing models imaginatively, including valid behavior profiles. Large or dispersed datasets may include useful information buried amid enormous volumes of irrelevant data, which may be analyzed by human analysts. In the pre-processing and post-processing stages, a more in-depth analysis should be carried out, including developing theories, hypotheses about new contexts or action models, filtering out irrelevant information, and finding clues on topics that require a high degree of professional knowledge. However, automation is superior to humans: use judgment in certain situations and eliminate human error or prejudice in such situations (Figure 16.1).

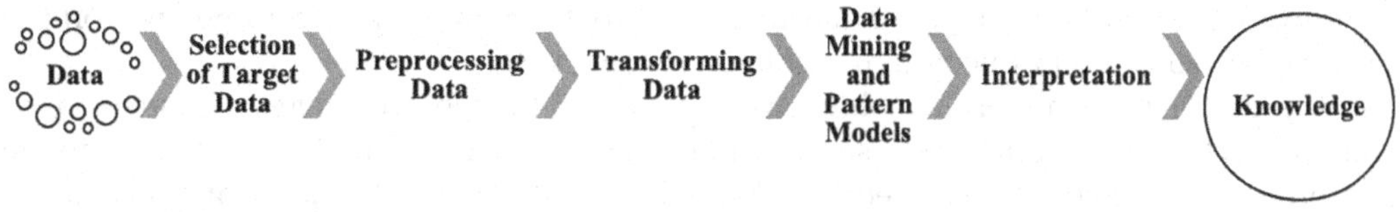

FIGURE 16.1
Represents the data mining process from the data to the conversion of knowledge.

Existing Affairs of Data Mining in Conflict with the Law

Mining applications include huge quantities of data that may have come from many sources, possibly external to them. The quality of the information is thus not guaranteed. In addition, while pre-processing information takes place before executing a data quality application, people transact unpredictably, leading to the fast expiry of personal data. This might lead to an improvement in data quality. Inaccurate patterns will most likely be found when mining takes place over expired data [8].

If personal data is gathered, it is usually planned strategically and removed from the individual, thereby increasing confidentiality but causing abuse and error. There has recently been a tendency towards the treatment and selling of personal information as a resource. The data may be copied and re-sold easily. The sentence data mining employs the metaphor of natural resource exploitation, which further helps to see data as goods. Furthermore, there was little intellectual and legal discussion on the topic of whether it was permissible with regard to human rights to trade in personal data. The negative implications of this kind of business are comparable to data mining: privacy violation and the harmful repercussions of incorrect information. However, when the potential for legal responsibility is created, incorrect data have an increased consequence for organizations trading personal data. The practice of data trading or data mining might create errors and so lose severely in the courts.

Any company determined to have harmed (or failed to prevent injury) a person for whom it has a duty of care may be compensated. The plaintiff can seek compensation financially for any subsequent damages produced by the negligent conduct after culpability has been proven (tort of negligence). The size and accuracy of the losses are mostly unique for each claimant, although the limits of carelessness are never closed. A mining exercise may mistakenly condemn a person as bad credit risk, and actions on this premise may be taken to the detriment of that person. Algorithms might in some circumstances be properly classifiable, but they can be based on problematic characteristics (i.e., morally sensitive), for example, gender, race, religion, or sexual orientation.

Another legal problem is whether organizations that manipulate personal data are permitted to defame a person whose data they have extracted. Because data mining creates previously new material, it is highly feasible that the organization which uses the data mining technology may be regarded as the author for the purposes of defamation legislation. In addition, organizations trading in personal data may be considered to be comparable to publishers, as collections of data are issued for sale and dissemination. Hence, if the material may be judged by courts to be defamatory, data mining firms might be held responsible for damages.

Market experts are frequently unrealistic in their opinion. Confidentiality is a barrier to consumer comprehension and improved product provision. Each year, 200 super offices sell hundreds of millions of personal information to direct marketers, commercial companies, researchers, and public organizations in America [8].

In the era of information, absolute privacy cannot be achieved. Due to their shared everyday activities, people disclose data like e-government, ATM transactions, mobile communications, e-commerce, credit card, email, etc. [9]. The most important data sources are cookies and server logs, customer data, smart Internet operators, and central resident records and official registers as well. Due to the lack of information about the social and ethical impacts of data, reliable technology for processing, storing, and transmitting data guarantees unrestricted use of this data by other people and institutions and is inappropriate in most cases. The result is not necessarily causal. Data mining does not provide any information, so it is mainly for miners to decide how and where to use the obtained data. This method will have a negative impact on people's privacy and rights [10].

Crime prevention data mining is based on individuals recording acts such as traveling, mailing, telephone conversations, trading, and meetings between persons. Data mining in such applications might provide erroneous and defective findings that are typically privacy infringements and damaging to people. For instance, an individual might be categorized as a suspect while he or she is unrelated to the business. In such a situation, the repercussions may be quite significant, and the entire procedure is undoubtedly unethical, if not criminal. If the data are inaccurate, defective, or even manufactured, negative consequences, including litigation, unfavorable publicity, loss of reputation, discrimination, etc., may further worsen. This is where a procedure that is intrinsically immoral may also become illegal.

Legal concerns relating to data mining utilization are complicated and hard to assess and cannot be placed under specific laws as such. Furthermore, the legal theories and legal systems of countries have not yet correctly addressed the issues of the digital era [11]. Most countries decide to enhance their privacy by a special statute. However, such legislation cannot be expected to predict every conceivable breach and sort of crime. In fact, legal disputes are typically settled by the application of some other codes or legislation available to safeguard the rights of the individual in general. Here, the main difficulty is the expansion of IT technology, which over a relatively short period makes norms, regulations, and even laws outdated. It is exceedingly hard to identify legal or private, let alone establish coherent court rules in many cases. In many cases, the key reason is that the majority of physical limits of technology hardware and software are fast getting newer and more versatile. This issue affects the methods and settings used for an application, including data mining. Diverse approaches and practices of privacy conservation are designed to resolve such ethical and legal questions by minimizing the chances of happening data mining applications.

Data Mining and Indian Legal System

The Puttaswamy verdict is a significant legal advance in the privacy discourse and, in particular, in information privacy; previous legislative initiatives have been made in various sectors of India to guarantee information privacy. This covers the general data protection requirements in accordance with IT Act, 2000 (IT Act) and numerous sectoral data protection legislation [12].

Information Technology (Amendment) Act, 2008 (IT ACT)

As per the "Information Technology Act of 2008", civil and criminal penalties can be imposed in cases of improper disclosure, abuse, and breach of contractual obligations pertaining to personal data. If a body corporate negligently did now no longer preserve safety practices at the same time as keeping non-public information or records could be held liable for damages to the affected man or woman beneath Section 43A [13] of the *IT Act, 2008*. It is crucial to highlight that the compensation that can be sought by the aggrieved party in such situations has no maximum limit.

The Act also prescribes computer-related offenses and punishment for those computer-related offenses [14] like dishonestly stealing computer resources [15], identity theft [16], cheating by personation [17], violation of privacy [18], transmitting obscene material [19], publishing depicting children in sexually explicit [20], abetment of offense [21], and attempt to commit offenses disclosure of information [22].

IT Rules, 2021

The "Ministry of Electronics and Information Technology" promulgates the "Information Technology (Guidelines for Intermediaries and Digital Media Ethics Code) Rules, 2021" on February 25, 2021. The new provisions supersede the old ones from 2011 and are published under the "Information Technology Act, 2000". The new rules require companies operating in India to create new positions on the ground, add particular terms to companies' policies, adopt new technology for data removal, and execute a new plan of action. The new rules have several parts, sections that contain essential due diligence for intermediaries and ethical codes for digital media. This rule focuses on the new due diligence essentials for all intermediaries or Parts I and II of the rules. The rules define three types of intermediaries: "intermediaries" [23], "social media intermediaries" [24], and "significant social media intermediaries" [25].

The rules also require distributors to publish rules and regulations, privacy policies, and terms of use for their users on their website or mobile application once a year [26]. If any user fails to comply with the rules, then the intermediary has the right to terminate the user [27]. Even after termination or cancelation of registration, the intermediary can retain their information for not more than 180 days [28]. The intermediary must stipulate in its policy that no user may publish, modify, disclose, view, transfer, upload, update, share, or store any other person's data that is harmful to children or violates intellectual property rights, deceive citizens, threaten national security, software viruses [29]. All intermediaries must authorize a grievance officer [30]. After receiving the complaint, the grievance officer must confirm receipt of the complaint within 24 h and 15 days in order to resolve the issue [30].

The Information Technology (Reasonable Security Practices and Sensitive Personal Data or Information) Rules, 2011 (SPDI Rules)

The SPDI Rules have been put out under Section 43A [13] of the Information Technology Act, 2000. The OECD guiding guidelines, in particular, collection restriction, purpose definition, restrictive use, and personal engagement, are included in the SPDI Rules to a limited extent. The SPDI Regulations prescribe specific information collecting requirements [31] and urge that this only be done for a legitimate reason related to the operation of the organization [32]. In addition, a thorough privacy policy is necessary for each organization

[33]. The SPDI Rules also layout directions that information may be maintained for the duration of time [34] and allows persons to update their information [35]. Disclosure shall not be allowed without the agreement of the supplier or without the permission of such disclosure by contract [36]. The provider is not needed to consent in the exchange of information with government agencies, and such information can be distributed for the objectives of identification verification, preventing, detecting and investigating events, prosecution, and the punishment of crime [37]. The regulations of the SPDI only apply to companies [38] and leave government and public bodies out of their remit; the rules are limited by sensitive personal details, including traits such as sexual orientation, health records and history, biometrics, etc. [39]. In addition, its latest ruling was given in 2011 by the "Cyber Appellate Tribunal", which handles appeals in accordance with IT Act. Therefore, the lack of an efficient implementing mechanism raises issues over the application of the SPDI Rules. Therefore, an extensive law must be established to safeguard personal data in all its aspects sufficiently and to guarantee that they are implemented effectively.

The Aadhaar (Targeted Delivery of Financial and Other Subsidies, Benefits, and Services) Act, 2016 (Aadhaar Act)

The *Aadhaar Act* allows the government, in accordance with such biometric [40] information, to gather people [41] identifying information, including its biometrics, to provide a unique identification number or Aadhaar number, and then to offer subsidies, benefits, and services for them in a targeted form [42]. The Aadhaar Act also provides authentication services based on Aadhaar whereby an applicant entity (public/government or private/agency) may request the Indian Unique Identifying Authority (UIDAI) to verify that identification information provided by persons is correct to enable them to extend services [43]. The applicant shall seek the individual's agreement before obtaining identification information for authentication purposes and shall use its identity information exclusively for authentication purposes [44]. The Aadhaar law provides for the administering of the UIDAI Act, specifically the *Aadhaar Act* [45]. It also sets up a "Central Identifies Data Repository" [46] database, including Aadhaar numbers with related demographic and biometric information [47]. In accordance with the Aadhaar law, collecting, storing, and using sensitive personal data is a requirement for receiving a grant, service, or benefit [48]. Although the *Aadhaar Act* does not, per se, require the use of an Aadhaar number (specifically as an entitlement pursuant to Section 3) other than for certain subsidies, benefits and services financed under the Indian Consolidated Fund, the use of the Aadhaar number is in practice compulsory for the benefit of most services through a number of cognate laws [49].

Different data protection standards are recognized via the Aadhaar Act and its rules to safeguard the safety and privacy of Aadhaar number holding companies. Firstly, UIDAI has an obligation to make sure safety and sensitive information of the identification data and the individual's authentication data, including taking all steps necessary to prevent illegal access, use, or divulgation of such data and accidentally or intentionally destroying, losing, or damaging them [50]. Furthermore, the Aadhaar Act bans the sharing and use of basic biometric information for purposes other than generating and authenticating Aadhaar numbers [51]. Under some situations, the exchange of information other than basic biometric information is allowed. The Aadhaar Act also enables individuals to request the UIDAI to give them access to their identification data [52] and authentication documents [53] (except their basic biometric information) [52]. She may also seek to amend their demographics if they are changed or inaccurate, and if they are lost or modified her biometric information [54]. The Aadhaar (Data Security) Regulations 2016 also set forth the Data Protection Standard for sensitive personal data gathered under the *Aadhaar Act* (Aadhaar Security Regulations). In order to ensure information is secured, the Aadhaar Security Regulations place a duty on the UIDAI to develop a protection policy laying forth the technological and organizational methods it would use [55].

Personal Data Protection Bill, 2019

The bill was submitted to Lok Sabha by the "Ministry of Electronics and Information Technology" on December 11, 2019. The aims of the bill [56] include the safeguarding of the data related to any information which is sensitive, proper mechanisms of using private information and its subsequent circulation, developing a foundation of security and trust between people and organizations regarding the process of private data, the due rights of the people whose private information is being processed must be safeguarded, and finally, a proposal for the structure of measures that are related to the organization and other technical measures.

Criminal Law (IPC)

Data privacy is not expressly addressed in Indian criminal law. Liability for such violations must be inferred from related offenses under the Indian Penal Code (IPC). If you are stealing or converting "movable property" for your personal use, you might be facing criminal charges under Section 403 [57] of the Penal Code of India. So, if it falls within the responsibility of another, the question arises of whose rights are to be safeguarded? Whoever misappropriates another person's property is penalized under criminal breach of trust, according to Sections 405 [58] and 409 [59]. No one can dishonestly remove any movable property from the custody of another person without that person's agreement. If someone does this, they are considered to have committed theft and are penalized. As a matter of fact, the state is the sole victim of the offense. As a result, maintaining law and order is a major problem. Punishments are included in the Penal Code, while damages are assessed by a jury in civil cases. In addressing the right problem, it is important to bring this up. A suitable link exists between the Indian Penal Code and data protection laws when it comes to addressing the right. In this context, the state is likewise tasked with protecting the personal data of individuals.

Telegraph Act

Many laws apply to the telecommunications industry, like "Indian Telegraph Act, 1885 (Telegraph Act)", "The Indian Wireless Telegraphy Act, 1933", "The Telecom Regulatory Authority of India Act, 1997 (TRAI Act)", and several regulations were issued. However, data protection rules in the telecom sector are mainly run by the "Unified License Agreement (ULA)" issued to "Telecom Service Providers (TSP)" by the "Department of Telecommunications (DoT)". DoT prescribes the layout and categories of the data to be gathered from the person [60]. TSP is obliged to take appropriate measures to protect the confidentiality and confidentiality of the data and information about the personnel it provides services, and it receives this information from these personnel due to the services provided [61]. As a result, the TSP is required to keep a record of all commercial transactions for at least one year in order to be examined by the Department of Telecom [62]. When it comes to security measures, the TSP is bound by a number of duties, including the requirement to induct into its telecom network only those network parts that have been tested in accordance with current Indian or International security standards [63], among others [64]. If the subject has consented to the disclosure, it must be done in line with the conditions of consent [65]. This privacy invasion is justified in large part by national security. TSPs must also oblige with the Telegraph Act, which allows the government to intercept communications in the event of a crisis. Interception is subject to certain procedural guarantees [66]. "Telecom Regulatory Authority of India (TRAI)" has also issued the "Telecom Commercial Communication Preference Regulations, 2010 (TRAI Regulations)" to tackle unsought business communications [67]. This facility was put up in accordance with the provisions of TRAI's Regulations [68] by TSP, allowing customers to opt-out of receiving business communications; however, these guidelines are limited to messaging and other communications over the phone and do not apply to advertisements displayed in email applications or web browsers.

Credit Information Companies (Regulation) Act, 2005 (CIC Act)

It is a legal requirement in the financial business to maintain customer confidentiality and adhere to data protection laws. As far as data privacy laws go, the CIC Act and CIC Regulations are among the most extensive in the financial industry, if not the most complete.

CIC laws mainly apply to credit reporting agencies and treat them as information collectors [69]. All credit information [70] collected, used, and disclosed by credit reporting agencies must be accurate, complete, and protected from loss or unauthorized use, access, or disclosure in accordance with CIC and data protection laws [71]. They must also comply with widely recognized data protection standards, such as data collection restrictions, data use restrictions, data accuracy, data storage, and data access and alteration [72].

RBI Regulations

"Know your customer" (KYC) standards restrict the types of information banks can request from their customers. After collection, the bank is obliged to keep this information confidential. Bank and credit card issuer NBFC's rupee cooperative debit and prepaid card transactions, the 2009 Basic Customer Service Announcement and Customer Obligation Guidelines, all of which stipulate the confidentiality and confidentiality of various agencies within the federal scope.

Health Sector Regulations

The nature of health information, which is inherently vulnerable, seems insufficient in terms of the legislative framework for data protection in the health industry. The "Clinical Establishments (Central Government) Rules, 2012 (Clinical Establishments Rules)" required medical care providers to keep documents and provide electronic "medical records/electronic health records" and to store medical records electronically.

The "SPDI Rules" recognize health information as confidential and thus govern the acquisition, use, and disclosure of sensitive personal data. As noted previously, however, the SPDI Regulations only apply in the commercial sector, therefore keeping the entire public health sector out of reach.

The "Regulations 2002" issued by the "Indian Medical Council (Professional Conduct, Etiquette and Ethics)", established in 1956, require physician–patient confidentiality except when a patient's data is needed by law to be disclosed or if an individual or a community is a serious and identified risk, or the condition is notifiable [73]. This current legislation and regulations need to be analyzed and, if necessary, amended in conjunction with the implementation of new data protection laws.

Consumer Protection

In 2019, the "Consumer Protection Act" was passed, which defined the psychological or emotional pain caused by property damage as loss and regarded the disclosure of sensitive personal data as unfair business practices [74].

Examples of Data Mining

In May 2016, it came out that the Indian Railways e-ticketing website had been hacked and that the personal information of about 10 million users had been taken. Reports say that a CD with IRCTC customers' phone numbers, dates of birth, and other personal information was sold for Rs 15,000. The Indian Railways Catering and Tourism Corporation said it wasn't true that their website had been hacked.

It is alleged that cybercriminals stole financial information from Axis Bank, Visa, ICICI Bank, YES Bank, and MasterCard, in October 2016 after malware was injected into the Hitachi Payment Services system. This attack affected approximately 3.2 million cards.

According to reports, in May 2017, an Indian food delivery service, Zomato, was hacked, and 17 million customers' email addresses and passwords were stolen. A Zomato spokesperson verified that no payment information had been stolen in the alleged hacking incident.

In July 2017, information of consumers of cell phone provider Reliance Jio was exposed on the Internet. Compromised data includes mail ID, full name, SIM activation dates, and Jio cell phone number. Reliance Jio did not initially confirm the leak, saying it is conducting an internal investigation into the situation. The company subsequently reported "illegal access" to its network to the police.

Judicial Position of India

The Indian Judiciary indirectly stated about data mining in any of the cases, but the judiciary showed concern about privacy matters from the very beginning, and it has been reflected in their judgments from time to time. The below mentioned cases discusses about the data privacy which can be broken if safety precautions are not taken. These cases are the example for an individual that how they prevent themselves from data breach. Judiciously the privacy issues can only be inspected by data mining which is the starting point. The interpretation of every decision can be turned in a high-tech way, and the source is the information or data, which is prime in the legal arena.

In a case M.P Sharma v. Satish Chandra [75], the Supreme Court ruled that the allegations of search and seizure violated Article 19(1)(f) of the Constitution because a simple search itself does not affect property rights, Although its seizure affects them, these effects are temporary and reasonable restriction in the right to privacy. Therefore, the right to privacy is established in the Indian Constitution in accordance with Article 19(1)(a) and

Article 21. Hence, the interpretation of physical property and digital property also has the same value.

In a similar angle, "R. Rajagopal v. State of Tamil Nadu" [76], also known as "Auto Shanker Case", the Supreme Court specifically pointed out that the right to privacy or soli- tude is protected by Article 21 of the Constitution. Citizens have the right to protect private life, family privacy, education, marriage, motherhood, and procreation from infringement, among other things. It is also accepted that the idea of protection of personal information needs to be respected, and it needs not to be breached through data mining.

In "State of Maharashtra v. Madhulkar Narain" [77], the court held that no one can vio- late the privacy of a woman of easy virtue. It is evident that the exposure of very personal affairs needs to be checked with proper legal setup and control of data mining.

In a similar way, tapping phones and extracting information are also a blatant violation of the right to privacy. In "People's Union for Civil Liberties v. Union of India" [78], widely known as "Phone Tapping Case", the Supreme Court ruled that wiretapping is a serious violation of personal privacy, which is part of the right to life and personal freedom stipu- lated in Article 21 of the Constitution. The state should not resort to wiretapping without public interest or urgently needed for public emergency or safety. The Indian Telegraph Act of 1885 and the Information Technology Act of 2000 allow the government to con- duct surveillance activities based on certain standards conducive to India's sovereignty and integrity, national security, friendly relations with foreign countries, public order, or incitement to obstruct crime. These reasons are based on the reasonable restrictions of the Indian Constitution on freedom of speech.

Access to the database and extraction of the information which is available to the public needs to be assessed before sharing. In "Unique Identification Authority of India (UIA) & Anr. v. Central Bureau of Investigation" [79], as part of an investigation, the "Central Bureau of Investigation" requested to access the database of (UIA). However, in an interim ruling, the Supreme Court ruled that India's unique identification agency should not dis- close the biometric information of the person assigned the Aadhaar number to other agen- cies without prior written authorization from an individual.

Proper address to data mining aspects cannot be supported without acknowledgment of the "right to privacy". Over a year of discussion in the apex court, India finally recognized the right to privacy as a fundamental right. In "Justice K.S. Puttuswamy (Retd.) & Anr. v. Union of India & Ors." [80], the Supreme Court, privacy issues are dealt with based on a unique identity system. The court argued whether this right is protected by the constitu- tion, and if so, where the right has come from since its existence in India. The Attorney General stated that the "right to privacy is not a basic right of Indian people" till date, but in the end, the court left because of the previous ruling of the bigger bench rejection of the right to privacy. It became a major case in which the "right to privacy was recognized as a basic right" and therefore questioned the broader constitutional discussion; in the end, the case was submitted to a larger court for final judgment. In 2017, the Supreme Court of India ruled that the "right to privacy is a constitutional right".

Conclusion and Suggestions

It is accepted that the privacy breach is the starting point of data mining. The idea to min- gle data mining with different aspects of privacy, ethics, business practice, and consumer practice in this research has given priority to the demand for legal phenomenon. When it comes to data mining, it is particularly important to be aware of what is going on with the data mining element on a global and domestic scale. An enormous amount of Internet activity is taking place on the domestic level today. For example, when you are speaking about Internet messages in this way, it means that there are significant numbers of messages that go across the Internet that are being intercepted by someone other than the sender or the intended receiver. Companies must allow consumers to access and manage their personal data without being overly prescriptive. The international and national laws, as well as international instruments, implement mechanisms that prohibit the transfer of data outside of countries where the standards of fair information practices are needed to be maintained adequately. Data export bans are likely to follow the failure of the global community to agree on standards of fair information practice.

All throughout history, banks have stored people's trust in the financial system. Consumers use banks because they need secure payment systems that guarantee privacy. Using legal, technological, and societal approaches, regulatory flexibility can be increased. Once the vision of the appropriate techniques is adopted in combining various regulatory schemes in a harmonious manner, then the justification for legal aspects of data mining can be seen for societal benefit.

References

1. Chen, H., Chiang, R., & Storey, V. (2012). Business intelligence and analytics: From big data to big impact. *MIS Quarterly, 36*(4), 1165–1188. https://doi.org/10.2307/41703503
2. Schroeck, M., Shockley, R., Smart, J., Morales, D. R., & Tufano, P. (2013). Analytics: The real-world use of big data, how innovative enterprises extract value from uncertain data, IBM global business services business analytics and optimization executive report, IBM institute for business value. In collaboration with Saïd Business School, The University of Oxford, 1–19. Retrieved from https://www.ibm.com/downloads/cas/E4BWZ1PY
3. Manyika, J., Chui, M., Brown, B., Bughin, J., Dobbs, R., Roxburgh, C., & Byers A. H., (2011). *Big data: The next frontier for innovation, competition, and productivity.* McKinsey Global Institute, 1–137. Retrieved from http://www.mckinsey.com/insights/business_technology/big_data_the_next_frontier_for_innovation
4. McAfee, A., & Brynjolfsson, E. (2012). Big data: The management revolution. *Harvard Business Review*, 1–19. Retrieved from https://www.utoledo.edu/library/help/guides/docs/apastyle.pdf
5. TNS Opinion & Social. (2011). Attitudes on data protection and electronic identity in the European Union. Retrieved from https://www.utoledo.edu/library/help/guides/docs/apa-style.pdf
6. Jetten, L., & Sharon, S. (2015). Selected issues concerning the ethical use of big data health analytics, 1–8. Retrieved from https://bigdata.fpf.org/wp-content/uploads/2015/12/Jetten-Sharon-Selected-Issues-Concerning-the-Ethical-Use-of-Big-Data.pdf
7. Taipale, K. K. (2003). Data mining and domestic security: Connecting the dots to make sense of data. *Columbia Science and Technology Law Review, 5*, 1–83.
8. Wahlstrom, K., Roddick, J., Sarre, W., Estivill-Castro, V., & de Vries, D. (2009). Legal and technical issues of privacy preservation in data mining. In John Wang (ed.), *Encyclopaedia of data warehousing and mining* (2nd ed., pp. 1158–1163). Hershey, NY: Information Science Reference.
9. van Wel, L., & Royakkers, L. (2004). Ethical issues in web data mining. *Ethics and Information Technology, 6*, 129–140. Retrieved from https://pure.tue.nl/ws/files/1901768/612259.pdf

10. Okur, M. C. (2008). On ethical and legal aspects of data mining. *Journal of Yasar University*, 3(11), 1455–1461. Retrieved from https://dergipark.org.tr/tr/download/article-file/179196
11. Seifert, J. W. (2007). Data mining and homeland security: An overview. CRS Report for Congress. Retrieved from https://fas.org/sgp/crs/homesec/RL31798.pdf
12. Ministry of Electronics and Information Technology. (2017). White paper of the committee of experts on a data protection framework for India. Retrieved from https://www.meity.gov.in/writereaddata/files/white_paper_on_data_protection_in_india_171127_final_v2.pdf
13. Section 43A, Information Technology Act, 2008
14. Section 66, Information Technology Act, 2008
15. Section 66B, Information Technology Act, 2008
16. Section 66C, Information Technology Act, 2008
17. Section 66D, Information Technology Act, 2008
18. Section 66E, Information Technology Act, 2008
19. Section 67, Information Technology Act, 2008
20. Section 67B, Information Technology Act, 2008
21. Section 84B, Information Technology Act, 2008
22. Section 84C, Information Technology Act, 2008
23. Section 2(i), Information Technology (Intermediary Guidelines and Digital Media Ethics Code) Rules, 2021
24. Section 2(w), Information Technology (Intermediary Guidelines and Digital Media Ethics Code) Rules, 2021
25. Section 2(v), Information Technology (Intermediary Guidelines and Digital Media Ethics Code) Rules, 2021
26. Section 3(1)(c), Information Technology (Intermediary Guidelines and Digital Media Ethics Code) Rules, 2021
27. Section 3(1)(c), Information Technology (Intermediary Guidelines and Digital Media Ethics Code) Rules, 2021
28. Section 3(1)(g), Information Technology (Intermediary Guidelines and Digital Media Ethics Code) Rules, 2021
29. Section 3(1)(b), Information Technology (Intermediary Guidelines and Digital Media Ethics Code) Rules, 2021
30. Section 3(2), Information Technology (Intermediary Guidelines and Digital Media Ethics Code) Rules, 2021
31. Rule 5(2), The Information Technology (Reasonable Security Practices and Sensitive Personal Data or Information) Rules, 2011
32. Rule 5(2), The Information Technology (Reasonable Security Practices and Sensitive Personal Data or Information) Rules, 2011
33. Rule 4, The Information Technology (Reasonable Security Practices and Sensitive Personal Data or Information) Rules, 2011
34. Rule 5(4), The Information Technology (Reasonable Security Practices and Sensitive Personal Data or Information) Rules, 2011
35. Rule 5(6), The Information Technology (Reasonable Security Practices and Sensitive Personal Data or Information) Rules, 2011
36. Rule 6, The Information Technology (Reasonable Security Practices and Sensitive Personal Data or Information) Rules, 2011
37. Rule 6(1), The Information Technology (Reasonable Security Practices and Sensitive Personal Data or Information) Rules, 2011
38. Section 43A, Information Technology Act, 2008
39. Rule 3, The Information Technology (Reasonable Security Practices and Sensitive Personal Data or Information) Rules, 2011
40. Section 3, The Aadhaar (Targeted Delivery of Financial and other Subsidies, Benefits and Services) Act, 2016, 2016
41. Section 30, The Aadhaar (Targeted Delivery of Financial and other Subsidies, Benefits and Services) Act, 2016, 2016

42. Section 7, The Aadhaar (Targeted Delivery of Financial and other Subsidies, Benefits and Services) Act, 2016, 2016
43. Section 8, The Aadhaar (Targeted Delivery of Financial and other Subsidies, Benefits and Services) Act, 2016, 2016
44. Section 8(2), The Aadhaar (Targeted Delivery of Financial and other Subsidies, Benefits and Services) Act, 2016, 2016
45. Section 11, The Aadhaar (Targeted Delivery of Financial and other Subsidies, Benefits and Services) Act, 2016, 2016
46. Section 10, The Aadhaar (Targeted Delivery of Financial and other Subsidies, Benefits and Services) Act, 2016, 2016
47. Section 2(h), The Aadhaar (Targeted Delivery of Financial and other Subsidies, Benefits and Services) Act, 2016, 2016
48. Section 7, The Aadhaar (Targeted Delivery of Financial and other Subsidies, Benefits and Services) Act, 2016, 2016
49. Gupta, K., & Roy, S. (2017, March 25) Aadhaar to be mandatory for mobile phone verification, Mint. Retrieved from http://www.livemint.com/Industry/wyGskI48Ak73ETJ5XW0diK/Aadhaar-now-amust-for-all-mobile-phone-connections-after-ta.html
50. Section 28, The Aadhaar (Targeted Delivery of Financial and other Subsidies, Benefits and Services) Act, 2016, 2016
51. Section 29, The Aadhaar (Targeted Delivery of Financial and other Subsidies, Benefits and Services) Act, 2016, 2016
52. Section 28(5), The Aadhaar (Targeted Delivery of Financial and other Subsidies, Benefits and Services) Act, 2016, 2016
53. Section 32(2), The Aadhaar (Targeted Delivery of Financial and other Subsidies, Benefits and Services) Act, 2016, 2016
54. Section 31, The Aadhaar (Targeted Delivery of Financial and other Subsidies, Benefits and Services) Act, 2016, 2016
55. Regulation 3, Aadhaar Security Regulations
56. The Personal Data Protection Bill, 2008
57. Section 403, Indian Penal Code, 1860
58. Section 405, Indian Penal Code, 1860
59. Section 409, Indian Penal Code, 1860
60. Clause 39.17, Unified License Agreement
61. Clause 37.2, Unified License Agreement
62. Clause 39.20, Unified License Agreement
63. Clause 39.7, Unified License Agreement
64. Clause 39, Unified License Agreement
65. Clause 37.2, Unified License Agreement
66. Rule 419A, Telegraph Act
67. Regulation 2(i), Telecom Regulatory Authority of India Regulations
68. Regulation 3, Telecom Regulatory Authority of India Regulations
69. Regulation 2(b), Credit Information Companies Regulations
70. Section 20, Credit Information Companies (Regulation) Act, 2005
71. Section 19, Credit Information Companies (Regulation) Act, 2005
72. Chapter VI, Privacy Principles, Credit Information Companies Regulations
73. Regulation 2.2, Professional Conduct, Etiquette and Ethics) Regulations, 2002
74. Section 2(22), Consumer Protection Act, 2019
75. AIR 1954 SCR 1077
76. 1995 AIR 264, 1994 SCC (6) 632
77. AIR 1991 SC 207, 1991 (61) FLR 688
78. AIR 1997 SC 568
79. SLP (CRL) 2524/2014
80. WP (Civil) 494/2012; (2017) 10 SCC 1; AIR 2017 SC 4161

17

Peaceful Neo-Luddism: How the Employee and the Employer Get on a Digital Diet?

Dulatova Natalya Vladimirovna and Ofman Elena Mikhailovna

CONTENTS

Introduction

A certain part of people believe that technological progress is the most important factor in the development of human society. But it also has the other side of the coin: the generation of social problems, a threat to fundamental human rights and freedoms, widespread surveillance, and the emergence of new addictions. Today, we can clearly see how society is rigorously engaging in 4.0 industry and transitioning to smart traffic lights, cars, smart buildings, smart health care, and even smart cities. However, the digitalization and transformation of society come with a number of challenges related to security, privacy, and accountability. Many questions arise related to artificial intelligence: What is the place of artificial consciousness in the big picture of the world and the role of legal regulation in such a digital world? Which decisions without human involvement can a machine make on its own? How to define the decision-making process of artificial intelligence [7]? What is the basis for evaluating the use of data needed in the context of global data policy [2. pp. 754–772]? What data needs to be protected?

The International Labor Organization's 2019 centennial report, Work for a Better Future, includes the following wording:

> The world of work is being transformed by new forces. … Technologically driven production processes can make labor unnecessary, ultimately alienating workers and oppressing their development. Automation can reduce worker control and autonomy and impoverish the content of their work, resulting in loss of skills and lower satisfaction … New technology generate large amounts of data on workers. This poses risks for workers' privacy.

Robots are likely to replace humans, A. Levitan and Clifford M. Johnson believed back in 1982. The scientists cited research at Carnegie at Mellon University. They believed that today's generation of robots could technically replace nearly 7 million existing jobs in

DOI: 10.1201/9781003215998-17

existing factories, equal to one-third of all manufacturing jobs [5. pp. 10–14]. In 1988 in Russia, the famous scientist G. S. Pospelov [3] wrote: "Artificial Intelligence is a complex scientific and technological problem, the solution of which requires collaborative work of mathematicians, electronic engineers, programmers, knowledge engineers, philosophers, psychologists, sociologists and other specialists in a wide variety of aspects of human society".

Today humanity observes robots writing poetry, music, or paintings instead of humans. However, modern scientific and technological society has not yet reached the necessary level to realize these ideas. The relationship between artificial intelligence and the legal sciences is extremely complex. In the near future law, philosophy, technical research, sociology, ethics, and other fields will develop together, all of which will provide a wide scope for research. Nevertheless, at all times there has been a part of the human community that has denied the role and importance of artificial intelligence or has been limited by technical progress. Today, denial or voluntary self-limitation of modern technology can be expressed in the absence of a human gadget or phone app. Such people call themselves "neo-Luddites" or "digital anarchists" (nihilists). They are critical of modern conditions, especially in the field of computer technology.

Historically, the term "neo-Luddism", that is, "new Luddism", is related to the historical legacy of the introduction of machine production during the Industrial Revolution in England. Luddites believed that machines had displaced humans, and this led to technological unemployment. As production increased, the situation of workers in England became even more difficult. They worked at least 14–16 h. The owner of the manufactory did not abide by the law by exploiting his workers. Wages were extremely low, and the workers did not have enough bread. Children were also brought to work in the manufactories, the youngest helpers being only 5 or 6 years old. The Luddites believed (and not without reason) that the mass introduction of machines would worsen the situation of the workers. They were serious about destroying the fruits of progress and invented a legend about their ideological leader, a certain Ned Ludd. The legend was that Ludd had destroyed the stocking machines and made the common people think that these machines had taken jobs away from the women who made their living by knitting stockings. In an instant, no one needed the skilled labor of stocking makers.

Modern neo-Luddites believe that technological processes have a negative impact on the environment, people, and society as a whole. Among the modern theorists of neo-Luddism is Martin Heidegger. He reflected his negative attitude to technological progress in the following words: "The artificial production of life is logically balanced by the artificial production of death". In one of his lectures, Heidegger compared the motorization of agriculture with the murder of people in gas chambers.

The fears of neo-luddites are well-founded: job loss is not a pleasant situation. According to the World Economic Forum's 2018 report, 75 million jobs will disappear from the economic sector by 2022. But forum experts promise 133 million new jobs that will require specialized skills. Economic sectors are disappearing because of the adoption of digital technology. For example, unmanned trucks could replace America's popular truck driving profession, which allows someone without a college degree to have a middle-class salary – 3.5 million drivers will be out of work.

Some countries are trying to develop strategies against possible unemployment in cases of mass introduction of innovative technologies.

> More than 50 billion machines in the world are expected to be connected in the next
> five years. The introduction of artificial intelligence into the workplace in the era of the

fourth industrial revolution is different from the third revolution and is that the work environment will be changed, which will entail changing every human life.

Some authors have argued that "not believing in the potential of artificial intelligence is like not believing in the potential of humans" [6].

The authors conclude that the model of digital labor organization today is aimed at protecting the rights of employers (business) and the state. Labor legislation in this area is inconsistent and imperfect, and the adopted normative legal acts not only do not regulate these relations but also mislead their participants, which can ultimately lead to serious systemic violations of human rights.

The authors believe that despite the massive penetration of media technologies in all spheres, it is necessary to keep in mind their main purpose – to simplify and regulate life, and not to become an object of fanaticism.

Monitoring and Control: Digital in the Workplace

Employers actively and arbitrarily use digital technology. This seems to be done for the purpose of controlling and monitoring employee behavior. This raises the problem of readiness of labor legislation to effectively regulate labor relations and to observe the balance of rights and interests of employees, employers, and the state. In modern conditions, the mechanism of exercising the subjective rights and obligations of employees and employers is transformed: The interaction of subjects of labor law is becoming more indirect; a striking manifestation of the digitalization of labor law is the use of numerous and diverse technical devices by employers to monitor the behavior of employees, which often leads to surveillance (actually – to spying) on the latter and invasion of his private/personal life.

A striking manifestation of the digitalization of labor law is the ability of employers to control and monitor employees. Researchers point to various technologies and technical devices: body-worn equipment, exoskeletons, collaboration software, virtual personal assistants, comier networks, and facial recognition systems. From the perspective of labor relations, these tools constantly collect, produce, exchange, and integrate information, which is used by the employer for various purposes and leads to the transformation of labor relations, to its so-called "genetic change" [1. pp. 95–121].

In today's environment, it is extremely easy for employers to monitor employee behavior. Forms of monitoring are becoming especially unconventional through the use of various devices and technologies that have become ubiquitous in the digital economy: (1) monitoring phone calls; (2) monitoring of email messages; (3) Internet monitoring; (4) monitoring of social networks of employees posts on various events and facts, both related and unrelated to the performance of their work duties; (5) monitoring the employee's activity at the computer during working hours; (6) video surveillance of the employee and wiretapping of his conversations; (7) GPRS, GPS monitoring; (8) monitoring of psycho-physiological condition of an employee during work performance.

In the US, there is a specific term "electronic monitoring". Within the framework of said monitoring, the employer may collect information on employees' activities during working time by any possible means (using a computer, telephone, radio, camera, electromagnetic, photoelectronic, photo-optical systems), except direct observation. It is unacceptable to collect information about employees on the employer's public property or information, the collection of which is prohibited by state or federal law. The employer can implement any type of electronic monitoring. He or she must follow the rules: notify all interested employees in advance in writing and inform them about the types of monitoring to be applied. Exceptions to advance notice to employees are when the employer has reasonable grounds to believe that employees are committing misconduct, violating the rights of the employer or other employees, and creating a hostile work environment by their behavior, and when electronic surveillance can provide evidence of such unlawful behavior. In these exceptional cases, the employer may conduct surveillance on an employee without his or her prior written notice.

The European Court of Human Rights believes that the state authorities can and must create an effective legal mechanism giving the employer the right to control the labor activities of employees through electronic and other means of communication. Only the manipulation of employees from the workplace is subject to control. The purpose of this regulation is to analyze the activities of employees, which may be conventionally called "non-workplace communications". Nevertheless, the discretion afforded by the state in this regard cannot be unlimited. The national authorities must guarantee that the employer's application of measures to control correspondence and other means of communication is supported by reasonable and sufficient safeguards against abuse. In the famous September 5, 2017, the ruling of the European Court of Human Rights, "Barbulescu v. Romania" (Complaint No. 61496/08) lists such guarantees: (1) the employee must be made aware of the possibility that the employer may use measures to control employees, as well as the implementation of such measures; (2) the employer's control and privacy intrusion into the employee's personal space must not be unlimited; (3) the employer must have legitimate reasons to justify controlling the employee's electronic monitoring and indicate (if necessary) these legitimate reasons; (4) the employer has the right to choose a certain form of monitoring. In doing so, the employer must assess whether it is possible to monitor in a less aggressive manner without intruding on the employee's privacy; (5) the employer must use the consequences of the control to achieve the objective stated in advance.

For example, in America, employees have the right to demand that their employer respect their right to the confidentiality of information. Employers should determine the form of monitoring necessary to protect their business interest; analyze the law to be applied; inform employees of the monitoring policy; set reasonable monitoring limits (employer monitoring policies and practices should be designed solely to protect their business interest and should not be overly broad: employers should not monitor employees' private communications, install video cameras in locker rooms, restrooms, toilets, etc.); apply measures to ensure the confidentiality of information obtained as a result of monitoring; appoint an official who is responsible for the storage, processing, and transmission of the information received.

It is important to note that the Labor Code of the Russian Federation does not differentiate such concepts as "control" and "monitoring". Moreover, this law does not even contain a definition of the concept of "control"; nothing is said about "supervision".

Meanwhile, the European Foundation for the Improvement of Working and Living Conditions (Eurofund) in its December 2020 study "Employee monitoring and surveillance: The challenges of digitalization" established that there are important differences between control and surveillance by the employer over the behavior of the employee.

> Compared to control (monitoring) of employee behavior, which is usually limited to collecting information about work-related activities, surveillance is more intrusive and aggressive because it uses technology by the employer to collect a broader range of information about both the employee's work activities and his (the employee's) non-work-related activities ... Although the practices of monitoring and surveillance overlap, the distinction between the two suggests that the resulting observation All this suggests that observation violates a person's autonomy.

Indeed, control should most often be understood as a one-time or repeated (but not systematic!) verification by the employer of the employee's activities to determine whether his actions comply with labor law and other regulations containing norms of labor law (local regulations, job descriptions, regulations, orders of the employer, technical rules, etc.), with the right to take necessary and possible legal responsibility for the guilty employee (bringing to disciplinary and/or material responsibility).

Supervision is essentially the observation of an employee, a group of employees, or all employees, which involves a system of regular and prolonged inspections not of a single action but of a set of actions of the employee(s). As a rule, the employer conducts surveillance throughout the working day/shift, as a result of which the employer becomes aware of information about the private/personal life of the employee or such information that the latter did not want to share (e.g., health or illness).

Digital surveillance has advantages: The technologies used by employers allow to ensure compliance of employees with the established rules (including occupational safety rules), minimizing accidents at work (for example, technology can monitor exposure to dangerous materials on the human body; employee health, blood pressure, blood sugar levels, fatigue); help to increase work intensity, reduce downtime; are an effective verification mechanism for the commission of an accident.

But experts point to the side (negative) effects of such constant and pervasive surveillance: Work becomes a source of stress (reduced autonomy and increased work intensity; increased stress and anxiety levels); reduced confidence in the employer, supervisor, and other workers; psychological discomfort in performing work; increased competition between employees, which increases the likelihood of gamification of the work process.

Russian courts state:

> The employer's use of video recording equipment (in fact – any digital technology) does not violate the basic constitutional rights of employees, because video recording is not disclosure of personal data of an employee and is not used to establish the circumstances of his private life or his personal and family secrets ... Setting of video surveillance at the employer's checkpoint is not a change in terms of the employment contract, as they are related to ensuring.

When courts consider cases to restore the violated rights of employees in the application by the employer of various types of control and surveillance of their behavior using digital technology, they usually establish: in order to qualify the employer's behavior as lawful he (the employer) must comply with a number of procedural rules: (1) adopt a document regulating the issue of the employer's implementation of digital monitoring of employees; (2) define the purpose of digital monitoring and the period of storage of records of digital monitoring of employees; (3) familiarize employees with the relevant rules and notify them of the introduction and implementation of such surveillance; (4) place information signs on the implementation of observation by the employer in the workplace; (5) appoint a specially authorized person who will have access to the employees' personal data; (6) conclude an agreement with employees on the gathering and processing of personal data or include appropriate provisions in the employment contract, specifying the purposes of processing, intended sources of personal data, methods of obtaining such data, consequences of employees' refusal to consent to their receipt.

The authors of the article believe that the above measures are not enough to protect the rights of employees from the intrusion of employers into their private lives of employees. Not always the information that became available in the course of monitoring and controlling the behavior of employees is related to the production activities of the employer. Probably, today the issue should be resolved at the legislative level, if not a ban, then at least a significant restriction of the employer's right to perform digital control and digital surveillance of employees, if this monitoring of employees' behavior can be made in the classical form, that is, directly by the employer himself, his representative or other employees, to whom the controlled person is directly subordinated.

Anti-Machine Rebellion: Combating Artificial Intelligence and Robotization in Russia

Today, the need for corporate digitalization has become most acute in the midst of the global pandemic of the new coronavirus, when the need for remote working, remote sales, and automated reporting arose. As the practice of many companies showed, the faster the digitalization of the business, the more the rank-and-file employees of such an enterprise resisted it. This phenomenon has been called "digital neo-Luddism". Employees are both a prerequisite for a successful digital system and one of the main reasons why digital innovations fail. People dislike change to such an extent that they often deliberately sabotage innovations in the work environment. In the first quarter of the 19th century, Luddites smashed stocking machines; nowadays, "neo-Luddites" are quietly "breaking" the digital environment. It is noteworthy that in both cases workers react to change according to the model of psychologist Elizabeth Kübler-Ross: "Denial – Anger – Bargaining – Depression – Acceptance". That is, the person almost never loses hope of returning to the previous conditions. And only at the last stage – acceptance – does he become convinced that everything is over and he has to live with it. But before that, at the stage of "Denial" employees sabotage innovation, find reasons to bypass the system or break it. In some cases, the confrontation between the customer relationship management (CRM) or other digital system and its users turns into an unspoken competition with the company's management: who is who. No matter how the digital system has made life more convenient, this is primarily a change of habits and new rules, and also – additional external control. Therefore, employees are bound to resist. As evidence of this thesis, the following examples can be given.

The case was in an African country. The team accompanying the digitalization project from Moscow began to receive reports that the monitoring system of one of the excavators regularly malfunctioned and stopped working. After a while, however, it would return to work for no apparent reason. The engineer who serviced the system always found it to be working properly. Diagnostics and replacements did nothing. It seemed as if some sort of magic was going on: the system would occasionally shut down and then start up again. The situation was saved by the engineer. He asked to see the excavator operator's work schedule. He compared the equipment shutdown times to the machine names and saw that all the shutdowns were occurring during the same person's shift. It turned out that when the employee came to his workplace, he would neatly disconnect the power wires and then reconnect them at the end of his shift. According to him, he just didn't like anyone controlling his work.

Not all workers are as peaceful as this African worker. Some don't just disconnect equipment, they try to break it. This is even safer for the perpetrator because it is enough to damage the equipment once, and with regular shutdowns, it is much easier to catch and find the culprit.

Another question arises: How to break equipment so that it looks like it broke itself? On production sites, it is possible to simulate a power surge or a short circuit. It turned out to be no problem at production sites: One of the workers took a welding machine and used it on the power terminals. In doing so, something like a short circuit occurred in the equipment with a tremendous amount of current. If the equipment is not specially protected, it is almost guaranteed to burn out. And it will be almost impossible to understand that someone did it on purpose. Nevertheless, by video surveillance, the actions of such a worker have been traced.

But what to do if the equipment is reliably protected? If it is protected against water and dust and you can drop it and destroy it as much as you want? And that, as it turns out, isn't a problem for neo-luddites with a knack for it, at least for one worker in a field outside the Urals. One worker decided to use sulfuric acid to disable it. Using a syringe, he poured acid from the battery into the computer through a connector that was not protected against such manipulation. When the worker was caught, he was asked why he was trying to disable the circular vision system on the excavator, which was useful to him. He said he thought he had been listening through it to discuss his superiors with his friends.

In some cases, automation eliminates inefficiencies that employees are already used to. For example, a driver used to deliver goods to seven points and spend the rest of his time dealing with personal issues. Now the system builds routes for him from 15 points instead of seven. From the driver's point of view, this will not look like optimization, but like an attempt to give him extra work for the same salary.

Another example: When introducing robots to a video surveillance installation company, a very interesting situation occurred. The employees pretended to accept the robots and work with them, but decided to circumvent one system. They didn't like the program's strict regime of timekeeping, where they couldn't be a minute late. Eventually, the three programmers who worked with the robots decided to "negotiate" with the machines and created an additional feature that allowed a call from a cell phone to a specific number to show the system the time of arrival to work.

In most cases, the introduction of artificial intelligence in any field is a tool to optimize processes and overall benefit, not to replace employees. But not all Luddites understand this, so they start to gently block the process: They fear that they may be replaced by algorithms. Any attempts at a direct confrontation between the luddite and the system are fairly easy to detect, so employees try to question the results of artificial intelligence in the eyes of management.

The strategy for implementing a new digital system depends on the answer to the question of what position the employer takes. In some cases, total control over the actions of employees in the enterprise is established and management consciously accepts the attitude "my employees can only be managed with a whip. If you comply, you get a bonus; if you don't comply, you get a reprimand or are fired". In other cases, complete freedom and trust on the part of the employer dominates. In such companies, the means of control are not needed at all. That is, automation tools are implemented not so much to collect data and control, but to facilitate the work of employees, giving them a tool to increase efficiency.

V.V. Putin said at a conference on artificial intelligence that he did not rule out the risk of machines rebelling against humans.

> Is it possible for machines to revolt? There is a fear that they will control humans. But people will control these machines. That's the first thing. The second is, are there any dangers and risks in this regard? Yes, there are,

said the President of Russia. Perhaps one of the important problems facing the state in the current period (the era of digitalization, algorithmization, and robotization) is the coming unemployment. Reasonable questions arise: "How prepared is the legislature to deal with the universal unemployment of people who do not work in traditional jobs?" "What is the future of humanity without the ability to realize their abilities to work?"

It is very important to remember that modern peaceful neo-Luddism and its associates must clearly distinguish between aggressive hostile technology (i.e., the means of total government surveillance or corporate control of people) and friendly digital technology (i.e., necessary work automation tools aimed at ensuring worker compliance (including workplace safety rules), minimizing workplace accidents (for example, technology can monitor exposure to hazardous substances on the human body; employee health, blood pressure, blood sugar levels, fatigue, and overwork); technology contributes to increasing the intensity of work, reducing downtime; is an effective and objective mechanism to confirm the commission of disciplinary misconduct by employees).

Conclusion

Technological advances have made possible what employers were previously deprived of: control and observation of employees are allowed in real time, access to employees (to their thoughts, actions) is possible at any time and in any place. Employers can establish employees committed a disciplinary offense by various means: video recording of behavior and misconduct that caused material damage to the employer's property, the audio recording of conversations, listening to phone calls made by employees, GPS and GPRS monitoring, monitoring email, records (posts) of social media accounts, interaction with employees by sending them legally significant documents via email, and employee biochipping.

Employers' use of digital technology engenders the problem of invasion of the private (personal) life of the employee, as there are processes that worsen the legal situation of the latter: the boundaries between work and personal life are blurred, the employer may become aware of such information about the employee that the latter does not want to advertise and which may affect his status (up to and including termination).

Russian legislation on the protection of personal data is not elaborate and incorrect, and this manifests itself, for example, in relation to the collection, storage, use, and transfer of employees' personal data obtained through various technological means. The employees' negative evaluation of the changes is understandable.

Information about the employee obtained using information technology refers to personal data, which under the law is not biometric personal data, because its processing is necessary for the performance of the contract (employment contract). At the same time, biometric personal data is data on the physiological and biological characteristics of an individual by which his or her identity can be established. The employee must consent to the treatment of such data. But Roskomnadzor emphasizes that biometric personal data is related to the purpose of its processing. If the purpose is to establish the identity of the subject of personal data, processing must take place with the consent of the employee, and the information must be classified as biometric personal data (for example, a photographic image and other information used to ensure single and/or multiple entries into the protected area and identification of a citizen). Processing employee information for other reasons does not give the employee's personal data the status of biometric data, since they are aimed at confirming their belongings to a specific individual whose identity has already been determined and whose personal data is already available to the operator. Information about the employee that has become known to the employer does not belong to the specified category of personal data (a photographic image contained in the employee's personnel file). In the second case, the processing of the employee's personal data takes place without the employee's consent.

Thus, according to Roskomnadzor, the main thing for recognizing personal data as biometric is the purpose of its handling, not its content.

This approach is flawed. Obviously, it is more correct to say that the main characteristic of biometric personal data is the scope of its content, the fact that they allow the identification of the employee, rather than the objectives of processing. In addition, with the use of modern means and methods of monitoring and control (audio and video recording), the employer can determine the identity of the employee – the violator of labor discipline. All this allows us to argue that personal data obtained by the employer in the process of monitoring (surveillance) of the employee with the help of information technology (voice, retina, fingerprints, image of the employee) should be classified as biometric personal data. The employer must obtain the employee's consent to process such information. Exceptions are possible in cases prescribed by law.

European countries use the results of monitoring and control to improve labor productivity, as well as to control employee behavior and increase labor productivity; monitoring solely for the purpose of controlling employee behavior is very rarely used by employers. The authors of this article conducted a survey of employees of a large employer in Russia, the data of which (the survey) formed the basis for several conclusions. Firstly, according to the survey data provided, employees could not answer the question about the purpose of employer monitoring of their behavior. Three purposes prevailed: improvement of the labor process; evaluation of labor efficiency; and control over the observance of labor protection rules in the performance of work duties. In addition, 91% of respondents indicated that they did not know what data the employer collected about them. Surveillance is carried out frequently (36%) or sometimes (55%), and the answer "never" was obtained in 9% of cases. The attitude towards observation is indifferent (55%) or even positive (36%); negative was 9%. The most common ways to monitor employee behavior are software surveillance (40%); performance monitoring (20%); video monitoring (20%); and Internet usage monitoring (20%). Access control is only 10%. There is no control of phone calls and no control over employee spending.

The answer to the question of whether the employer monitors/monitors the behavior of employees outside of working hours meets the requirements of international, Russian and foreign legislation (in the opinion of employees, the employer does not control their behavior during the specified period of time). However, 80% of respondents indicated that they "do not know" or "are not sure" whether the employer monitors the behavior of employees in their free time or not.

Thus, there is an imbalance in the respondents' response. It can be assumed that a significant part of employees do not trust the employer in the question of whether he controls their behavior during non-working hours. Moreover, when answering the question about the consequences of digital monitoring of employee behavior, the answer "digital technologies give the employer the opportunity to monitor what I do in my free time" prevails; "digital technologies are a source of stress when doing work". This indicates a negative attitude of employees to digital monitoring of their behavior.

The results of the survey on the consequences of the employer's use of digital means of monitoring employees for certain provisions correspond to the results obtained in European countries ("provides security in my relations with customers, buyers, users, etc."; "gives the employee less flexibility in the working day"); in some cases, the indicators in the Russian Federation for this employer are higher ("is a source of stress when performing my work") or lower ("I feel uncomfortable because of digital monitoring").

The authors of this article believe that the employer should not constantly monitor the behavior of all employees throughout the working day with the help of digital technologies, since such behavior is very aggressive and acquires the features of surveillance. The use of digital technologies for the purpose of monitoring employees should be limited by the following restrictions:

The inescapable conclusion is that the employer should not continuously monitor the behavior of all employees through digital technology throughout the work day (shift), because such behavior is very aggressive and takes on the traits of surveillance. Digital surveillance of workers should be limited to the following limits:

1. Monitoring is possible only in the process of employees performing their work duties; the behavior of employees during the rest period (the so-called "out-of-work behavior of an employee") cannot be monitored by the employer and cannot be considered as a basis for bringing an employee to responsibility.

2. Constant (throughout the working day) monitoring of employees should be prohibited.

3. Processing of biometric personal data of employees is possible in exceptional cases: for the purpose of monitoring attendance and access to the employer's premises; for employees of certain fields of activity or industries (remote workers, IT workers, employees who become aware of restricted access information in connection with the performance of their work duties, financially responsible employees, senior employees, etc.).

4. Control/supervision is possible if the employer has reasonable suspicions about the legality of the employee's behavior. In this case, only offenses that are the basis for bringing an employee to disciplinary and material responsibility are subject to investigation; crimes and administrative offenses should not be the subject of electronic surveillance of the employer.

Classical labor relations, representing permanent, lifelong work for the breadwinner-worker, with decent wages for the family, guaranteed rights, and social protection, are the most developed form of wage labor. But the emergence of the digital economy tends to accelerate the erosion of these traditional ("classical") labor legal relations: labor is becoming increasingly diverse – from traditional employment, it is transformed into self-employment and even into various forms of casual labor ("labor contract on call", "zero employment contract"). Another manifestation of this process is that the employer becomes "invisible", he sort of "disappears", while retaining the possibility of bringing workers to labor and legal responsibility (disciplinary, material).

The peculiarity of monitoring employee behavior with artificial intelligence (as compared to electronic monitoring) is that the robot manages most of the employer's processes: it sets the start of work, tracks the speed of work, monitors the performance of each individual employee, and automatically generates any warnings and even layoffs without the participation of managers. Such an automated system registers an employee's downtime, and if he or she takes enough time (according to the algorithm set by the machine) to get off the job, the machine will signal the employee about the inefficiency of his or her productivity. Thus, now it is no longer a person (employer) who decides on the lawful or, on the contrary, unlawful behavior of another person (worker), but a machine, a robot, an automaton that cannot understand a person from the point of view of physical and psycho-emotional experiences.

It is unacceptable to automate labor processes from hiring employees to making decisions about dismissal solely with the help of artificial intelligence. The robot's lack of ability to analyze the data it collects leads to the conclusion that human (employer) participation in deciding on the legality (or, on the contrary, the illegality) of employee behavior is still necessary today. A machine cannot yet fully replace human labor for any task, at least not until artificial intelligence comparable to human intelligence is developed.

The International Labor Organization notes that "rules governing the use of data (i.e., 'big data on workers') and the introduction of algorithmic responsibility" (meaning the responsibility of the robot, the algorithm).

Industrial revolutions are never painless. Labor automation saves time and costs for workers, but it puts many professions in the "museum of history" (the telephone operator, the lamplighter, the cashier, the truck driver, and many other low-skilled professions). What will solve this problem is not yet clear: An adequate system of retraining and constant training/retraining throughout a worker's "working life"; the establishment by the state of an unconditional income for citizens, regardless of whether or not the subject in question is in labor relations, or perhaps another way will be found. But the state must strike a balance between increasing flexibility in the legal regulation of labor, observing the standards of employment established in labor law, and protecting the rights of workers as the weaker side of labor legal relations. What will solve this problem is not yet clear: An adequate system of retraining and constant training/retraining throughout a worker's "working life"; the establishment by the state of an unconditional income for citizens, regardless of whether or not the subject in question is in labor relations, or perhaps another way will be found. But at the same time, the state must strike a balance between increasing flexibility in the legal regulation of labor, observing the standards of employment established in labor law, and protecting the rights of workers as the weaker side of labor legal relations. For example, it seems necessary that the employee's right "to be off-line" ("the right to silence") should be fixed in a codified act; the default of the employee's right not to be constantly in touch with the employer is illegal because it can lead to a violation of the employee's right to rest and to various abuses of the employer (bringing the employee to disciplinary responsibility; not paying overtime work).

To reduce risks, humans need to remain competitive even after the introduction of artificial intelligence into the workplace. There are two main areas of work: identifying opportunities for workers to be competitive with artificial intelligence systems and creating an ethical framework for shaping the trajectory of artificial intelligence [8. pp. 69–77]. The state needs to develop an effective regulatory framework. "In the digital age, governments and employers' and workers' organizations will need to find new ways to effectively enforce nationally enshrined maximum hours, such as by establishing the right to be digitally disconnected".

We would like to finish the article with the words of the Russian philosopher V.A. Emelin:

> Being aware of the inalienable cultural-historical need of people for technological expansion, let us limit ourselves to the hypothesis: technology should be commensurate with natural capabilities and abilities in the objective conditions of our world and serve not simplification, but development of personality.

[4. pp. 103–115]

References

1. Aloisi, A. and Gramano, E. (2019). "Artificial Intelligence is Watching You at Work. Digital Surveillance, Employee Monitoring, and Regulatory Issues in the EU Context", *Special Issue of Comparative Labor Law & Policy Journal, "Automation, Artificial Intelligence and Labour Protection"*, edited by Valerio De Stefano, Vol. 41, No. 1, pp. 95–121. Available at SSRN: https://ssrn.com/abstract=3399548. (accessed 2021.01.07).
2. Mantelero, A. (2018, August). "AI and Big Data: A blueprint for a human rights, social and ethical impact assessment", *Computer Law & Security Review*, Vol. 34, No. 4, pp. 754–772. https://doi.org/10.1016/j.clsr.2018.05.017. (accessed 2021.01.07).
3. Pospelov, G. S. (1988). *Artificial Intelligence: A Foundation of a New Information Technology* [in Russian]. Moscow: Nauka.
4. Emelin, V. A. (2018). "From Neo-Luddism to Transhumanism: Singularity and Vertical Progress or Loss of Identity?" *Philosophy of Science and Technology*, No. 1, pp. 103–115. https://doi.org/10.21146/2413-9084-2018-23-1-103-115. (accessed 2021.01.07).
5. Levitan, S. A. and Clifford, M. J. (1982). "The Future of Work: Does It Belong to Us or to Robots?", *Monthly Labour Review*, No. 9, pp. 10–14. https://www.jstor.org/stable/41841878
6. Kyriakidou, M. (2017). "Ethical Challenges to Survival With and From Robots: Cyborgs and Expiry Dates. Manuscript". http: //www.worlds cientific.com/; https://doi.org/10.1142/9789813149137_0074. (accessed 2021.07.07).
7. Krupiy, T. T. (2020, September). "A Vulnerability Analysis: Theorising the Impact of Artificial Intelligence Decision-Making Processes on Individuals, Society and Human Diversity from a Social Justice Perspective", *Computer Law & Security Review*, Vol. 38, pp. 105429. https://doi.org/10.1016/j.clsr.2020.105429. (accessed 2021.07.07).
8. Filipova, I. A. (2019). "Artificial Intelligence, Labor Relations and Law: Influence and Interaction", *State and Law*, ISSN: 0132-0769, pp. 69–77. http://doi.org/10.31857/s013207690007472-1; https://gospravo-journal.ru/s013207690007472-1-1/?sl=ru. (accessed 2021.06.07).

18

International Child Sexual Exploitation Database to Fight Child Sexual Abuse and Artificial Intelligence

Sayan Das and Paramita Choudhury

CONTENTS

Introduction

While the Internet has helped foster economic growth, education, and creativity, it has also created a more vulnerable society by allowing those who would harm children to access, create, and distribute child sexual abuse materials; to locate like-minded offenders; and to lower their chance of detection. Child sexual exploitation has become popular with the development of the Internet and may be found in various forms such as grooming, live streaming, consumption of child sexual abuse material, and coercing and blackmailing minors for sexual purposes. Advances in technology spawn new varieties of crime. Prior to this, offenders have had to put in significant effort to meet children, exchange pictures of abuse, disguise their identity, and generate money, and also to encourage each other to do further crimes. Despite appearing to be only pixels, digital images and videos include a corresponding actual kid who is being sexually exploited. Online abuse is similar to other types of sexual abuse in that it may leave victims with mental and physical scars for the rest of their lives. Although other types of abuse may only victimize the kid once, every time a picture is seen, emailed, or received, the youngster might be revictimized (Ali et al., 2021).

Since the criminals frequently use technology like Darknet portals and other anonymous methods to help them escape discovery, they are typically hard to identify and investigate. A person being sexually exploited online is typically in a different country from their abuser, which is something that happens on a global scale.

Online child sexual exploitation includes online child sexual abuse material, grooming of children for sexual purposes, live streaming sexual abuse of children, sextortion: coercing and blackmailing children for sexual purposes. The ECPAT network is engaged in combating the sexual exploitation of children online in all industries and countries across the world, collaborating with law enforcement agencies like The International Criminal Police Organization (INTERPOL), The European Union Agency for Law Enforcement Cooperation (EUROPOL), and police authorities of various nations; creating and building partnerships with member countries, its government and leaders so that it can tackle such crimes more efficiently. ECPAT also advocates for stronger domestic legal frameworks in consonance with internationally agreed conventions. To get a whole picture of this crime, ECPAT does research to gather information on its victims, offenders, and tactics. In nations all around the world, members assist law enforcement in educating officers on victim identification and treatment. And pertinent to mention is artificial intelligence (AI) in which ECPAT urges and promotes the use of technical tools based on AI to locate and delete contents of online child sexual abuse. An example of this may be seen in YouTube's video recommendation system, which recommends videos to viewers on a sidebar. Videos on the platform are prioritized according to users' pasts and contexts, and fresh uploads are typically favored. *The New York Times* has revealed that YouTube's algorithm is using partly dressed youngsters to encourage sexual predators to watch films after watching videos containing sexual material. Though the reports stated that they were intended for a different purpose, these films were frequently domestic and depicted youngsters just at the swimming pool or even on vacation. A research report published by three Google workers explained the deep neural networks used to deliver more relevant suggestions for YouTube users and stressed the necessity of a two-stage strategy to remain assured that only personalized and interesting movies show on a device. Browsing history will decide what we are encouraged to see, and it might lead to an increase in the chance of seeing improper or illegal information for a number of people (Quayle, 2020).

As of July 29, 2021, The International Criminal Police Organization (INTERPOL) on its website has mentioned its International Child Sexual Exploitation (ICSE) Database to Fight Child Sexual Abuse, which enables global experts in the identification of child sex abuse victims to analyze and compare photos. A global database of images and videos relating to child sexual exploitation (ICSE) is an intelligence and investigation tool that allows specialist investigators to share information about child sexual abuse cases. Investigators can utilize picture and video comparison tools to find links between victims, perpetrators, and locations immediately. The database saves investigators time by avoiding duplication of work and identifying comparable photos for them to compare in other countries or regions. It facilitates information-sharing across the world's 64+ nations, providing a forum for global experts to communicate and collaborate. As digital, visual, and auditory evidence are all included in pictures and films, victim identification professionals may combine their efforts and discover child sexual assault victims. More than 2.7 million photos and videos are contained in INTERPOL's Child Sexual Exploitation database, and the resource has helped identify 23,564 victims of child sexual exploitation globally. More than 3,800 identifiable victims were documented in 2019 and 10,752 offenders were identified till October 2020. ECPAT International and INTERPOL study indicated that boys and young children were at a higher risk of serious online sexual exploitation.

Despite the fact that most people are unaware of this that child sexual abuse includes acts of abuse committed against children who are just a few years old and even newborns. INTERPOL and ECPAT International collaborative report on unidentified victims of child sexual exploitation material was published in February 2018 following the analysis of a random sample of videos and photos in the ICSE Database, namely "Towards a Global Indicator on Unidentified Victims in Child Sexual Exploitation Material". The research revealed a number of troubling trends: the more severe the abuse was in the case of younger victims. Approximately 84% of the pictures had explicit sexual behavior (ECPAT, INTERPOL, 2018). Nearly three-quarters of those who could not be recognized were prepubescent, including babies and toddlers. About two-thirds of unknown victims were female. Children were the primary victims of the horrendous abuse photos. The vast majority of the offenders were male (INTERPOL, 2021).

INTERPOL and International Child Sexual Exploitation

This International Child Sexual Exploitation (ICSE) picture and video database is managed by INTERPOL and serves as a useful investigation tool for expert detectives. Investigators and victims' identification professionals analyze the digital, visual, and streamed content of child sexual abuse content to identify victims and locating wrongdoers (child sexual abuse material, CSAM). ICSE database employs state-of-the-art image and video analysis tools to find similarities between various pictures and videos, which assists in the matching process. This helps detectives uncover links across instances, in which the same victim, offender, or location is involved. The INTERPOL Child Abuse Image Database (CAID) was superseded in 2009 by the International Child Sexual Exploitation (ICSE) database. The ICSE database is a unique software designed for use by law enforcement officials to examine CSAM and child sexual exploitation material (CSEM), which are forms of pictures, videos, and hash codes.

INTERPOL develops partnerships with a range of stakeholders to improve our chances of capturing all sex offenders, namely ECPAT, Human Dignity Foundation, INHOPE, International Justice Mission, Internet Watch Foundation, WeProtect Global Alliance, Virtual Global Taskforce, the National Center for Missing and Exploited Children, Regional law enforcement organizations, THORN. Private-sector partners, including financial institutions, internet service providers (ISPs), and software companies, play a vital role in monitoring and shutting down unlawful distribution routes for child sexual abuse content. Their contribution is critical and has a major impact on our synchronized plan.

ECPAT International

The mission of ECPAT is to put an end to child prostitution, human trafficking, and sexual exploitation of minors that occurs in the tourist and travel industries. It engages with governments and the international community to effect change and carry out research to understand the problem better and addresses the victims' and survivors' needs. Children through trafficking and prostitution and online hazards such as caregivers, sextortion, and the spread of pictures of child sexual abuse are at risk of sexual exploitation around the world. As a result of international travel, child sex offenders have had millions of new victims during the past several years.

Efforts to stop these crimes must be based on accurate information and proof so that choices and actions may be made with certainty. ECPAT conducts primary research and collects information from a range of trustworthy and professional academic sources from diverse industries and nations all over the world. In addition to aiding vulnerable children of sexual exploitation and abuse, ECPAT also helps them comprehend their rights, the judicial system, and available medical, psychiatric, and social assistance. Our service helps survivors collaborate to recover as a global network of advocates.

Child Sexual Abuse Material: Challenges

Child sexual abuse (CSA): When a child is manipulated or coerced into participating in sexual acts, it is considered child sexual abuse. These might entail either physical or non-physical interactions and can occur in either the real world or in cyberspace. Child sexual exploitation (CSE): One of the forms of child sexual abuse is child sexual exploitation. It happens when an individual or group abuses a power imbalance to coerce, manipulate, or deceive a child or children below the age of 18 into sexual encounters: by (a) giving the victim something they need or want in exchange; and/or (b) making the perpetrator or promoter wealthier or more powerful. Even if the victim was sexually active, the assault might have still taken place if it was forced. The exploitation of children for sexual purposes is not limited to physical contact; it may also happen through the use of technology. Child sexual exploitation material (CSEM) to include all sexualized depictions of children under the phrase "child sexual exploitation material" and its variant child sexual abuse material. Wherein, child sexual abuse material (CSAM) refers to pornographic materials depicting children refers to any image of a child participating in sexual acts, whether actual or simulated, or a child's sexual organs, all for the goal of sexual stimulation (UNICEF, 2020). Knowing access to child pornography has been increasingly called CSAM (child sexual abuse material), and similarly, the acronyms Online Child Sexual Exploitation (OCSE) and CSAM have been coined to replace the word "child pornography."

CSAM offenses are different, change often, and the language used by academics and practitioners is inconsistent, all of which make it difficult to compare study findings. In the worldwide fight against data collecting of child sexual abuse and exploitation, this has had a detrimental effect, and online CSAM is no exception. There is limited ability to measure the scope of the issue by the amount of material on the open and hidden Internet, as well as the fact that content may be deleted but not erased. There are indications of a rise in CSAM offenses based on both data from public records and research questionnaires. The age of the kid is also difficult to ascertain in youth-produced photos, as is the likelihood that the pictures are depicting coercion.

It has been observed that youth victims are more likely to be involved in sexual violence, especially within a family setting, and their spread is more likely to be broad. It is necessary to investigate more since Internet connection is growing via hand-held devices, and many are worried about live broadcasting of abuse (ECPAT International, 2018). If a photo or video features one child, the other similar pictures and videos are clustered around it. When viewed by themselves, these pictures may look harmless, but when utilized in the context of the series, they explain it better. Grouping related pictures and videos makes it easier for investigators to build a single case.

In addition to the information-sharing on investigations that the ICSE database makes possible, the technology also allows investigators in over 64 countries to work together on shared data utilizing existing data. By providing information on whether or not similar photographs have been discovered in other countries, the organization may prevent redundant investigations. An investigation can be expedited by detectives collaborating to find victims of child sexual abuse (INHOPE, 2021).

Using Artificial Intelligence to Protect Children

Artificial intelligence (AI) is the ability of computers to make choices or decisions using algorithms without human interaction and at high speed, and as they are able to make lots of judgments quickly from vast data and decisions, it can make a big difference in supporting and protecting children online. Fast processing of large volumes of information is critical. Without specialized tools, law enforcement agencies need to comb through every file found on a suspect's computer to see if it contains CSAM. This would be very hard on investigators and would take a long time for a human team to complete. AI algorithms or AI-integrated systems may be used in tandem with other specialized devices to efficiently sift through this data and locate child pornography videos and photos. It is this that helps them develop a stronger case more quickly, identify victims, and give aid to those who want justice and assistance. Meanwhile, Internet sites and applications such as YouTube, Facebook, TikTok, Snapchat, and many more are using this type of technology to detect child sexual abuse content and delete it, in addition to reporting it to police enforcement.

Website blocking: This is one essential aspect of the battle against child sexual abuse in preventing people from accessing websites that display child pornography. Access is blocked to help protect minors who have been victimized and as a teaching tool for individuals who may be considered a criminal act. Police can send ISPs a list of domains, or web addresses, to ban on their networks in order to limit access to websites that distribute child sexual abuse material. After a failed attempt to visit the page, visitors will be sent to a stop page that gives details on the rationale for the block, links to the legislation in question, places to lodge complaints, and more.

Baseline: The baseline system assists business and governmental sectors in locating, report, and delete known child pornography. To do this, they can run photos and videos through INTERPOL's baseline list, which has some of the worst child abuse images and films' digital signatures. The content's distribution is restricted when network operators inform the police and delete the matching material.

The following criteria must be met in order to be included in the baseline list: being recognized as child abuse imagery or video by our investigators and having content that is either known to show minors under the age of 13 or looks like a young child and because of the stringent standards, the baseline list is limited to unlawful photos and films, which are regarded such in any nation (INTERPOL BLOCKING, 2021).

AI and Identification

The International Child Sexual Exploitation Database (ICSE Database) is a crucial tool in helping investigators, and the ICSE Database includes picture comparison software that assists in the identification of victims. The act of identifying victims is a challenging task due to the complexity of the activity. It involves a broad array of diverse specialties. Civilian analysts are often recruited to help law enforcement to pinpoint the origin of a series of pictures or videos. People that specialize in victim identification cooperate with their peers throughout the world in order to prevent the signals that are specific, consistent, or easily identifiable in one nation from being ignored in another.

Identifying victims of child sexual abuse requires a thorough examination of both still and moving photos to find and rescue the victims. Police investigators need to start with the evidence and work their way back to the crime site when dealing with online child sexual assault. The photos can also be discovered through child exploitation research; proactive online platform monitoring; forensic examination of confiscated mobile devices, computers, digital storage units, etc.

When victims are discovered, forensic professionals take control. They meticulously scrutinize each photograph in order to protect the youngster and catch the offender. The majority of cases of sexual abuse are committed by someone the kid knows, such as a relative, a neighbour, or a daycare provider.

> **Image scanning**: Images are thoroughly examined, and standard investigation procedures are used for victim identification. Officers can glean information about the victim or place by looking at pictures and videos' digital, visual, and aural material. Victims' identities can be discovered utilizing a variety of resources and approaches.

> **Task force to identify victims**: Although databases and secure communications networks are of primary importance to us, on occasion we must rely on having everyone physically there at the same time to solve difficult issues that span many nations in a specific region.

> **PhotoDNA**: Many European online technology platforms employ hash fingerprint technologies like PhotoDNA to automatically identify previously identified (known) child sexual abuse material (CSAM) that is uploaded on their sites. This gives them the ability to find data about the services that they might use for CSA without invading their consumers' privacy (INHOPE e-Privacy, 2021).

AI, in the end, is essential for protecting children, which therefore becomes a crucial and pertinent application for ECPAT International. The organization believes that businesses need to be held accountable for their usage of algorithms. Thus, ECPAT lends its support to the rising movement of human-centric AI over data-centric AI in the EU. It is important to design and use AI to improve human and societal well-being, making sure there are sufficient safeguards to protect against adverse consequences like human and data-driven bias (ECPAT, 2021).

Pandemic and Challenges

INTERPOL evaluation of the COVID-19 pandemic shows that its effect includes children being sexually abused and a rise in the amount of child exploitation material shared over peer-to-peer networks. The study illustrates current trends and dangers, as well as what immediate effects they have, as well as what is expected to happen in the long run if

COVID-19 limitations are loosened. A wide range of key environmental, social, and economic factors change that has taken place as a result of COVID-19. These have impacted child sexual exploitation and abuse (CSEA) across the world. Such changes include the closing of schools and an increased reliance on virtual learning environments; a rise in the amount of time children spend online for entertainment, social, and educational purposes; a reduction in international travel and the repatriation of foreign nationals; a decrease in the availability of community support services, child care, and educational personnel, who are often key to identifying and reporting cases of child sexual exploitation. There are worries that some crimes may never be reported following a considerable wait, which may be attributed to the increased barriers that victims of crime now face in reporting offenses or receiving help (INTERPOL, 2020).

Communities on the Darknet

Another discovery was that there was a spike in forum conversations on the Darknet related to CSEA. Sexual offenders with technical competence in managing forums took extra time to build new ones, while users took advantage of more online time to arrange their CSEA collections. A rise in child exploitation via live streaming for monetary compensation has been seen in recent years, and there is reason to believe that travel limitations will further fuel this desire. As the number of CSEA victims is expected to grow, so too will the quantity of live-streamed content. One factor contributing to this is that more people will be relying on facilitators who, in turn, will have less money to spend. Areas where COVID-19 is very prevalent will likely have a greater chance of children being abused when parents are hospitalized, and the children are under the care of others or without any care at all.

COVID-19 impact on policing includes:

- Campaigns to raise awareness and teach citizens how to identify and avoid the risk of CSEA through online gaming, messaging, and social media platforms should be created.
- Conduct exercises that map occurrences in many schools to pinpoint growing concerns with CSEA, for example, instances like "zoom bombing".
- Also, make sure that hotlines stay open and manned, consider implementing free texting services, and integrated reporting channels for minors through gaming, social media, and messaging services.
- Share relevant CSEA information with INTERPOL to aid and support cross-jurisdictional investigations.

Conclusion

It is imperative that the strategies to safeguard online safety improve both education and health, as well as limit the dangers of exposure to violence, exploitation, and abuse, and privacy breaches. United Nations Children's Fund (UNICEF) is committed to ensuring that the Internet is a safe environment for children to learn, socialize, and grow. UNICEF tackles the issue of child sexual exploitation in countries and worldwide.

The objective of accurately measuring the scale of CSAM/CSEM worldwide is difficult to achieve, and it will involve lengthy discussions and consultations between researchers and the gatekeepers of international CSAM/CSEM archives, whether they be law enforcement or helpline officials. A common agreement across all countries on standards and categories of case data, and even a database on unidentified CSAM/CSEM crimes would be extremely challenging given that many law enforcement agencies do not keep official databases or archives of confiscated CSAM/CSEM. Instead, even though what is followed, a different standard of administration, record-keeping, and data categorization is required. Nonetheless, it is apparently necessary both for harmonious categorization and sharing approaches for unidentified CSAM/CSEM victims among databases worldwide and to facilitate, as far as possible, consolidation of a database based on common data standards and enable analyzing the current state of affairs of unidentified children.

Research needs to be done to provide a framework for the creation of insights into the situations of CSAM/CSEM victims who have not been identified. It also offers a categorization technique that may be further improved and altered to help create detailed profiles of unidentified victims in future research combined with a number of anonymous data collection tools and the exchange across information gatekeepers of standardized data. There is an urge to use and upkeep the ICSE Database and to continue to encourage new members to join. It should be completed within the limits of law enforcement's ability and funding and should prioritize further use of technology with AI integration to help and improve investigators and analysts throughout the world.

Further, to create and build harmonious approaches to CSAM and CSEM categorization across countries and jurisdictions, and to share case-related relevant data on child victims of sexual abuse and exploitation, and to cooperate to identify victims through the ICSE Database or more broadly by holding meetings of experts and specialized investigators, build in mechanisms that allow for the regular analysis of trends and data, which will assist in planning and executing future research programs that look at child sexual exploitation and abuse. Additionally, experiment with technology partners to see how best to leverage existing and future technologies to identify victims.

CSAM has changed in the eyes of police and hotline analysts. Although expert judgments may provide a foundation, global metrics need significant statistical evidence to support them. To better comprehend and counter-terrorism, governments must analyze and follow CSAM data patterns at both national and international levels.

Since investigations, arrests, and covert operations are utilized to deal with child sexual abuse material, prevention is also employed. It is worth noting that no criminal charges have been created because of a person being sent to a website with child pornography. In order to deal with child sexual exploitation, a multi-pronged approach should be utilized, including the usage of access restrictions. Need to keep a registry of websites that publish the most graphic child abuse content. The "Worst of" list includes websites distributing content on child sexual abuse and which have been confirmed by at least two separate countries/agencies.

Furthermore, it is high time to amend the local definition of "child pornography" to include "virtual child pornography" in order to criminalize it. To encourage ISPs to disclose instances of child pornography, it is imperative to include a need for them to report situations where their resources have been used to carry out activities linked to child pornography. People who commit child pornography-related crimes while overseas and who are habitual residents of any state should be prosecuted under state laws in a special territorial jurisdiction. And for offenses that constitute "child pornography", ensure that extradition is possible even if no particular extradition treaty exists. It is imperative to determine whether the matter should be transferred to national prosecution if extradition is denied.

Bibliography

Ali, Sana, Youssef, Enaam, & Hyakal, Hiba. 2021. *Child sexual abuse and the internet-A systematic review.* Human Arenas. 4. https://doi.org/10.1007/s42087-021-00228-9.

Crimes Against Children Research Center. Accessed July 27, 2021. http://www.unh.edu/ccrc/.

ECPAT. 2021. *The role of artificial intelligence in protecting children in the digital space: Why does it matter?* Posted on 17/08/2021. Accessed August 21, 2021. https://www.ecpat.org/news/ai-digitalspace/.

ECPAT International. 2018. *Trends in online child sexual abuse material.* April 2018, Bangkok: ECPAT International. Accessed July 05, 2021. https://www.ecpat.org/wp-content/uploads/2018/07/ECPAT-International-Report-Trends-in-Online-Child-Sexual-Abuse-Material-2018.pdf/.

ECPAT, INTERPOL. 2018. *Towards a global indicator on unidentified victims in child sexual exploitation material.* Accessed July 12, 2021. https://www.interpol.int/en/Crimes/Crimes-against-children/International-Child-Sexual-Exploitation-database/.

INHOPE. 2021. *What is the international child sexual exploitation (ICSE) database?* Article on published on 09.06.2021 by INHOPE. Accessed July 24, 2021. https://inhope.org/EN/articles/what-is-the-international-child-sexual-exploitation-icse-database/.

INHOPE e-Privacy. 2021. *e-Privacy derogation passes & new regulations.* Accessed August 23, 2021. https://inhope.org/EN/articles/e-privacy-derogation-passes-new-regulations/.

INTERPOL. 2020. *INTERPOL report highlights impact of COVID-19 on child sexual abuse.* Accessed July 12, 2021. https://www.interpol.int/en/News-and-Events/News/2020/INTERPOL-report-highlights-impact-of-COVID-19-on-child-sexual-abuse/.

INTERPOL. 2021. *International child sexual exploitation database.* Accessed July 12, 2021. https://www.interpol.int/en/Crimes/Crimes-against-children/International-Child-Sexual-Exploitation-database/.

INTERPOL BLOCKING. 2021. *Blocking and categorizing content.* Accessed July 11, 2021. https://www.interpol.int/en/Crimes/Crimes-against-children/Blocking-and-categorizing-content/.

Quayle. 2020. *Prevention, disruption and deterrence of online child sexual exploitation and abuse.* ERA Forum. https://doi.org/10.1007/s12027-020-00625-7.

UNICEF. 2020. *What works to prevent online and offline child sexual exploitation and abuse?* Review of national education strategies in East Asia and the Pacific 2020. Accessed July 1, 2021. https://www.unicef.org/eap/media/4706/file/What%20works.pdf/.

19

EU and American Charter on Use of Artificial Intelligence in Judicial Process: Model Charter for India?

Taruna Jakhar and Surya Saxena

CONTENTS

Introduction

Artificial intelligence (AI) is a broad field of computer science that focuses on creating intelligent machines that can accomplish activities that would normally need human intelligence. Although AI is a multidisciplinary subject with several techniques, advances in machine learning and automated reasoning cause a paradigm shift in nearly every sector of the computer industry.

The Joint Technology Committee under the National Centre for State Courts, USA, released a resource bulletin titled "Introduction to AI in Courts". The report discusses the similarity in the decision making of humans and AI decisions as follows:

    i. Humans and AI make a decision based on the data they possess or are provided.

    ii. Humans and AI come across extensive data and sources and thus their assumptions and conclusions keep evolving based on the data they come across.

    iii. They have biases and intuitions that develop based on their experience.

DOI: 10.1201/9781003215998-19

However, the report also states that AI-based decisions are made quickly because of their computational capacity, whereas human decisions develop at a slow pace with inputs that are difficult to evaluate.

The advent of AI in the field of law has brought revolutionary changes. The biggest challenge plaguing the Indian judiciary is the pendency of cases. The statistics reveal that as of January 2021, there were 37,251,615 pending cases in India.

Integrating AI with judiciary will enable and aid the decision-making process and simultaneously increase efficiency in justice delivery. The inception of AI in judiciary started in November 2019 when the Apex Court injected a tool called SUPACE (Supreme Court Portal for Assistance in Court Efficiency) to expedite and catalyze judicial processes by translating judicial orders from English to respective vernacular languages. Furthermore, Justice S.A Bobde in his speech highlighted the use of AI in the judiciary. However, he was also of the opinion that penetration of AI in the judiciary should be limited to assisting and aiding judges and the final discretion should be left upon the judges.[1]

A research paper by Vidhi Centre for Legal Policy[2] reveals that the Indian judiciary has grown by leaps and bounds to fully harness the potentials that AI technology has to offer on our platter. The paper sheds some light on "Task-Specific narrow AI tools" which can aid judges to focus less on administrative matters. The judiciary is broadly a mirror to the society that reflects the standards which have to be adhered to by the society as a whole. The courts are premised on the information. They are tasked to collect the information and process the information in the course of the procedure with the outcome as information. However, not all information processing requires complex customization. Many cases are settled with a simple assessment and only a few cases are complex and contradictory. Hence, the need for information technology is not similar across the spectrum. The degree of complexity of information and predictability of outcome are decisive factors that determine the handling of cases in administrative and civil matters. A large proportion of cases generally have a predictable outcome, such cases are settled via a court ruling which is a document produced through automation based on the data available. However, in complex cases, where the knowledge of a human judge has to be deployed to settle the case, information technology ensures that a digital case file is readily available to process a large amount of information.

AI is helpful in innumerable ways to address difficulties. There have been several contentions and debates about whether robots can act as a judge to perform the functions of organizing information, advising, and predicting the outcome.

Article 6 of the European Court of Human Rights (ECHR) emanating from the Ethics Guidelines establishes a proper procedure for ensuring transparency and well-founded judgment. The vast legal information should be first made machine-processable for effective understanding. Anomalies in data can lead to flawed decisions, consequently reducing the quality of judgments. Statistical data and analysis of data are not enough to substantiate a good judgment. The information must be properly framed and structured to provide a rational meaning. The addition of legal meaning with well-framed terminologies can increase the effectiveness of AI in legal proceedings. AI should be able to posit rational reasoning that made it reach a particular conclusion. The explanation is expected to be as convincing as a human explanation. However, humans can effectively articulate some aspects of AI. So, there is still a long way to go for AI and machine learning to permeate courts and legal proceedings.

Comparative Analysis of European and American Ethical – Principles on the Use of Artificial Intelligence in the Judicial System

Position in the European Union

Application of AI in the realm of judicial systems across the globe aims to enhance the efficiency, speed, and quality of justice delivery. However, we stand to lose as much as we may gain from this integration. For instance, open data systems make it possible to compare a gigantic volume of files within the same database. These files can then be processed by an algorithm to draw a series of observations and conclusions, which can benefit all stakeholders during trials. However, as explained in an interview[3] by Mr. Georg Stawa, President of the European Commission for the Efficiency of Justice (CEPEJ), by referring to a car accident case, the AI tool always decides in favor of the white man with a red car who prefers chocolate ice cream as it is the most probable outcome. To avoid such scenarios, it is essential to lay down certain fundamental principles that must be followed while the development and subsequent use of different AI applications within judicial systems.

Thus, the CEPEJ endorsed the first "European Ethical Charter on the Use of Artificial Intelligence in Judicial Systems and Their Environment" at its 31st plenary meeting.[4] The novel framework was developed with the following objectives:

1. To present a scientific and unbiased outlook on the limits and prospects of the integration of AI with the judiciary.
2. To draw attention to the major issues and to help to identify constructive solutions.
3. To advise upon governance and ethical aspects.

With this, the charter effectively guides and helps the 47 member states of CEPEJ in developing their systems of application of AI in the arena of justice. Furthermore, these guidelines address policymakers, legislatures, justice professionals, and private and public companies as they grapple with the development or use of algorithms while integrating AI in their national judicial systems. This emphasizes the fact that all stakeholders must be conscious of the risks of technology and be aware of the repercussions of their actions. As put by Mr. Stéphane Leyenberger, Executive Secretary of the CEPEJ, "the said charter is crucial to avoiding the Frankenstein Syndrome."

Against the aforementioned background, the CEPEJ believes that AI must be applied to the realm of justice with utmost prudence and responsibility. Hence, through this charter, the CEPEJ formally adopts five fundamental ethical principles, which drive their validity from the ECHR and the Convention on the Protection of Personal Data apart from the other core principles of justice. These fundamental principles are as follows:[5]

1. **Basic rights principle:** Guarantee that artificial intelligence products and services are designed and implemented in a way that is consistent with fundamental rights.
2. **Non-discrimination principle:** Particularly prohibit any discrimination between people or groups of individuals from developing or intensifying.

3. **Quality and security principle:** Employ verified sources and intangible data with models developed in a multidisciplinary approach in a secure technology environment when processing court decisions and data.

4. **Transparency, impartiality, and fairness principles:** Make data processing processes transparent and intelligible and allow external audits.

5. **The principle of "under user control":** Prevents a prescriptive approach and assures that users are informed actors who are in charge of their decisions.

These principles are crucial as they form a *system of checks and balances* at each phase of the process of developing AI in the field of justice – at the level of legislatures, judicial practitioners, and legal techs. It emphasizes the contention that AI is only a helping tool and never the final objective as it can never go beyond the judgment and discretion of a rational, reasonable, and thinking human judge.

The application of these principles is going to be a colossal challenge for judiciaries across Europe. The integrity of judicial data and information systems will have to be protected. Many advantages come with digitization, but the change is fraught with difficulties, including how to effectively comply with international ethical norms.[6]

Status Quo of the Use of AI in Europe

The Charter also enshrines the prevalent and the possible application of AI in the judiciary while analyzing the potential and limits of AI. It highlights that the wave of digital transformation in the judiciary has undergone development across the Council of Europe member states. *Many European countries appear to have evolved an exceptionally advanced strategy to using practical applications (in terms of both technology and legal backing), while others are still grappling with the issue and focusing mainly on effective IT management.* However, through a survey in 2018, CEPEJ found that overall, the use of AI in the judiciary across its member states is still very low. It has been speculated *that the use of predictive tools in criminal trials is very rare in Europe "as many other algorithmic instruments are, and there is room to set up a legal discussion before the market rules overarch any effective possibility for it".*[7] Thus, apart from establishing ethical guidelines for the use of AI I judiciary, the Charter also provides a comprehensive roadmap to its member nation for the same by enshrining upon the possible uses to be encouraged or discouraged, necessary precautions, and impending research.

The following is the present extent of the use of AI in the judiciary in two European countries: France and Germany.

France

France is not so in favor of using AI in the judiciary, and this can be seen from its decision of passing legislation that banned the use of AI in the judiciary.

Though the possible reason for France banning the use of AI can't be found, many theories have been proposed by scholars and academicians such as anonymity of judiciary, and the possible fear that predictive data analytics used in AI may depict the true distinction between actual judiciary rules and regulations and the reality that whether they were being followed or not.

They instead wished to have a statistical analysis done of the decisions that they made so that their performance could be constantly monitored and they could find the deficiencies if any in their conduct. But there is opposition to the use of statistical analysis done by AI as they are mostly seen as tools that are imbibed with a particular set of biases and notions that could provide error-prone decisions and results, and this is a possible reason that France is not so in favor of using AI in their judicial systems.[8]

Germany

It is not very clear in Germany whether judges can be permitted to take the help of AI and use it to improve and further enhance the quality of their decisions. However, it is seen as possible for the lower courts in Germany to use AI as the cases they handle are comparatively easier and less ambiguous in their language.

With regard to lawyers using AI services, some regulations need to be complied with if they wish to use those services. Bankruptcy restoration firms could potentially employ technological systems to examine legality concerns but only if they are correlated to the claim, according to the current law in place there.

In Germany, lawyers can use AI only if they are providing the legal services themselves. What this means is that lawyers can use AI in their daily functioning to speed up their work and give them a better perspective. This may save them the research work as it would be done by AI. But this may downgrade their professional efficiency and thinking ability as they would become dependent on AI for their research work.

At present there is no legislation in Germany to regulate AI in the field of law, only the general regulations that are in place for human liability are applied as there is absence of specific statutory regulation for use of AI.[9]

Position in the United States of America

The USA is IT hub and birthplace of several technological advances, which are transforming the world. The inclusion of AI in their courts is something which was at a natal stage when the world was being acquainted with the same. Thus, it makes sense to look into their model for a better perspective and then modify and make it according to our uses. Judicial decisions need to be precise and accurate mostly so there is no room for learning from mistakes, as that mistake can take a count for someone's life, so a better stance would be learning from the one already learned.

They have instilled AI in their very basic work too, their AI bot GINA works to solve queries for the people visiting the site, pay fines, and several other things like the date for proceedings and help police in their investigation for quick search of previous criminal records and gauging the probability of crime that person will commit in future. This is a feature that every big corporate giant is providing to its people even railways and telecom biggies. This does not dilute human intelligence or intrudes into it but tries to provide another arena and reduces the workload to a great extent. Another example is Florida uses AI software to classify and docket e-filed documents making the process automatic. If the bot makes any error, it is rectified so that it does not repeat the same mistake.[10]

The USA has also used its AI in delivering judgments too by different calculations and algorithms, which has been into the quite contentious debate in the Bar Association and for the same they have raised queries and will issue certain guidelines for the attorneys. The queries raised by the Bar Association, USA, are worth going through. The first query is about the vendor details, the institution that will be providing the services, or even if they sell it to the system, how will it ensure that the data stays protected and there is no further relation with creator or vendor. Second, bias and explanation of the same, as it is a system that consists of some ruled out inputs and it has a fixed output, thus every decision will be ex parte and the concept of justice will devolve. Third, the Bar has asked lawyers to be aware of such things and should use AI only to that extent that does not breach the ethical code of conduct.

Judges have been quite vigilant and have taken some serious observations of the threats by the AI if used for giving judgments, one judge in his ratio decidendi writes,

> The court found that using COMPAS (Correctional Offender Management Profiling for Alternative Sanctions), a "risk-need assessment tool to provide decisional support for judges in criminal situations," did not violate the defendants' due process rights. However, an important independent report revealed that the risk assessment computer algorithm utilized by judges was prejudiced towards black prisoners. COMPAS was "far more likely to designate black offenders as likely to re-offend – wrongly labeling them at nearly twice the rate (45 percent to 24 percent) than white defendants."[11]

There have been many cases in the USA which talked about the issues that would start arising constantly due to the frequent use of AI in everyday lives. In the case of *Inc.* V. *Robins*, it had been held by the court that a "search engine" which portrayed false information about the person's job was a serious violation. This case highlighted the fact that the judges needed to consider the impact AI would bring on people's lives, how their lives would be changed, and what would be the consequences of relying on AI to do most of the tasks.[12]

Now, there are both shortcomings and benefits as a result of using AI. The most important issue that can be corrected and proximity can be reduced to a large extent is that of errors made by humans. Lawyers are generally less competent, inept, and sluggish when compared to AI for some particular errands, such as reviewing documents at a fast pace. AI is sounder and more precise when sufficient data is provided to it, and this has proven to be true in envisaging the consequences of the decisions made by the US Supreme Court. Some law officials had provided data to the AI software on the rulings made by US Supreme Court. Using the data provided, the software was successful in forecasting the judgments made by the Court with an efficiency rate of 70%.

The USPTO is also in the process of augmenting its skills in patent searching. A call for AI had been issued by it, which would have taken it further in its goal of achieving it as USPTO was highly interested in using modern technology that would have complemented and enhanced its search proficiencies with unique and distinguished solutions for its overall development. The USPTO was scanning for a mechanism that would assess and evaluate patent filings in detail to understand the source code that was pertinent to the revelation undergoing scrutiny.[13]

However, with such use, it is necessary to follow ethical principles to ensure that AI is used in an ethical and unbiased manner. It is pertinent to note that, unlike the European Union, the United States does not have a separate charter for the ethical use of AI in the judicial system. Most big tech companies that are based in the US have formulated their ethical charter. However, any public sector organization did not have such a charter until recently. The Department of Defense (DoD) in the US adopted ethical principles for use of AI, which is the first such development in the US public sector. These principles are as follows:[14]

1. **Responsible:** While maintaining responsibility for the development, deployment, and use of AI capabilities, DoD employees shall apply appropriate judgment and care.

2. **Equitable:** The Department will take intentional steps to ensure that AI capabilities are free of unintended prejudice.

3. **Traceable:** The Department's AI capabilities will be developed and deployed in such a way that relevant personnel have a thorough understanding of the technology, development processes, and operational methods that apply to AI, including transparent and auditable methodologies, data sources, and design procedure and documentation.

4. **Reliable:** The Department's AI capabilities will have specific, well-defined uses, and their safety, security, and effectiveness will be tested and assured across their entire lifecycles within those stated uses.

5. **Governable:** The Department will design and engineer AI capabilities to perform their intended activities while detecting and avoiding unintended outcomes, as well as the capacity to disengage or deactivate deployed systems that exhibit undesired behavior.

It is pertinent to discuss the Joint Technology Committee under the non-profit – National Centre for State Courts here. The NCSC is made up of several chief justices, lawyers, and administrators who are approached by Court for authoritative knowledge and information, and the JTC is a committee under the NCSC which was established to develop and promote technology standards for courts, educate the court officials in technology, etc. The JTC released a resource bulletin "Introduction to AI in Courts" acknowledged the ethical principles adopted by DoD. The bulletin also states that these principles are not unique only to the DoD and apply to all public sector organizations such as Courts.

Application on Indian Judiciary

Indian judiciary stands in a troublesome situation as the number of pending cases is mammoth and constantly rising. This has the potential to reduce people's access to justice and the efficiency of the judiciary.

Thus, the adoption of AI in the domain of the justice delivery system depends firstly on the identification of legal processes where technology can improve efficiency. In an attempt towards the same, a tool named SUPACE (Supreme Court Portal for Assistance in Court Efficiency) was launched by the Supreme Court of India. The tool understands the judicial procedures that require automation and then assist the court to reduce the pendency.

Further, with this initiative, the Indian Supreme Court has become a frontrunner in the globe in terms of the adoption of AI in the judiciary. With the adoption of AI, there are several issues reported over biasness, privacy, data protection, data storage, etc. The complex algorithms and correlations followed by a huge database make it very difficult to establish if the originator and the building block cause harm. However, in a recent paper, Justice Bobde has noted that a futuristic judiciary is not an impossible dream now. He further noted that the determination of a judicial decision of the final consideration must be that of humans and not the machine.

Added to that, NITI AAYOG has also released a national strategy on AI (NSAI), highlighting the potential of AI in solving various challenges across sectors. The paper focuses on India-specific challenges in areas of AI implementation along with the impact and opportunity. The report also describes the growing ethics, privacy, and security-related issues. Also, this is followed by the added advantage that the pandemic has given in the form of virtual courts (Jauhar, 2020). Also, the potential of alternative dispute resolution in an online medium provides for the capacity building of stakeholders through training and skill development. This can be a step ahead, but it will certainly face the challenge of the performance of technology with the role of judges and obligations that are requisite for humans.

The proper and ethical rollout of AI has been addressed by several think tanks as well like Vidhi Center for Legal Policy has noted that it involves the expansion of supreme courts AI committee, publishing openly available database and involvement through PPP model (Public–private partnership). Thereafter, the stance of the Indian court is a phase-wise acceptance of change by the Supreme Court. This will ultimately have the potential to transform the Indian justice system, some of the tools will also aid in the development of accurate tools for adding predictive justice.

As already discussed above, AI has great potential in almost every industry and the judiciary is no exception although there have been various criticisms of AI technology as a judge and that it cannot replace a human mind where AI can be used to make the process of litigation time efficient and more transparent. Inclusion of AI to the fullest extent would mean from filing, notarizing to uploading evidence, and giving judgments. Partial inclusion would mean that AI would be strictly limited to the procedural work like registering, filing, and uploading of documents to assist in case laws or specific sections for research, making it a less cumbersome job to file a case or a petition.

In few steps of litigation, the use of AI will help the judicial system to dispose of the cases, namely efficiently organizing information, adjournments, advising, and predictions [e.g., COMPAS (Correctional Offender Management Profiling for Alternative Sanctions) used in some states of USA], decision-making assistance.

Across the globe certain nations have allowed AI tools in the legal system for assistance although no country has allowed an AI tool to be a judge, countries like Estonia have allowed robot judges in small claim cases, China has allowed AI tools for legal advice and assistance, in Abu Dhabi AI tools are being used for predicting the scope of the settlement, in US COMPAS is being used, in the UK H.A.R.T is being used to determine recidivism in a criminal case. Now when we see the Indian legal system the two major drawbacks are, loopholes in the procedural aspect of the cases which people use for their benefits and the disposal rate of the cases is very low. From the above discussion, these two problems can be resolved with the use of AI tools in Indian court as proper documentation and evidence review.

The American and European models on ethics can be applicable in India by the Indianization of their concept. As discussed earlier, India is at the first stage of the application of Artificial Intelligence in the judicial context. The models will give a sense as to the necessity of human intervention in our country. The principle can be applied from the eDiscovery theme of the USA, thereby making it feasible for India to bring down the backlog of cases. The EU Charter of Ethics about AI established the importance of fairness. Fairness is a concept based on the naturality of common sense. It is a concept that is dynamic and cannot be molded in any shape as such. There are several situations where humans have failed to be fair; therefore, expecting a programmed machine to be fair is too idealistic. The Indian Supreme Court, for example, is considered as a polyvocal court as some judges are well known for accepting far more cases for regular hearing than other judges, believing the court should leave its doors more widely open. Hence, it might cause algorithmic biases. The word justice will not hold to its meaning if there won't be any fairness. However, as mentioned in the NITI Aayog report, the Indian model can be first applied to few courts thereby taking feedback and in the second phase unveiling the same at the national level.

The use of technology in courts will be the most significant development in justice delivery system in the coming decade. AI will not only assist in the organization of cases, but it will also speed up the process of bringing references into the judgment.

Our lives are influenced by the law and judicial system in two ways. First, in the realm of our daily lives: transactional events. Second, in the realm of litigation: lawyers, judges, and other fora for conflict and dispute resolution. Major reforms are required and will be implemented at both levels. These constitutional objectives will be realized in the coming decade. Furthermore, citizens must be taught in general, as well as about their essential human rights. They must be given the authority to demand that society fulfill these basic rights. They also require instruction on how society may be forced to provide them with the essentials of life. In the 2030s, these will also grow.

Notes

1. N Joshi, "Artificial Intelligence can supplement but not supplant a judge; cannot be allowed to determine the outcome of the case: Chief Justice of India S.A Bobde," *Bar and Bench*, April 16 2021.
2. A Sengupta, "Responsible AI for the Indian Justice System – A Strategy Paper" *Vidhi Centre for Legal Policy*, April 15, 2021, Accessed 16 April, 2021, https://vidhilegalpolicy.in/research/responsible-ai-for-the-indian-justice-system-astrategypaper/#:~:text=%20Responsible%20AI%20for%20the%20Indian%20Justice%20System,sections%20of%20the%20paper%20elaborate%20some...%20More%20.
3. Counsel of Europe, "CEPEJ European Ethical Charter on the use of artificial intelligence (AI) in judicial systems and their environment" June 6 2018, Accessed on 20 April, 2021, https://www.coe.int/en/web/cepej/cepej-european-ethical-charter-on-the-use-of-artificial-intelligence-ai-in-judicial-systems-and-their-environment.
4. Strasbourg, European Commission for the Efficiency of Justice, *"European Ethical Charter on the use of Artificial Intelligence in judicial systems and their environment"* 3–4, July 2018.
5. Ibid.
6. A. D. (Dory) Reiling, "Courts and Artificial Intelligence" *International Journal for Court Administration*, 11(2) (2020) Accessed April 22 2021.
7. S. Quattrocolo, "An introduction to AI and criminal justice in Europe." *Rev. Brasileira de Direito Processual Penal*, 5 1519 (2019) Accessed April 20 2021.
8. G. Bufithis, "Understanding the French ban on judicial analytics" December 21, Accessed on 20 April 2021, https://www.gregorybufithis.com/2019/06/09/understanding-the-french-ban-on-judicial-analytics/.
9. Schirmbacher. "Insights on Public Policies to Ensure AI's Beneficial Use As A Professional Tool" (IBA Alternative and New Law Business Structures Committee). Accessed on April 30, 2022, https://www.ibanet.org/MediaHandler?id=df7ac29b-7cc9-43e1-94df-799ea4b00017.
10. Joint Technology Committee, "Introduction to AI for Courts" *JTC Resource Bulletin* March 27 2021 Accessed on 20 April 2021, https://www.ncsc.org/__data/assets/pdf_file/0013/20830/2020-04-02-intro-to-ai-for-courts_final.pdf.
11. J. Willie, "Artificial Intelligence: Now Being Deployed in the Field of Law" *American Bar Association*, February 3 2020, Accessed on 26 April 2021, https://www.americanbar.org/groups/judicial/publications/judges_journal/2020/winter/artificial-intelligence-now-being-deployed-the-field-law/.

12. B.K. Newman, "Recent Developments in Artificial Intelligence Cases" *American Bar Association,* June 16 2021, Accessed 27 April 2021, https://businesslawtoday.org/2021/06/recent-developments-in-artificial-intelligence-cases/.
13. Caroll J. Timothy, "Pros and Pitfalls of Artificial Intelligence in IP and the Broader Legal Profession" *Landslide American Bar Association* 2019, Accessed 30 April 2021, https://www.americanbar.org/groups/intellectual_property_law/publications/landslide/2018-19/january-february/pros-pitfalls-artificial-intelligence-ip-broader-legal-profession/.
14. US Department of Justice, Accessed on April 18 2021, https://www.justice.gov/opa/speech/assistant-attorney-general-kristen-clarke-delivers-keynote-ai-and-civil-rights-department.

Index